Mesoporous Materials for Advanced Energy Storage and Conversion Technologies

Mesoporous Materials for Advanced Energy Storage and Conversion Technologies

Editors

Jian Liu
Department of Chemical Engineering
Faculty of Science and Engineering
Curtin University
Perth, WA
Australia

San Ping Jiang
Fuels and Energy Technology Institute
& Department of Chemical Engineering
Curtin University
Perth, WA
Australia

CRC Press
Taylor & Francis Group
Boca Raton London New York

CRC Press is an imprint of the
Taylor & Francis Group, an **informa** business
A SCIENCE PUBLISHERS BOOK

CRC Press
Taylor & Francis Group
6000 Broken Sound Parkway NW, Suite 300
Boca Raton, FL 33487-2742

Printed on acid-free paper
Version Date: 20170119

International Standard Book Number-13: 978-1-4987-4799-8 (Hardback)

Library of Congress Cataloging-in-Publication Data

Names: Liu, Jian (Chemical engineer), editor. | Jiang, San Ping, editor.
Title: Mesoporous materials for advanced energy storage and conversion technologies / editors, Jian Liu, Department of Chemical Engineering, Faculty of Science and Engineering, Curtin University, Perth, WA, Australia, San Ping Jiang, Fuels and Energy Technology Institute & Department of Chemical Engineering, Curtin University, Perth, WA, Australia.
Description: Boca Raton, FL : CRC Press, Taylor & Francis Group, 2017. | Series: A science publishers book | Includes bibliographical references and index.
Identifiers: LCCN 2016042509| ISBN 9781498747998 (hardback : alk. paper) | ISBN 9781498748018 (e-book)
Subjects: LCSH: Electric batteries--Materials. | Fuel cells--Materials. | Solar cells--Materials. | Mesoporous materials.
Classification: LCC TK2901 .M47 2017 | DDC 621.31/24240284--dc23
LC record available at https://lccn.loc.gov/2016042509

Visit the Taylor & Francis Web site at
http://www.taylorandfrancis.com

and the CRC Press Web site at
http://www.crcpress.com

Preface

Mesoporous materials have attracted a great deal of attention of scientists and engineers for reasons well beyond the scope of their appealing porous structures and tunable physical and chemical properties. In the past decades, rapid advances have been made in the synthesis, morphological and structural control, functionalization and applications of mesoporous materials. Benefitting from their high surface area, large pore volume, controllable morphology and tunable porosity, mesoporous materials can offer new opportunities for advanced energy storage and conversion technologies. The aim of this book is to systematically summarize the important developments in the area of mesoporous materials for energy related applications, such as supercapacitor, Li-ion battery, Li-S battery, fuel cells, solar cells and photocatalysis.

The seven chapters of the book are grouped into four parts. We will bring all the pieces of mesoporous materials for energy related application together under one umbrella, using the unifying concepts as glue and key materials as histories. Chapters 1 and 2 cover mesoporous carbon for energy storage and conversion. Chapter 1 systematically summarizes the functionalization, synthesis strategies, heteroatom doping, morphological control of mesoporous carbon. The research on the mesoporous carbon has broad applicability beyond the electrochemical energy storage and conversion. Following Chapter 1, Chapter 2 reports the specific application of mesoporous carbon for electrochemical capacitors (ECs). From the viewpoint of fundamental physical properties of capacitor materials, Chapter 2 summarizes the synthesis, structure-properties relationships of carbon materials and their electrochemical performance in ECs. Chapter 3 and Chapter 4 report on mesoporous materials for fuel cells. Chapter 3 details advances in the development of mesoporous structured proton exchange membrane for fuel cell applications. This is supporting the potential of developing next generation of proton exchange membrane fuel cells (PEMFCs) operating at high temperature and low humidity conditions. In Chapter 4, the methods researched to design and synthesize mesoporous precious metals and metal alloys are reported. The fuel cell applications of these mesoporous electrode materials are focusing on the reactions including methanol oxidation reaction (MOR), oxygen reduction reaction (ORR), formic acid oxidation and ethanol oxidation reactions. Chapter 5 and Chapter 6 describe the utilization of mesoporous materials as electrodes for rechargeable batteries. By selecting mesoporous metal oxides, mesoporous perovskites, mesoporous carbon composites as examples, Chapter 5 is devoted to the mesoporous materials for rechargeable batteries, from Li-ion batteries, Na-ion batteries, to Li-S batteries. The extraordinarily porous structure and high surface area of these mesoporous materials are critical in the development of performance and durable next generation batteries. The importance of engineered porosity of mesoporous materials in developing high performance lithium batteries with higher capacity and longer cycle life is highlighted in Chapter 6. Chapter 7 reviews the development of dye-sensitized solar cells and perovskite-sensitized solar cells by using mesoporous materials as electrodes. The impact of the mesoscopic properties and morphologies on device performance are highlighted to elucidate the advantages of enhanced light harvesting and charge collection for such mesostructured nanomaterials.

This book highlights the significant advancement in mesoporous materials for advanced energy storage and conversion technologies. In this book, we have attempted to present to the readers the fundamental theories, maturing synthesis methods and recent development in the field. We also make a balance between concepts and the vividness of examples in this book. Therefore, we hope that this book will be a good reference for researchers, scientists, graduate and undergraduate students in chemistry, chemical engineering, materials science who are interested in porous materials and nano energy. We, the editors, would like to take this opportunity to thank all the authors for their valuable contributions and the team from the Editorial Department of CRC Press for their professional assistance and help. We also sincerely hope that readers could provide feedbacks and comments to us.

Jian Liu
San Ping Jiang
Perth, Australia

Contents

Preface v

List of Contributors ix

1. **Functional Mesoporous Carbons from Template Methods for Energy Storage and Conversion** 1
Hao Tian, Shaomin Liu, San Ping Jiang and *Jian Liu*

2. **Nanoporous Carbon for Capacitive Energy Storage** 51
Clement Bommier and *Xiulei Ji*

3. **Mesoporous Structured Materials as New Proton Exchange Membranes for Fuel Cells** 97
Jinlin Lu, Jin Zhang and *San Ping Jiang*

4. **Mesoporous Structured Electrocatalysts for Fuel Cells** 152
Kamel Eid, Liang Wang and *Hongjing Wang*

5. **Mesoporous Materials for Batteries** 221
Kuang-Hsu (Tim) Wu, Hamid Arandiyan, Ju Sun, Cui Yanglansen and *Da-Wei Wang*

6. **Recent Advances in Mesoporous Materials for Lithium Batteries** 261
Yimu Hu, Hoang Yen and *Freddy Kleitz*

7. **Controlling the Photoanode Mesostructure for Dye-sensitized and Perovskite-sensitized Solar Cells** 292
Wu-Qiang Wu, Dehong Chen and *Rachel A. Caruso*

Index 325

List of Contributors

Hamid Arandiyan
School of Chemical Engineering, University of New South Wales, Sydney, NSW, Australia.

Clement Bommier
Department of Chemistry, Oregon State University, Corvallis, Oregon, United States.

Rachel A. Caruso
Particulate Fluids Processing Centre, School of Chemistry, The University of Melbourne, Melbourne, Victoria, Australia & CSIRO Manufacturing, Clayton South, Victoria, Australia.

Dehong Chen
Particulate Fluids Processing Centre, School of Chemistry, The University of Melbourne, Melbourne, Victoria, Australia.

Kamel Eid
College of Chemical Engineering, Zhejiang University of Technology, Hangzhou, Zhejiang 310014, P.R. China & State Key Laboratory of Electroanalytical Chemistry, Changchun Institute of Applied Chemistry, Chinese Academy of Sciences, Changchun, Jilin 130022, P.R. China & University of Chinese Academy of Sciences, Beijing 100039, P.R. China.

Yimu Hu
Department of Chemistry, Université Laval, Québec (QC), Canada G1V 0A6.

Xiulei Ji
Department of Chemistry, Oregon State University, Corvallis, Oregon, United States.

San Ping Jiang
Fuels and Energy Technology Institute & Department of Chemical Engineering, Curtin University, Perth, WA, Australia.

Freddy Kleitz
Department of Chemistry, Université Laval, Québec (QC), Canada G1V 0A6.

Jian Liu
Department of Chemical Engineering, Curtin University, Perth, WA, Australia.

Shaomin Liu
Department of Chemical Engineering, Curtin University, Perth, WA, Australia.

Jinlin Lu
School of Materials and Metallurgy, University of Science and Technology Liaoning, Anshan, China.

Ju Sun
School of Chemical Engineering, University of New South Wales, Sydney, NSW, Australia.

Hao Tian
Department of Chemical Engineering, Curtin University, Perth, WA, Australia.

Da-Wei Wang
School of Chemical Engineering, University of New South Wales, Sydney, NSW, Australia.

Hongjing Wang
College of Chemical Engineering, Zhejiang University of Technology, Hangzhou, Zhejiang, China.

Liang Wang
College of Chemical Engineering, Zhejiang University of Technology, Hangzhou, Zhejiang, China.

Kuang-Hsu (Tim) Wu
Institute of Metal Research, Chinese Academy of Sciences, Shenyang, China.

Wu-Qiang Wu
Particulate Fluids Processing Centre, School of Chemistry, The University of Melbourne, Melbourne, Victoria, Australia.

Cui Yanglansen
School of Chemical Engineering, University of New South Wales, Sydney, NSW, Australia.

Hoang Yen
Department of Chemistry, Université Laval, Québec (QC), Canada G1V 0A6.

Jin Zhang
Fuels and Energy Technology Institute & Department of Chemical Engineering, Curtin University, Perth, WA 6102, Australia.

CHAPTER 1

Functional Mesoporous Carbons from Template Methods for Energy Storage and Conversion

Hao Tian,[1] *Shaomin Liu,*[1] *San Ping Jiang*[2] and *Jian Liu*[3,*]

ABSTRACT

In order to make high-performance electrodes in the field of energy storage and conversion such as fuel cells, supercapacitors and batteries, a great deal of efforts has been devoted to the functionalization of nanoporous carbons. In this chapter, the classification of nanoporous carbon materials according to pore sizes will be briefly introduced first. In the following part, three main common strategies including hard template method, soft template method and hard-soft template method will also be presented to prepare ordered porous carbon materials in energy-related fields. The emphasis of this chapter will be focused on the control of porosity, morphology, surface modification and framework composition. Finally, the outlook of nanoporous carbon materials towards optimizing the electrochemical behaviour in fuel cells, batteries and supercapacitors will be discussed.

[1] Department of Chemical Engineering, Curtin University, Perth, WA, Australia.
[2] Fuels and Energy Technology Institute & Department of Chemical Engineering, Curtin University, Perth, WA, Australia.
[3] Department of Chemical Engineering, Faculty of Science and Engineering, Curtin University, Perth, WA, Australia.
* Corresponding author: jian.liu@curtin.edu.au

Keywords: Porous carbons, template methods, doped carbons, morphology control, energy storage and conversion

1. Introduction

Since the industrial revolution, energy consumption has been increasing exponentially in the whole world. Energy, which plays a vital role in advancement of the world's economy, continues to be dominantly derived from fossil fuels (natural gas, oil and coal). However, considering the future global energy requirement and environmental problems sourced from the depletion of natural resources, global warming and pollution from burning fossil fuels, the dependence on fossil fuels has caused urgent energy crisis. Therefore, alternative technologies for using clean and sustainable energy in the field of energy storage and conversion such as fuel cells, supercapacitors and batteries have been developed impressively.

Two main fuel cells including direct methanol fuel cells and proton exchange membrane fuel cells consists of Oxygen Reduction Reaction (ORR) at cathode and hydrogen or methanol oxidation at anode. Electrocatalysts are usually used in fuel cells to facilitate electron transfer to optimize electro-chemical performance. However, there are still two major problems including high cost of commercial electrocatalysts such as platinum particles loaded activated carbons or carbon blacks and the low durability which need to be dealt with for industrial production. In addition, commercial materials usually suffer from some deactivation problems such as the susceptibility to carbon monoxide and methanol poisoning [1]. To address these problems, there is a growing social and scientific interest in metal-free porous carbon catalyst with excellent electrical conductivity and high specific surface area [2–5]. Porous carbon materials have also been used in the area of energy storage such as batteries and supercapacitors, which have received worldwide concern and increasing research interests [6]. These are attributed to the advantageous properties of porous carbons such as porous structures, low cost, excellent biocompatibility, high resistance to acid and basic environment, large surface area, good mechanical stability, high conductivity [7].

There are several review papers on the nanoporous carbon for energy storage and conversion. Carbon dots, fullerene, amorphous carbons, active carbons and carbon nanotubes are not included in this chapter. In this context, functional nano-porous carbons from template methods have attracted a lot of attention for energy storage and conversion. Three main common strategies including hard template method, soft template method and hard-soft template method have been used to prepare ordered porous carbon materials in energy-related fields. This chapter aims to summarize

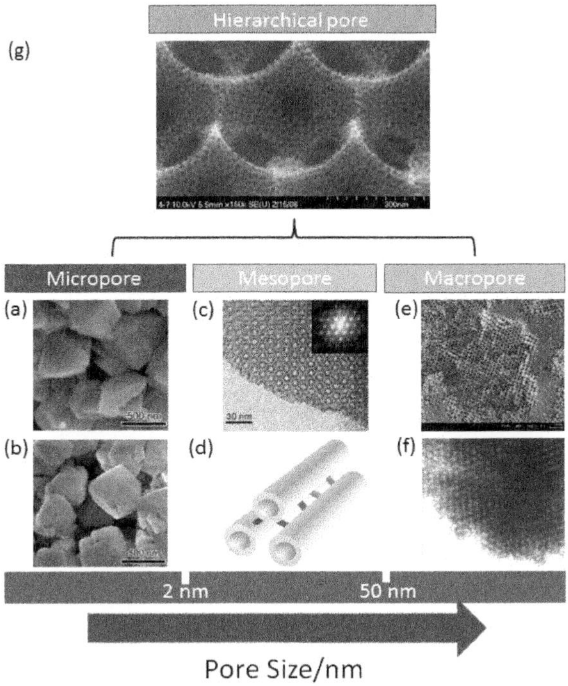

Figure 1.1. IUPAC classification of porous carbon structure: micropore (SEM image of zeolite Y (a) and A8(2) carbon (b)) [8], mesopore (TEM image (c) and illustration model (d) of ordered meso-nanoporous carbon) [9], macropore (SEM image (e) and TEM image (f) ordered macro-nanoporous carbon) [10] and hierarchical pore (SEM image (g) and pore size distribution (h) of hierarchical porous carbon [11].

the recent development of template carbons towards optimizing the electrochemical behaviour in fuel cells, batteries and supercapacitors.

2. Classification of Nano-porous Carbon Catalyst

There are three types of porous carbon materials according to pore sizes based on the International Union of Pure and Applied Chemistry (IUPAC) regulations: micro-porous < 2 nm, 2 nm < mesoporous < 50 nm and macro-porous > 50 nm (as shown in Fig. 1.1). Pore size of the nano-porous carbon materials has been tailored from micro-pore to macro-pore. Micro-porous carbons could allow high surface area to enhance reaction rate, but macro-porous carbon materials could facilitate mass transfer and diffusion of reactants and products in the reaction. Hierarchical structural carbon materials with combined properties will provide the opportunity for the large-scale production of carbon catalyst. Recent developments of the synthesis porous carbon materials with uniform pore sizes were summarized and this part was classified into three sections with regard to pore sizes: micro-porous, mesoporous and macro-porous.

2.1 Micro-porous carbon materials

Micro-porous carbon materials have been applied in many fields such as gas separation, electrochemical capacitor and shape-selective catalysts. Rigid inorganic templates are usually necessary for synthesizing micro-porous carbon materials with uniform pores and regular pore arrays. Zeolites with ordered and uniform pores are commonly considered as the inorganic template and utilized in molecular sieves and catalysts [12].

Kyotani et al. prepared porous carbon materials by using Y zeolite as a template and reported the template synthesis method. Poly(acrylonitrile) and poly(furfuryl alcohol) were used as the carbon precursor and were incorporated into the pores and channels of Y zeolite. Then after carbonization of the complex, the resultants were treated with acid to remove zeolite template to produce micro-porous carbon materials. Due to its high thermal stability, the Chemical Vapour Deposition (CVD) was also investigated as a method to introduce the carbon precursor (propylene gas) into the channels of zeolite, resulting in higher surface areas (over 2000 m^2/g) and larger pore volume than those of polymer carbons. Transmission Electron Microscope (TEM) and Scanning Electron Microscope (SEM) results showed the similar morphology of carbon/zeolite and original zeolite template, indicating carbonization occurred in the channel of zeolite. But irregular carbon framework was formed because the carbon layers were stacked together after acid washing with zeolite [13].

In addition, they also achieved a nitrogen-containing micro-porous carbon with well-ordered structure by using zeolite as an inorganic template. Two-step carbon filling process was operated through impregnation of furfuryl alcohol and the following chemical vapour deposition (CVD) of acetonitrile. The CVD process was necessary for the carbon-loading process in unoccupied channels after poly(furfuryl alcohol) carbonization. SEM image of A8(2) carbon in Fig. 1.1(b) showed smoother surface than zeolite Y (a) in Fig. 1.1(a), indicating the carbons were formed during the zeolite channels in acetonitrile CVD process and that the deposition on the external surface was not significant. The nitrogen-doped well-ordered micro-porous carbon material showed a stronger affinity to water molecular than a non-doped porous carbon material with similar pore structure, indicating nitrogen doping is responsible for the increase of hydrophobicity on the carbon surface [8].

Recently, an amine functionalized Al-MOF (metal-organic framework) was used to synthesize nitrogen-containing micro-porous carbon material with approximately 1 nm narrow pore size and large micro-porous surface area and volume through direct carbonization method by Zhu and co-workers. The doped nitrogen atoms in micro-porous carbon material derived from carbonation of amino groups from MOF. They reported that compared with commercial Pt/C (20 wt%), the synthesized materials showed higher surface area and specific nitrogen state, which induced the enhancement in the oxygen reduction performance [14]. Han and colleagues reported the one-step solvothermal carbonization of cyclodextrins to synthesize microporous carbon materials. Moderate sorption capabilities for hydrogen (about 1.07 wt%, 77 K and 1.0 bar) and carbon dioxide (approximately 12.7 wt%, 273 K and 1.0 bar) could be achieved by using these microporous carbon materials [15].

2.2 Mesoporous carbon materials

Many areas including dye adsorbents, catalysts support and electrodes have been involved in the use of mesoporous carbon materials [16–20]. Mesoporous silica materials were synthesized through the sol-gel method by using silica precursors with the help of a surfactant self-assembly by Mobil Corporation researchers in 1992 [20]. Changing the ratio of the surfactant to the silica precursor and the chain length of the surfactant could control the pore dimension and pore structure of mesoporous silica materials. These materials with uniform pore size, interconnected pore structure and high surface area have been considered as the inorganic templates for the preparation of mesoporous carbon materials. Ryoo's group reported that alumina silica material MCM 48 as template could be filled with carbon precursor such as sucrose and also could be polymerized with phenol resin. The following acid dissolution could lead to the removal of the template from the formed complex to produce mesoporous carbon material [21]. They also developed a general method synthesis for highly ordered mesoporous carbons with tuneable diameters by using ordered silica SBA 15 as templates. TEM image (Fig. 1.1c) and illustration model (Fig. 1.1d) of ordered meso-nanoporous carbon were also shown. Pt loaded ordered mesoporous carbon material showed superior electrocatalytic activity for oxygen reduction reaction than that loaded on carbon black due to the enhanced uniformity and the decreased size of Pt cluster [9].

The applicability of mesoporous carbon materials could be extended to biomolecular related area such as biosensing and separation because carbon could be used as an excellent support for nanoparticles [22]. For example, proteins could be adsorbed in the pores of mesoporous carbons and widening pore size could increase the protein capacitance [23]. Separation capability of biomoleculars could be improved by using well-ordered mesoporous carbons with high surface area [24]. To generate large pores (over 10 nm) in mesoporous material to reach biology-related goal and develop other applications such as electrodes of supercapacitors and pollutant adsorption, ultra-large mesoporous carbon material (about 27 nm) was synthesized by Hyeon's group by using meso-cellular aluminosilicate foams as template and phenol resin as carbon source [25].

In order to enhance electrical conductivity, mesoporous carbon materials with a large surface area and excellent graphitic crystallinity are needed. Fuertes and colleagues prepared mesoporous carbons with graphitic and large porous structures by using polypyrrole-incorporated mesoporous silica materials in the presence of $FeCl_3$. Iron salts not only played an important role in polymerization as an oxidant but also promoted the formation of a graphitic structure in carbonization stage. Graphic carbon materials as electrode in Electrical Double-Layer Capacitors (EDLCs) presented better charge-discharge capabilities than non-graphic materials, because graphic mesoporous carbon material provided large porosity and highly conductive framework to facilitate electron transfer [26].

Because mesoporous silica materials are usually of high-cost and large-scale unfeasible hard template, a facile, low cost and industrialized method has attracted a lot of attention from researchers. This method is sourced from the self-assembly of inorganic precursors and surfactants in a dilute surfactant water solution [27–30]. The formation of a periodically ordered organic-organic nanocomposite derived from thermosetting polymerized carbon material such as resol and the utilization of a thermally decomposable surfactant like Pluronic 123. For example, a well-ordered mesoporous carbon material with Ia3d symmetry was synthesized through direct assembly of resol and Pluronic 123 by Zhao et al. The formation in resol-block-copolymer mesophase in the dilute basic solution was caused by cooperative assembly. After carbonization, the remained mesoporous carbon materials had a uniform pore size (about 3 nm) and large surface area (nearly 1150 m^2/g) [31].

Additionally, hierarchically ordered mesoporous carbon materials have attracted a lot of interest and they could be synthesized by using hierarchically ordered mesoporous silica materials. Yoon et al. prepared carbon materials with hollow core and mesoporous shell structure by using solid core with mesoporous shell silica as hard template and resol or poly(divinylbenzene) as a carbon source. They also found that appropriate SCMS silica sphere templates could be used to control the diameter of the hollow core and the mesoporous shell thickness [32].

2.3 Macro-porous carbon materials

Macro-porous carbon materials with different structures could be achieved by using spherical silica particles as template. Yu et al. prepared well-ordered porous carbon materials with different morphologies through utilizing a colloidal crystal template and changing the acid sites for polymerization process of resol as a carbon precursor. Sulphuric acid was needed for silica particles to generate acid sites for polymerization, but strong acid sites existed on the surface of Al-impregnated silica particles. Therefore, the ordered nano-porous carbon formed through filling the whole void around the silica spheres, but Al-impregnated silica particles were considered as the template for polymerization of phenol and formaldehyde, leading to the formation of macro-porous carbons [33]. They also synthesized porous carbon-loaded Pt-Ru materials for methanol oxidation and found the prepared carbon materials showed enhanced activity than commercial E-TEK and Vulcan XC-72 supported Pt-Ru catalysts. This was derived from the high surface area of porous carbons to facilitate metal dispersion and the formation of 3D interconnected uniform macro-pores to improve product diffusion compared with commercial materials. Uniform macro-porous carbons with a pore diameter of 62 nm and BET surface area of 750 m^2/g were synthesized through using colloidal crystal template and resol as carbon precursor. SEM image and TEM image of macro-porous carbons were shown in Fig. 1.1(e) and 1.1(f), respectively. This material also presented type IV adsorption behaviour due to the well-regular macro-pores [10].

In addition, Anodic Aluminium Oxide (AAO) films could also be considered as the template to synthesize macro-porous carbon materials. Kyotani and coworkers synthesized carbon nanotubes by depositing propylene in the channels of AAO films and subsequently followed by thermal decomposition of propylene and acid etching. Adjusting current density and oxidation time could control the diameter and the length of the channels in AAO films. Additionally, the wall thickness could also be controlled by altering the carbon deposition period [34]. Hierarchically ordered macroporous carbon materials could also be achieved through hard template method with furfural alcohol as carbon sources. The macropores in the carbon materials, as shown in the SEM (Fig. 1.1g), were approximately 440 nm with connecting pores (about 160 nm) and mesopores in the wall (almost 20 nm).

3. Synthesis Methods of Nano-porous Carbon Materials

Nano-porous carbon materials with uniform pore sizes are very promising in the field of adsorption, super-capacitors, catalysts and biomedical applications. Most porous carbon materials were usually synthesized through carbonization of raw materials such as coal and wood [35]. Rigid structural inorganic materials such as silica, zeolite and metal-organic framework have been used as hard template to prepare porous carbons. The hard template method includes several steps (as shown in Fig. 1.2a [36]): (1) synthesis of hard template materials such as SBA 15, (2) preparation of

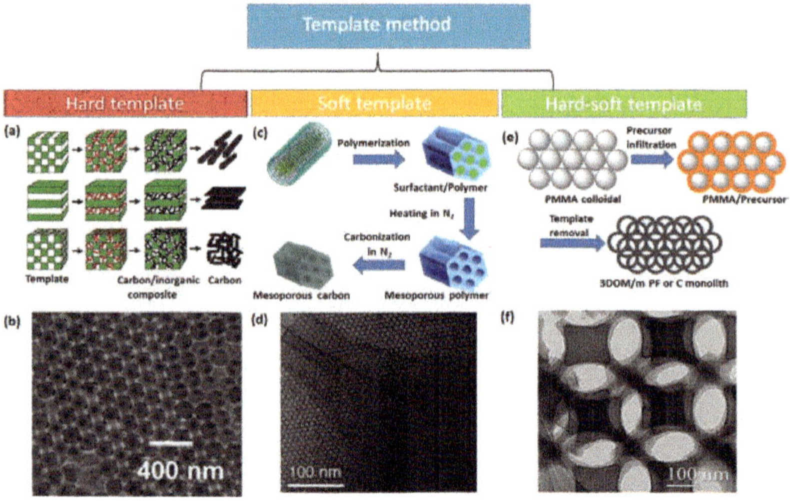

Figure 1.2. Comparison between hard template method and soft template method (a) illustration of hard template [36]; (b) TEM images of prepared porous carbon based on hard template method [37]; (c) illustration of soft template [38]; (d) TEM images of prepared porous carbon based on soft template method [38]; (e) illustration of dual template [39]; (f) TEM images of prepared porous carbon based on hard-soft template method [39].

carbon and hard template composite material, (3) subsequent carbonization of the formed material in an inert atmosphere, (4) the removal of hard template. A representative SEM image (Fig. 1.2b) showed three dimensionally ordered macroporous carbons (about 200 nm) with mesoporous walls [37]. In addition, preparation of porous carbons through carbonization of block polymers as soft template and resols has also been developed. Relatively fewer steps were needed for the soft template method (as shown in Fig. 1.2c): (1) preparation of the mixture of phenolic resin and block polymer, (2) the removal of block polymer (soft template), (3) carbonization of the formed mesoporous polymer in an inert condition. An example of well-ordered hexagonal carbons was shown in TEM image (Fig. 1.2d) [38]. Advantages and disadvantages between hard template method and soft template method are also summarized in Table 1.1. Recently, a combination technique between hard template method and soft template method was developed as well. Hard-soft template method usually involved several steps such as self-assembly of hard and soft template, incorporation with carbon precursors, carbonization of carbon/template complex and then removal of the template matrix as shown in Fig. 1.2e. The carbon materials derived from this method presented ordered macroporous/mesoporous structure (Fig. 1.2f) [39]. Here three main different synthetic template methods including hard template method, soft template method and hard-soft template method will be discussed and summarized.

Table 1.1. Advantages and disadvantages between hard template method and soft template method.

	Advantages	**Disadvantages**
Hard template	• Homogeneous and well-defined structure • Easy to control structure	• Multi-step • Template etching • Pore limitation • Mechanically unstable
Soft template	• Tailored architectures • Tuneable surface properties • Mechanically stable • Low cost • Convenient • Large-scale production	• pH-dependent • Temperature-control • Surfactant-directed

3.1 The hard template method

Ordered mesoporous carbons with controlled porous structure have been synthesized with the help from hard template such as mesoporous silica materials. This kind of material has attracted intense interests due to the advantageous functionalities such as high specific surface area, uniform pore diameters, large adsorption capacities and high thermal and mechanical stabilities [40]. Therefore, great attention has been paid on advancement of catalysts, hydrogen storage, gas separation, super-capacitors and fuel cell electrodes through utilization of well-regular mesoporous carbons [9]. Various structures and pore diameters have also been designed based on silica template (Table 1.2). Additionally, tailoring pore diameters become quite important to meet the application requirement. But this method is quite complicated, high-priced and difficult for large-scale production [20]. In addition, by using silica template, morphologies and nanostructures of carbon materials are usually restricted. Due to aggregation and cross-linking tendency of silica templates and carbon precursors, it is difficult to obtain small mesoporous carbon nanoparticles through the hard template method [41]. Mesoporous carbon materials synthesized through the hard template method are summarized below.

3.2 The soft template method

In addition to hard template method, mesoporous carbon materials with various structures could be synthesized through an organic-organic assembly method so called the soft template method by using amphiphilic triblock copolymers as templates and phenolic resols as carbon precursors [38, 50–53]. The preparation of mesoporous materials includes two ways: (1) Evaporation Induced Self-Assembly (EISA) method, (2) solvent thermal method [54, 55] (Table 1.3). The EISA method is used to synthesize

Table 1.2. Synthesis of nano-porous carbon materials through hard template method.

Template	Carbon precursor	Porous structure	Pore size (nm)	Pore volume (m^3/g)	Surface area (m^2/g)	Application	Reference
Monodisperse silica particles	Phenol and formaldehyde	Macroporous	62	1.68	750		[10]
Silica synthesized from monodisperse polystyrene (PS) spheres as a template	Furfuryl alcohol	Hierarchical meso-macroporous	17–23, 150–170, 430–450	1.93	962	Electrocatalytic oxygen reduction	[11]
SBA-15	Sucrose	Meso-porous	4.9–8.8	1.4–1.9	1378–1583		[22]
Colloidal silica crystal	Phenol and formaldehyde	macroporous	200		706	Methanol fuel cell	[33]
Colloidal silica	Monodisperse polystyrene	Three dimensionally ordered macroporous (3DOM) carbons with mesoporous walls	6–90	2.1–2.9	490–1450	Electric double-layer capacitors	[37]
SBA-15	Sucrose	Hexagonally meso-porous	2.2–3.3				[40]
SBA-15 and boric acid as pore expanding agent	Sucrose	Meso-porous	3.8–10.5	0.36–0.56	850–1340		[42]
SiO_2 particles (16.8 nm–39 nm)	Sucrose	Meso-porous	12.4–34.5	2.4–3.7	1130–1221		[43]
Meso-macroporous Silica monolith	Furfuryl alcohol	Meso-porous	4.3	1.27	1115		[44]
Micro-meso-macroporous zirconia	Sucrose	Meso-macroporous	10–17, and 25–50	0.44	950		[45]
Macroporous/mesoporous silica	Phenolic resin	Hierarchical macroporous	2.8, 286	0.93	1261		[46]

MgO	Poly(vinyl alcohol) (PVA), hydroxyl propyl cellulose (HPC) and poly(ethylene terephthalate) (PET)	Hierarchical meso-macroporous	2–25		300–1800	Electric double layer capacitor	[47]
$Ni(OH)_2$	Phenolic resin	3D hierarchical porous texture with macroporous cores, mesoporous walls, and micropores	1–2, 5–50 and 60–100	0.69	970	Electrochemical performance	[48]
Silica microspheres	Sucrose and cyanamide	Macroporous	150		97	Electrochemical oxygen reduction	[49]

Table 1.3. Synthesis of nano-porous carbon materials through soft template method.

Template	Carbon precursor	Porous structure	Pore size (nm)	Pore volume (m^3/g)	Surface area (m^2/g)	Application	Reference
Pluronic F127	Phenol/formaldehyde	Mesoporous	6.8	0.51	590		[38]
Polystyrene-block-poly (4-vinylpyridine)	Resorcinol monomers/ formaldehyde vapor	Mesoporous	33.7 ± 2.5				[50]
Pluronic F127	Resorcinol/formaldehyde	Mesoporous	5.9–7.4	0.74–0.81	624–1354		[51]
Pluronic F127	Phloroglucinol/formaldehyde	Mesoporous	5.4–9.5		378–569		[52]
Pluronic F127	Phenol/formaldehyde	Mesoporous	2.8	0.38	750		[53]
Pluronic F127	Resorcinol/formaldehyde	Mesoporous	5.0	0.46	675		[54]
Pluronic F127/P123	Phenolic resols	Mesoporous	3.0	0.36	620		[55]
Pluronic P123	Phenol/formaldehyde	Mesoporous					[57]
Poly(ethylene oxide)-block-poly(methyl methacrylate)-block-polystyrene	Resol	Mesoporous	18–24	0/28–0.86	500–1400		[58]
Poly(ethylene oxide)-block-poly(styrene)	Resol	Mesoporous	32–54	0.7	1000	Magnetization	[59]
Poly(ethylene oxide)-b-polystyrene and homopolystyrene as a pore expander	Resol	Mesoporous	26–90	0.45–1.1	350–1210		[60]
Pluronic F127	Phenolic resol	Mesoporous	2.6–3.0	1.1–1.5	940–1130	cell permeability and drug adsorption capacity	[61]
F123, CTAB	3-aminophenol, formaldehyde	A mesoporous interior with a microporous shell		0.19–0.23	343–394	Electrochemical oxygen reduction	[62]
FC4 and Pluronic F127	Resorcinol and formaldehyde	Mesoporous	3.5	0.34, 0.45	640, 857		[63]

mesoporous carbon films and monolithic and solvent thermal method usually produces carbon powders with various particle sizes. There are four important conditions for synthesizing mesoporous carbon materials through soft template method. Firstly, the soft template precursors should own the ability to assemble themselves into nanostructures. Secondly, carbon-generating and pore-forming precursors are necessary. Then, the pore-forming precursor should tolerate the temperature of the carbon formation, but also should be easily decomposed during the carbonization period. Lastly, the carbon-generating precursors should form a highly cross-linked polymeric material and have thermosetting capability to reach the decomposition temperature that the pore-forming precursors need [56]. Some of the related mesoporous carbon materials prepared through soft template method are summarized below and will provide some new thoughts for the future development of mesoporous carbon materials.

3.3 The hard-soft template method

In order to reduce mass transportation derived from long diffusion path lengths, carbon materials with a well-ordered and hierarchically structure containing macropores and mesopores were designed through the hard-soft template method [39] (Table 1.4). By using colloidal crystal template, macro-pores with a three-dimensionally ordered structure can be produced [64, 65]. At the same time, mesoporous structures could be achieved with the assistance of soft template such as Pluronic F127 and P123. Wang et al. reported that this method involved several steps: (1) incorporation polymer spheres with mono-dispersed silica particles, (2) calcination of silica/polymer complex to generate three-dimensionally interconnected and well-ordered macroporous structure with aggregated silica nanoparticles, (3) mixing porous-silica framework with carbon precursors, (4) carbonization of carbon/silica precursor and the subsequent etching of silica particles. They also found the electrochemical activity was enhanced in methanol fuel cell when using hierarchical porous carbons as catalyst support due to the addition of mesoporous structure into the three-dimensionally ordered macroporous structure [39]. In addition, Zhao et al. also used silica colloidal crystals and triblock copolymers PEO-PPO-PEO as dual templates to synthesize hierarchically ordered macro-/mesoporous carbon products [66]. Mesoporous carbon materials synthesized through dual template method are summarized below.

Table 1.4. Synthesis of nano-porous carbon materials through dual template method.

Template	Carbon precursor	Porous structure	Pore size (nm)	Pore volume (m^3/g)	Surface area (m^2/g)	Application	Reference
Poly(methyl methacrylate) (PMMA) colloidal crystals, Pluronic F127	Phenol-formaldehyde	Hierarchically ordered macro-/mesoporous	3, 340	0.35	~500	depth-sensing indentation	[39]
Monodispersed silica colloidal crystals, amphiphilic triblock copolymer PEO-PPO-PEO	Resol	Hierarchically ordered macro-/mesoporous	11, 230–430	1.25	760		[66]
SiO_2 opal, amphiphilic triblock copolymer PEO-PPO-PEO	Phenolic resin	Hierarchically ordered macro-/mesoporous	15–25, 300–400	0.92–4.80	650–1660		[67]

4. Factors for Catalytic Activity of Carbon Materials

4.1 Porous structure

Great efforts have been made on controlling pore size, pore length, pore orientation and pore morphology to enhance catalytic performance of carbon materials. While macropores can provide easy access to the active sites or larger molecules and fast mass transfer, size and shape-selectivity for guest molecules, large surface area and large pore volume can be offered by micropore materials [68]. Porous structure of carbon material can influence the energy of electro-catalytic reactions and electrochemical corrosion. The effect of porous structure was discussed in detail earlier.

4.2 Heteroatom-doped effect

In order to design cheap and stable nanoporous carbon materials with novel electrochemical and physical properties, heteroatom-doped porous carbons including monodoped atoms such as boron (B), fluorine (F), sulphur (S), nitrogen (N) and phosphorus (P) and dual doped atoms such as boron with nitrogen and nitrogen with sulphur have been reported. The synthesis methods of doped porous carbons and their catalytic performances will be summarized and discussed below.

4.2.1 Boron-doped

Boron is a unique element that has been reported as a replacement in carbon framework to reinforce electrochemical performance [73]. Because the trigonal sites in carbon framework can be substituted by heteroatom boron atoms which could be considered as electron acceptor due to three outer electrons, the electronic band structure of the carbon framework could be modified [74]. This kind of low-level boron doped material also facilitates oxygen chemisorption on carbon surface and assists redox reactions [75, 76]. Therefore, boron doping can be used to modify electrochemical behaviour and metal-free B-doped ordered mesoporous carbons have also been designed in the density of states field of oxygen reduction reaction [69, 77–79]. B-doped ordered mesoporous carbons could be achieved by impregnation of sucrose, 4-hydroxyphenylboronic acid and SBA-15 silica template and the followed carbonization and etching process. As shown in Fig. 1.3a, catalytic current for this material increased with rotation rate. Both the boron content and surface area affected the catalytic activity for oxygen reduction reaction in alkaline medium. This B-doped mesoporous carbon material also showed higher catalytic selectivity against CO poisoning, superior methanol tolerance and longer stability than commercial Pt/C

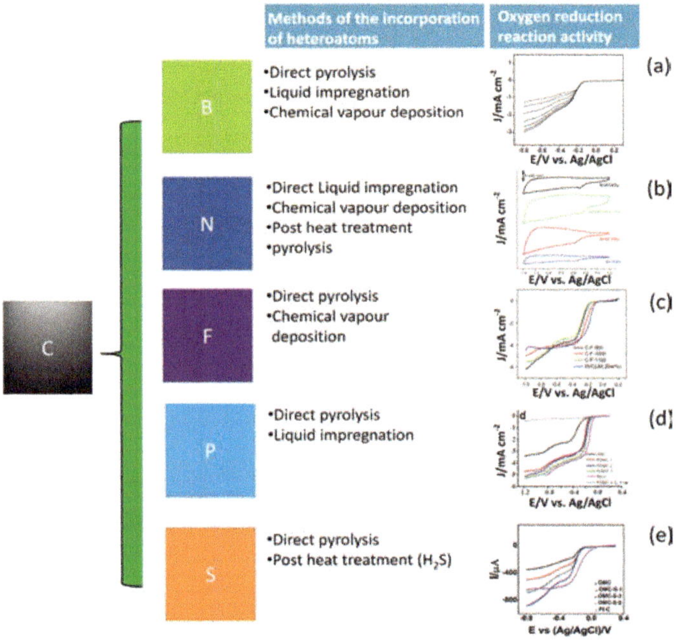

Figure 1.3. Methods of the incorporation of heteroatoms and the example of the correspondent heteroatom doped porous carbon materials for oxygen reduction reaction (ORR) activity. (a) Linear sweep voltammetric curves at different rotation rates for boron-doped ordered mesoporous carbon materials [69]. (b) Linear sweep voltammetric curves for various nitrogen-doped carbon spheres at 1600 rpm [62]. (c) ORR polarization curves for different fluorine-doped mesoporous carbon and Pt/C [70]. (d) Linear sweep voltammetric curves for phosphorus-doped and un-doped ordered mesoporous carbons at different rotation speeds [71]. (e) Rotating Disc Electrode (RDE) voltammetric curves for various sulphur-doped ordered mesoporous carbons [72].

[69]. He et al. reported the synthesis of ordered boron-doped carbon films with resol as carbon precursor and boric acid as boron precursor via soft template (tri-block polymer F123) method. Graphitization degree of the boron-containing carbon material was increased through the doped method. The prepared boron-doped carbons also presented more corrosion-resistant, hydrophobic and electrically conductive than unmodified carbons [80].

4.2.2 Nitrogen-doped

N-doped carbon materials have also attracted a lot of attention in recent years. Because nitrogen atoms have higher electro negativity and smaller atomic diameter than carbon atoms, large quantities of defects could be derived from nitrogen doping to provide active sites and further enhance the interaction between nitrogen-doped carbons with reactants. Three different ways are usually used to prepare metal-free N-doped carbon catalysts. The first method is direct pyrolysis of nitrogen-containing precursors such as melamine foam, carbon nitride and polymer framework. For example, Yang et al. synthesized N-doped mesoporous carbon spheres with tunable particle size from 40 to 750 nm through carbonization of nitrogen-containing phenol resin precursors in inert atmosphere. As shown in Fig. 1.3b, compared with undoped mesoporous carbons, higher specific capacitance was shown for N-doped mesoporous carbons because of the activation of Faradic process in carbons. N-doped mesoporous hollow carbon spheres presented the highest capacitance among these doped carbons due to mesoporous structure, highest surface area and pore volume [62]. The second approach is *in situ* incorporation of doped nitrogen atom into the carbon framework via liquid impregnation or chemical vapour deposition (CVD) [81–85]. Xia et al. reported the synthesis of N-doped mesoporous carbons via carbonization of SBA-15 in acetonitrile-saturated nitrogen atmosphere with ordered mesoporous structure, graphitic pore walls and hollow sphere morphology [81]. They also developed this chemical vapour deposition method to synthesize N-doped mesoporous carbon by using various porous silica such as SBA-12, SBA-15, MCM-48, MCM-41 and hexagonal mesoporous silica. Ordered mesoporous carbon materials could not be achieved by using MCM-41 and hexagonal mesoporous silica as templates because of the lack of pore connectivity in their pore channel system. The third way is post heat treatment of carbon materials in ammonia or polyaniline at high temperature [86–88]. Rafael et al. synthesized mesoporous nitrogen-doped carbons through *in situ* polymerization of polyaniline in the pore channel of SBA-15, followed by calcination of polyaniline and SBA-15 composite materials and etching process. Excellent electrocatalytic activity toward oxygen reduction reaction was also presented by using these obtained N-doped mesoporous carbons [87].

4.2.3 Fluorine-doped

Structural and chemical properties of mesoporous carbon materials can be modified through fluorination. The commonly-used method is chemical vapour deposition such as fluorine gas [89, 90]. For example, Dai et al. synthesized fluorinated carbon with an ordered mesoporous structure through reaction between carbon material and fluorine gas [89]. But the problem of non uniform modification and pore blockage may cause trouble for post-synthesis because of the inert carbon surface. Mesoporous structure may be destroyed under rigorous conditions [91]. Therefore, highly ordered fluorinated mesoporous carbons were synthesized by the one-pot soft template approach using phenol-formaldehyde/p-fluorophenol as carbon precursors. This material sustained well-ordered porous structure with large pore size (4.4 nm), large pore volume (0.7 cm^3/g) and high surface area (998 m^2/g) and C-F covalent bonds were maintained at 900°C. A higher electron transfer rate was also observed by using the synthesized fluorinated mesoporous carbon materials than pure unmodified carbons [91]. Electrocatalytic activity toward oxygen reduction reaction was also evaluated by using F-doped mesoporous carbon materials through hard template method. This material showed comparable activity to Pt/C (as shown in Fig. 1.3c) due to high surface area, large active sites and partly graphitization degree [70]. In addition, mesoporous carbons with higher F/C ratio (0.81) were prepared through soft template and subsequent fluorination method. The diameter of the prepared carbon material can reach 11 nm and the specific area can also approach 850 m^2/g after fluorination. Moreover, compared with commercial carbon fluorides with comparable fluorine content, the synthesized carbons showed better performance in Li/CFx batteries such as higher discharge potentials, energy and power densities and faster reaction kinetics when applied in high current densities [92].

4.2.4 Phosphorus-doped

Phosphorous atoms possess similar chemical properties as nitrogen atoms due to the same number of valence electrons, but larger atomic radius and higher electron-donating capability of phosphorous atoms could induce enhanced catalytic activity and durability [71]. Recently, the phosphorus-doped porous carbon materials were reported to exhibit superior electrochemical performances than non-modified materials as super-capacitor electrodes [71, 77, 93]. For instance, Yu et al. reported the synthesis phosphorus-incorporated ordered mesoporous carbons through co-pyrolyzing a phosphorus precursor and a carbon precursor with SBA 15 as a hard template. This metal-free catalyst showed superior electro-catalytic activities long term stability and resistance to alcohol crossover effects for oxygen reduction reaction (ORR) in alkaline conditions compared with Pt/C (Fig. 1.3d). This was attributed to the defects in carbon framework, leading to enhanced electron delocalization and formed active sites. By shortening the length of doped carbons, the increased ORR activity was observed due to larger surface area and reduced resistance of shorter channels [71].

4.2.5 Sulphur-doped

Sulphur atoms have similar electro negativity with carbon atoms and could also be incorporated into carbon matrix through *in situ* doping such as direct carbonization of sulphur-containing species (benzyldisulfide [72], 2-thiophenemethanol [94–97], p-toluenesulfonic acid [98], 4,4'-thiodiphenol

[99]) and post treatment such as calcination in H_2S [100] or SO_2 [101] at a high temperature. S-doped mesoporous carbon was prepared by infiltration and polymerization of 2-thiophenemethanol in SBA-15 template and the following carbonization and HF etching process. Compared with commercial available sulfur-impregnated activated carbon, this material presented higher mercury saturation binding capacity and faster sorption kinetic of mercury adsorption in aqueous solution. This was probably derived from high sulphur concentration, easy accessibility and lack of blockage in this carbon material. Additionally, Guo et al. reported the synthesis of sulphur-doped ordered mesoporous carbon materials by using SBA-15 as a hard template, sucrose as a carbon source, benzyldisulfide as a sulphur source. The RDE voltammograms (Fig. 1.3e) clearly showed that all of the S-doped ordered mesoporous carbons have higher currents than undoped materials, indicating that the enhanced electrocatalytic performance was attributed to the formation of covalent bond between carbon and sulphur [72].

4.2.6 Dual doped

Dual doped or multi-doped carbon materials have also been developed and the synergistic effect of co-doped heteroatoms (Fig. 1.4) has led to enhanced catalytic activity [77, 78, 102–104]. N and P dual-doped hierarchical porous carbon materials were first synthesized by using poly (vinyl alcohol)/ polystyrene hydrogel as template, cyanamide as nitrogen source and phosphoric acid as phosphor source. This material with hierarchical interconnected macroporous structure (Fig. 1.4b) exhibited excellent electrocatalytic activity towards oxygen reduction reaction in acid, basic and neutral media. Figure 1.4b showed that better performance of methanol tolerance and higher stability were also exhibited compared with Pt/C. Dual doping provided large asymmetrical spin and charge density and the macroporous/mesoporous structure also facilitated the electron and reactant transport and large surface area [103]. Huang et al. reported the synthesis of sulphur and nitrogen co-doped carbon foam with hierarchical pore structure (Fig. 1.4c) with sucrose as carbon source, thiourea as nitrogen and sulphur source through hard template (SBA) method. They found the mass ratio between sucrose and thiourea was not only responsible for the morphologies and structures of the final carbon materials, but also played a key role in the difference of electrochemical performance. Compared with commercial Pt/C, the carbon foam (sucrose:thiourea = 1:4) exhibited higher electrochemical activity and better methanol tolerance, as shown in Fig. 1.4c. This resulted from the synergistic effect via co-doping, high graphic degree caused by high temperature, premium reactant transport from hierarchical porous structure, excellent electron transfer by three dimensional carbon

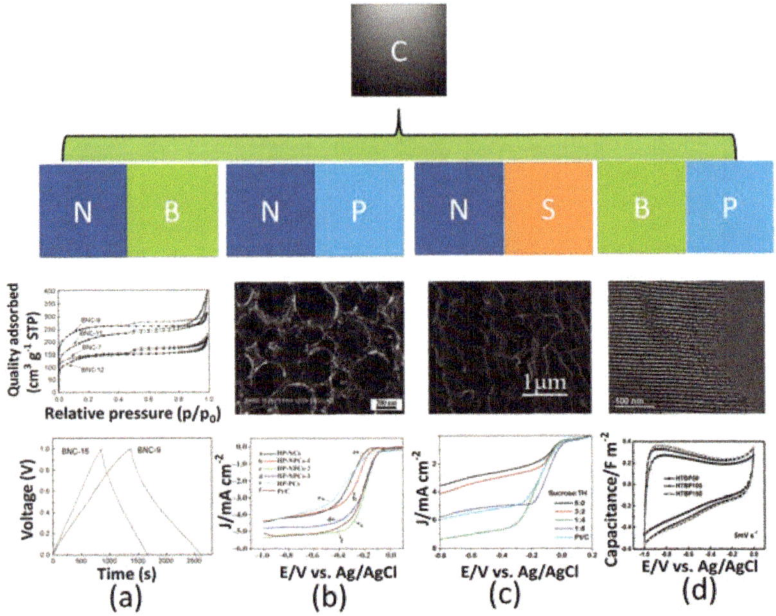

Figure 1.4. Summary of dual doped porous carbon materials. (a) Adsorption–desorption isotherms and voltage/time plots of the boron and nitrogen co-doped porous carbons [102]. (b) SEM image and linear sweep voltammograms of nitrogen and phosphorous doped porous carbons [103]. (c) TEM image and linear sweep voltammograms of nitrogen and sulphur doped porous carbons [104]. (d) TEM image and specific capacitance cyclic voltammetry (CV) curves of nitrogen and sulphur doped porous carbons [78].

frameworks and the bonding configuration [104]. Boron and phosphor dual doped ordered mesoporous carbon with high homogenous distribution of heteroatoms were achieved through a facile hydrothermal self-assembly method and the following carbonization process. The size of the pores in dual doped carbons, measured by the TEM image (Fig. 1.4d), was about 12 nm. When applied as supercapacitor electrodes, the supercapacitance increased with the hydrothermal temperature (Fig. 1.4d) due to the increased acidity and the improved surface polarity and the three materials showed similar symmetric shapes and CV curves.

4.3 Surface modification

The modification of the surface of porous carbon material is very significant because carbon materials are usually used as catalysts support to enhance application behaviour. There are two main ways to functionalize the surface of carbons: deposition nanoparticles and functional group modification.

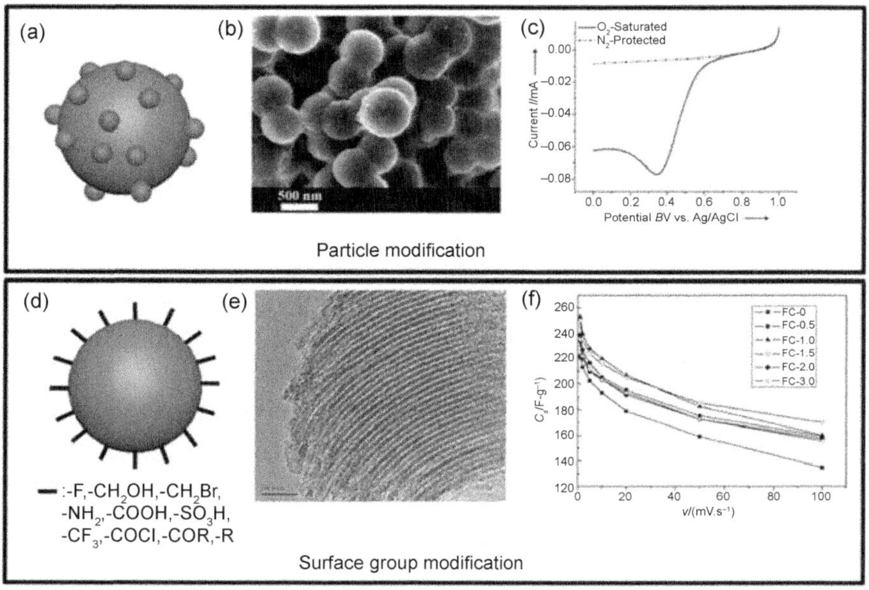

Figure 1.5. Surface modification of porous carbon materials (a) illustration of particle modified carbons; (b) TEM image of Pt/carbon spheres [105]; (c) linear-sweep voltammograms of Pt/carbon spheres for the oxygen reduction reaction (ORR) in an O_2-saturated and O_2-free (under N atmosphere) 0.5 m H_2SO_4 solution at room temperature at a scan rate of 10 mVs^{-1} [105]; (d) illustration of surface group modified carbons; (e) (f) TEM image and cyclic voltammogram of mesoporous carbon CMK-3 after oxidation in HNO_3 [106].

4.3.1 Nanoparticle modification

Nanoparticles deposition on the surface of carbons is often considered as a very effective method to adjust the interfacial properties and improve the interaction among molecules. An illustration of particle modified carbons is shown in Fig. 1.5a. Due to the reliability of surface modification in electrical process, electron conductivity and electrochemical activity can be improved through incorporation of nanoparticles on the surface of porous carbon materials.

Many methodologies have been developed to modify carbon materials through combining the secondary phase within the carbon structure. Two main ways are commonly used to incorporate secondary particles into carbon framework: impregnation metal salt precursors with carbon framework and incorporation hard or soft template with metal salt precursors with the following template removal process. The first technique is direct loading of nanoparticles on the surface of carbon framework through impregnation methods. Due to cheap and simple synthesis, impregnation method is widely utilized. After mixing metal salt precursors with carbon frameworks, drying and calcination process, metal oxides such as iron oxide, copper oxide, nickel oxide, cobalt oxide, manganese oxide and zinc oxide are loaded on the surface of porous carbons [107–111]. Zhao et al. synthesized SnO loaded ordered mesoporous carbon materials through deposition method by using $SnCl_2$ and phosphorus ester $OP(OCH_3)$. Compared with other hydrophilic phosphorus precursors such as PCl_3 and H_3PO_4, the use of hydrophobic $OP(OCH_3)$ facilitates precursor impregnation and uniform formation of SnO. The synthesized material presented better cycle-ability as negative electrodes for lithium-ion batteries than those discrete SnO_2 nanoparticles [112]. In addition, Pt nanoparticles (about 6 nm) could be loaded on the surface of carbon spheres (Fig. 1.5b) derived from monodisperse resorcinol and formaldehyde (RF) resin polymer. Figure 1.5c showed that an onset potential about 0.6 V in linear sweep voltammograms under the O_2 atmosphere for the oxygen reduction reaction was presented by using this composite material, indicating carbon spheres could become an excellent candidate of support materials [105]. The second method is a direct, one-port, co-assembly method through incorporation of carbon precursor, metal precursor and hard template such as SBA-15 or soft template such as block copolymer. Jaroniec et al. prepared mesoporous carbon materials with ultra-thin pore walls and highly dispersed nickel nanoparticles. They used SBA-15 as a hard template, nickel nitrate hexahydrate [$Ni(NO_3)_2{\cdot}6H_2O$] as a nickel source and 2,3-dihydroxynaphithalene (DHN) as a carbon source. Additionally, the nickel nanoparticles with uniform size about 3 nm were homogeneously distributed on the tubular carbon walls [113]. For example, Zhao et al. synthesized ordered mesoporous carbon materials with iron-oxide nanoparticles through a chelate-assisted multicomponent co-assembly method by using acetyl-acetone as a chelating precursor, iron nitrates as metal precursor and phenolic resol as a carbon source and copolymer Pluronic F127 as soft template. 2-D hexagonally pore structure with uniform pore size (about 4 nm) and high surface area (approximately 500 m^2/g) and highly dispersed iron oxide nanoparticles were generated after carbonization. They also found the amount of acetyl-acetone could control the particle size of iron oxide and the synthesized iron/carbon material showed excellent catalytic performance, high stability and high selectivity in Fischer-Tropsch Synthesis due to high porosity of carbon framework and semi exposed structure of iron-oxide nanoparticles [114].

4.3.2 Functional group modification

Functional groups on the surface of carbon materials are treated as anchoring sites for metal catalysts and can assist the metal adsorption on the carbon surface. This usually leads to high dispersion of metallic species on the surface and enhanced chemical performance in fuel cells [115]. Five main methods are usually used for functional group modification of carbon surface involving oxidation, KOH activation, sulphonation, halogenation, polymer decoration and grafting. An illustration of surface group modified carbons is shown in Fig. 1.5d.

Oxygen containing functional groups such as anhydride, carboxylic acid, ether, ketone, phenol are introduced to carbon surface through covalent, electrostatic and hydrogen bonding interactions, resulting in the enhanced wettability of the pores by polar solvents and electrochemical capacitance of porous carbon material [116]. Oxidizing gases such as air oxygen and ozone and oxidizing solutions such as nitric acid and hydrogen oxide are usually used for oxidation of carbon surface [117, 118]. Because excessive oxidation results in structural collapse of carbon framework, the oxidation conditions must be chosen carefully. For example, carboxyl-group and hydroxyl-group modified mesoporous carbons were synthesized through oxidation treatment of CMK-3 in nitric acid solutions. The ordered mesoporous carbon structures were maintained after acid treatment, as confirmed from TEM image in Fig. 1.5e. It could be also clearly seen from Fig. 1.5f that the capacitance of all the materials decreased with the increase of scan rate and FC-1.5 exhibited the highest specific capacitance among the test samples, indicating the optimum oxidation time is 1.5 hour [106]. In addition, functionalization by concentrated treatment acid induced the formation of oxygen-containing functional groups, resulting in strong binding between Pt nanoparticles and carbon surface and further improved electro-catalytic activity in methanol fuel cells [119]. Surface oxidation was also considered to be the cause of micropore generation in shorter oxidation time, leading to increased surface area, pore size and better capacitance [106]. But the electronic conductivity of porous carbons can also be decreased via surface oxidation, which is disadvantageous when porous carbons were used as electrodes in the area of fuel cell [46].

The carbon surface can be modified through functional group grafting method via organic chemical reaction. Firstly, after introduction of carboxyl groups, acyl chloride groups, amino groups, hydroxyl groups, bromine groups, thiol groups could be converted through organic chemical reaction such as substitution, esterification, reduction and hydrolysis reaction [120–123]. For example, nitric acid was used to treat microporous carbons and macroporous carbons to produce carboxyl groups. After that, the functionalized carbons were treated with thionyl chloride ($SOCl_2$), followed by immobilization of diamine compounds (ethylenediamine (EDA) and hexamethylenediamine (HMDA)). Although the BET surface area of amino-immobilized microporous carbon decreased largely compared with unfunctionalized microporous carbon, mesoporous carbon showed similar BET surface area and pore volume after functionalized diamine [122]. Secondly, in order to maintain carbon frameworks and surface smoothness in the process of carbon oxidation, reductive compounds such as diazonium salts, isoamyl nitrite or hydrophosphorus acid have been used directly to functionalize the carbon surface [124–128]. For example, Pinson et al. grafted aryl groups on the carbon surface through electrochemical reduction of diazonium salts and this method could be used to covalently attached a large number of diazonium salts. Alternation of the texture of carbon surface can also occur through KOH activation [129], sulphonation [130, 131], halogenation [89, 132], polymer decoration [133] method.

4.4 Graphitization degree

Graphitic carbon materials have various advantages such as high thermal and chemical stability, great electronic conductivity and excellent field emission behaviour [134]. Therefore, exceptional graphitization degree for nanoporous carbon materials can enhance their performances in direct methanol fuel cell, lithium ion battery and electrochemical catalysts [135–137]. Recently, a large number of porous carbon materials with a graphitic structure have been synthesized through four main different methods.

Firstly, to generate porous carbons with graphitic structure, carbon precursors such as mesophase pitch, acenaphthene, furfuryl alcohol have been employed to infiltrate the pores of mesoporous templates [138–141]. For example, Johnson et al. reported that mesoporous carbon prepared from Furfural Alcohol (FA) showed higher graphitization degree than that derived from Mesophase Pitch (MP). It indicated that FA-made carbons presented higher electrochemical stability in oxygen reduction reaction than MP-made carbons [138]. In addition, the effect of thermal treatment and different carbon precursors for graphitization degree were investigated. Heat treatment resulted in a gradual shrinking of structure and fracturing of pore walls and the carbons prepared from polyaromatic and acenaphthene and mesophase presented higher graphitic degree than that synthesized from furfuryl alcohol. But time-consuming and repeated infiltration and polymerization process is necessary to obtain ordered carbon materials. In addition, during high-temperature carbonization, a great deal of small molecules such as water will be emitted, deteriorating the pore structure of the template and the final carbon products [142, 143].

Secondly, the catalytic graphitization is an effective method to synthesize graphitic carbon with high crystallinity in mild conditions. With the help of catalysts (Fe, Ni and Co), relatively low pyrolysis temperature will be needed to prepare porous carbon materials with high graphitic structure by using mesoporous silica incorporated with metal salts as templates [26, 136, 144–151]. For example, a solid-state method by using metal pthalocyannines and SBA 15 was employed to prepare highly graphitic ordered mesoporous carbon materials. The degree of graphitization was enhanced by metal catalysts and the pore and graphitic structure remained complete during heat treatment, leading to high oxidative stability and capacitance [151]. In addition, the carbon materials synthesized from soft template method showed high graphitization degree in the presence of metal salts under 900°C heat treatment, resulting in superior capacitive performance (155 F/g) over a wide range of scan rates, even up to 200 mV/s. But the main disadvantage for this method is that synthesis of template is usually a time-consuming and high-cost process.

Thirdly, chemical vapour deposition is based on incorporating gaseous carbon precursors into silica template with the advantages of high degree of pore filling, simple control of pyrolytic carbon and easy formation of graphitic pore walls [152–154]. For example, two-dimensional hexagonally ordered mesoporous CMK-5 carbon materials were prepared by chemical vapour deposition of ferrocene in soft template SBA-15. Graphitization degree of CMK-5 was improved below the pyrolysis temperature of 850°C, but partial collapse of ordered CMK-5 occurred at above 850°C with the formation of entangled graphitic ribbons [153]. But the use of the template is still a shortcoming of this process.

Lastly, incorporation of ordered mesoporous carbon materials with other highly conductive carbon materials such as reduced graphene oxide can also enhance the graphitization degree [155–157]. For example, composite materials containing Ordered Mesoporous Carbons (OMC) and Reduced Graphene Oxide (RGO) were prepared through a solvent Evaporation-Induced Self-Assembly (EISA) method with Pluronic F127 as template. Ordered porous carbons were considered as supporting framework, avoiding the agglomeration of RGO and RGO could be connected by dispersed carbon nanoparticles to facilitate the electron transfer. Compared with the Pt loading on OMC-RGO, the amount of Pt loading on OMC was much less due to less functionalized groups for adsorption of the platinum ions on the surface of OMC [155].

4.5 Morphology effect

In addition to porous structure, heteroatom-doped effect, surface modification, graphitization effect, carbon nanoparticles with controlled morphologies also have great advantages on adsoption, catalysts, drug delivery and energy storage/conversion. These were derived from porous structure, exceptional properties of meso-channels and quantum effect on nanoscales. Many efforts have been made to design carbon spheres, hollow spheres, York-shell structure, fibres, rods and monoliths as shown in Fig. 1.6.

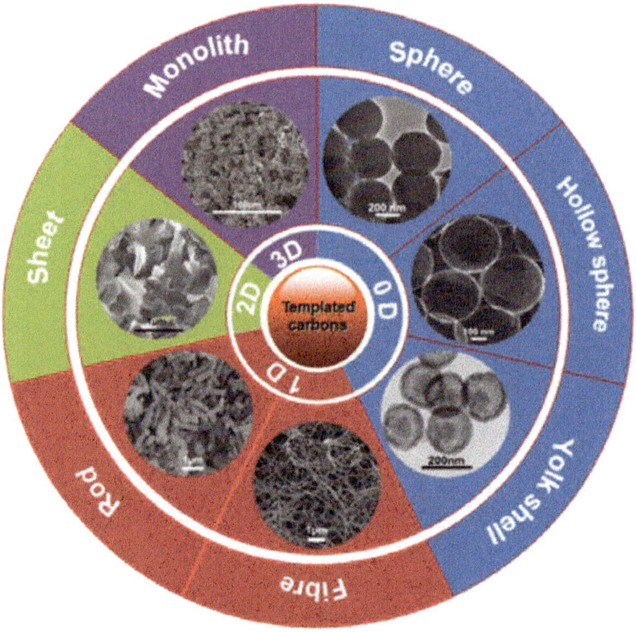

Figure 1.6. Tuned carbon nanostructures: (a) spheres [105], (b) hollow spheres [158], (c) york-shell nanoparticles [159], (d) fibres [160], (e) rods [22], (f) sheets [161], (g) monoliths [162].

4.5.1 Spheres

Hard template method by filling and etching silica template and soft template method by organic-inorganic self-assembly are two main techniques for preparation of carbon spheres. Compared with hard template method, fewer synthetic steps, less time and expenses were required for soft template method. Recently, the extension of the Stober method was used for the preparation of monodisperse resorcinol and formaldehyde (RF) resin polymer spheres with uniform and controllable particle size and the following carbonization induced the formation of mesoporous carbon [105]. This efficient, facile, general and environmentally-friendly method revealed that various different morphological carbons could be designed and synthesized. In addition, highly ordered mesoporous polymers with tuneable particle size were prepared through soft template method and porous carbon spheres could also be achieved through carbonization of the polymer spheres. This carbon material was further investigated as a cathode material for Li-S batteries and showed high initial discharge capability and good recyclability [63].

4.5.2 Hollow spheres

Hard template method was widely used to prepare hollow carbon spheres. A carbon precursor was usually coated on the surface of template core for the formation of core-shell structure and the following carbonization and core-removing process lead to the generation of hollow carbon spheres. Dai et al. has developed a facile synthesis method of hollow carbon with dopamine as carbon source due to the nature of dopamine polymerization and high carbonization yield of polydopamine [158]. Double-shelled hollow carbon spheres could be obtained with hollow SnO_2 sphere as the hard template. Due to the strong interaction between carbon and sulphur and the excellent encapsulation ability of double-shelled hollow carbon spheres, a large quantity of sulphur was encapsulated effectively in the double shell, suppressing the problem of polysulfides diffusion and volume variation. This also led to excellent electrochemical performance and great cycling ability as a cathode material for lithium-sulfur batteries [163]. Benzene or acetonitrile could also be used as a carbon source for the generation of various hollow carbon spheres through chemical vapour deposition (CVD) method. The relative large silica spheres, short CVD time and high CVD temperature favoured the formation of hollow carbon spheres [164].

4.5.3 York-shell nanoparticles

York shell structures are considered as promising functional materials in wide areas such as catalysts, drug delivery, lithium ion battery and biosensors because of easy fabrication of the composition of materials and simple tailored properties [165, 166]. York shell nanoparticles with a hollow space between the freely movable core and the protective shell could be synthesized by many methods such as selective dissolution, soft template, ship-in-bottle, Ostwald ripening or galvanic replacement process and kirkendall diffusion [167]. Recently, monodisperse Ag@C nanoparticles were prepared by carbonization of Ag, AgBr@RF york-shell nanoparticles which were synthesized by the extended stober method. Altering the concentration of the resorcinol and formaldehyde precursors and calcination atmosphere could tune the shape and thickness of the shell. Other type yolk shell structural material with metal core and carbon shell could also be synthesized by this simple one-pot synthesis [168]. Additionally, a hierarchical porous yolk-shell structured carbons with a mesoporous carbon core and a microporous carbon shell was achieved through facile stober coating method, showing the maximum specific capacitance of 159 F/g among the synthesized carbons [159].

4.5.4 Fibres

Carbon fibres have attracted tremendous attentions in the area of catalysis, adsorption, biomedicine, energy storage and conversion due to high thermal stability, good biocompatibility, and abundant electronic and excellent mechanical properties. Many methods such as chemical vapour deposition, electrospinning and the template techniques have been utilized to fabricate them [169]. Carbon nanofibers could be obtained through direct template-based synthesis method by using solvent-free infiltration of the mixture of the tri-block copolymer Pluronic F127 as a structure-directing agent and phloroglucinol as a carbon source into porous alumina. This method indicated that prior to the infiltration process the disposal of the solvents avoided presence of macroscopic phase separation and hydrodynamic instabilities when the solvents were evaporated from the pores. Etching alumina generated 60 nm carbon fibres with helical mesopores [170]. In addition, carbon nanofibres with large pore (> 10 nm) on tube wall could be achieved. Aluminium oxide (AAO) provided the template for the growth of carbon fibre and the self-assembly of block copolymer/carbonhydrates were responsible for the formation of nanopores inside AAO pore channels [171]. Recently, an effective and large-scale method for the synthesis of heteroatom-doped carbon fibres was reported by using bacterial cellulose as the carbon source. The large number of functional groups on the surface of bacterial cellulose made it easy for P, N/P and B/P doping through pyrolysis the mixture of bacterial cellulose and H_3PO_4, $NH_4H_2PO_4$ and H_3BO_3/H_3PO_4. The prepared carbon fibres showed superior supercapacitive behaviour (186.03 kW/kg) [160].

4.5.5 Rods

Hexagonally ordered porous carbon nanorods (CMK-3) were first synthesized with SBA-15 as hard template, sucrose as a carbon source and sulphuric acid as the carbonization catalyst. This ordered porous structure of CMK-3 showed an inverse carbon framework without structure transformation in the process of etching silica template compared with CMK-1 [172]. Ordered mesoporous carbon nanorods (MWCMK-3) with large pore size (about 9 nm) and narrow size distribution could also be formed by using microwave-assisted synthesis of SBA-15 (MWSBA-15) which needed much less reaction time (less than 5 hours) than the conventional hydrothermal method (72 hours). This temperature-programmed microwave system provided homogeneous heat to improve the thermal stability of mesoporous silica template, facilitating the synthesis of SBA-15 with large pore size and narrow pore size distribution [22].

4.5.6 Sheets

Carbon nanosheets were different from graphene or graphene-based sheets because very small amount of graphene were present in the obtained carbon sheets with a graphene inner layer and a carbon coating on both sides [161]. Ionic liquids were utilized for the synthesis of nitrogen-doped carbon nanosheets and changing the ratio of the raw materials could precisely control the thickness of the carbon layer. Excellent rate capability, high specific capacitance (341 F/g) and good cycling ability (over 35000 cycles) could be reached by using these materials when used as electrodes for supercapacitors [161]. An industrial production of porous carbon sheets could also be obtained through a graphene-directed and catalysis-free polymer synthesis and following thermal carbonization by using Schiff base-type polymer as precursors. The produced materials with high surface area and nitrogen content presented outstanding supercapacitive performance (424 F/g) [173]. Graphene oxides and poly(benzoxazineco-resol) were used as a shape-directing agent and carbon precursor for rapid fabrication of sandwich-type microporous carbon nanosheets with high surface area (1293 m^2/g), narrow pore distribution (0.8 nm) and controlled layer thickness (about 17 nm). Such behaviours provided the fast diffusion of organic electrolytes, leading to superior gravimetric capacitance (103 F/g) [174].

4.5.7 Monoliths

Due to escaping from the excessive pressure drop in fixed bed adsorber and dusting problems, carbon monoliths have attracted tremendous attention to reach daily needs for practical applications [175]. In order to enhance the performance of CO_2 capture and storage, amine groups were introduced into carbon materials through post-modification, copolymerization and direct pyrolysis of amine-containing precursors. Lu et al. designed and synthesized nitrogen-doped carbon monolith through direct pyrolysis of copolymer of resorcinol, formaldehyde and lysine as nitrogen source with higher CO_2-adsorption capacity (3.13 $mmolg^{-1}$) than melamine-derived mesoporous carbon (2.25 $mmolg^{-1}$) at room temperature. This was due to the acid-base interaction between basic nitrogen groups in monolithic carbons and CO_2 [176]. A rapid and large-scale synthesis of hierarchical carbon monoliths with interconnected macroporous and ordered mesopore structure through soft template method was reported. The presence of lysine proved to catalyze polymerization of resorcinol and formaldehyde in 15 minutes at 90°C and promoted the formation of mesostructured. Changing the amount of F127 copolymer could adjust the total pore volume and steam activation could also increase surface area up to 2422 m^2/g [177]. In addition, superior capture, separation ability, high mechanical strength and excellent selectivity of CO_2 and facile regeneration could be achieved by using nitrogen-containing carbon monoliths. They were synthesized through a rapid and scalable self-assembly of poly(benzoxazine-co-resol) and following carbonization process [162].

5. Applications for Carbon Materials

Nanoporous carbon materials have the advantages such as chemical stabililty, electronic conductivity and absorption capabilities, so they have become quite promising materials in the energy and environment-related area. Here applications of porous carbon will be concluded into four parts: catalysts, batteries and capacitors/supercapacitors, which will provide perspectives for future investigations.

5.1 Metal-free electrocatalyst for oxygen reduction reaction

Large attention has been focused on fuel cells due to direct conversion of chemical energy into electricity with no combustion process and high energy conversion efficiency. Among various types of fuel cells, direct methanol fuel cells and proton exchange membrane fuel cells have drawn much attention because of low operation temperature, abundant fuel supplies, small dimensions and little pollution during reaction [179]. These kind of fuel cells consist of oxygen reduction reaction (ORR) at cathode and hydrogen or methanol oxidation reaction at anode. The rate determining step is the process of the reduction of oxygen and the electrochemical reactions are usually accelerated with the assistance from electrocatalysts [180]. The ORR process occurs in two pathways: (a) four-electron reduction of oxygen to water in acid condition or to hydroxyl ions in basic condition; (b) two-electron reduction of oxygen to hydrogen peroxide in acid condition or HO_2^- in basic condition [181]. An illustration of oxygen reduction reaction with four-electron pathways by using cobalt–nitrogen-doped porous carbon material is shown in Fig. 1.7a.

Electrocatalysts are designed to facilitate electron transfer pathways to optimize the produced electricity [1]. However, high cost of commercial electrocatalysts such as platinum particles loaded activated carbons or carbon blacks and the low durability are still two major problems towards large scale productions. In addition, commercial materials usually suffer from some deactivation problems such as the susceptibility to carbon monoxide and methanol poisoning [1]. To address these problems, a lot of metal-free and non precious metal-doped carbons with high surface area, ordered structure and tuneable pores have been designed for oxygen reduction reactions. This was attributed to a large number of active sites

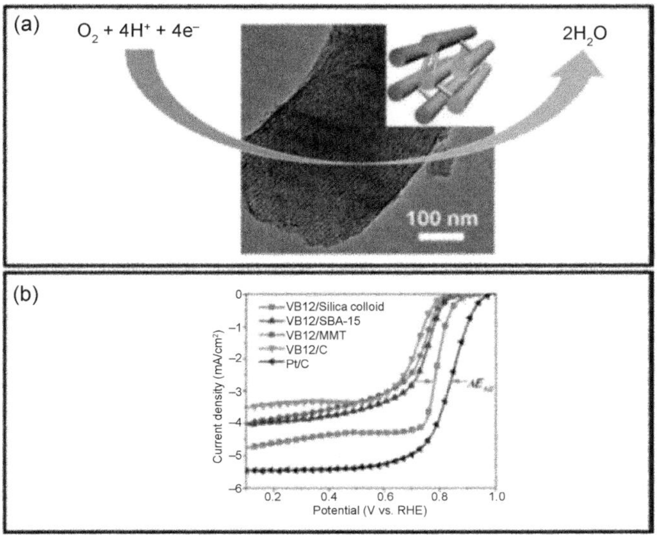

Figure 1.7. (a) TEM image with the inset of model illustration of cobalt–nitrogen-doped porous carbon material and illustration of oxygen reduction reaction (b) ORR polarization plots of various cobalt–nitrogen-doped porous carbon [178].

and enhanced mass-transport properties due to mesoporous structures in the carbons.

Heteroatoms doped porous carbons are quite promising because of the enhanced porosity and the tailored electronic structure through the incorporation of heteroatoms into the carbon matrix. B- [69], N- [62, 182–184], F- [70], P- [71], S- [72], NP [103], NS [104]-doped porous carbons with ordered mesoporous structure and high surface area have been prepared and showed superior electrochemical properties than undoped carbons and commercial Pt/C. For example, a series of nitrogen-doped porous carbon materials derived from the templates of silica nanoparticles, SBA-15 and montmorillonite were prepared and Fig. 1.7b also show all the three template carbons have better activity than Pt/C with regard to the onset potential and half-wave potential [178].

5.2 Batteries (Na-ion sulphur batteries, Na-ion battery, Li-air battery)

Energy storage has been recognized as one of most significant energy challenges and received worldwide concern and increasing research interest [6]. Batteries, as the most convenient form of energy storage, have played a very significant role in Electric Vehicles (EVs), Hybrid Electrical

Vehicles (HEVs), or car powered by other portable devices to provide abundant power and high energy density and at the same time could avoid the utilization of internal combustion engines. But several factors such as low energy and power densities, poor durability, high cost and poor safety have restricted the development of batteries [167]. Therefore, intensive efforts have been focused on designing next-generation batteries such as lithium ion batteries, lithium sulphur batteries, lithium air batteries, sodium ion batteries and sodium sulphur batteries with enhanced power density, cycling life and charge/discharge rate capability.

5.2.1 Lithium ion batteries

Because lithium is the lightest metal in the periodical table and thus can provide high energy density per electron, rechargeable lithium ion batteries are designed for the storage and conversion of electrochemical energy. The basic principle of lithium ion batteries is based on Li+ intercalation/conversion reactions. When the cell is discharged, lithium ions extract from the anode, pass through the electrolyte and intercalate into the cathode. In the meantime, the electrons pass through the external circuit to the cathode [185]. Carbonaceous materials are usually used as anode materials in the commercial rechargeable lithium batteries. Various carbon structures such as carbon nanofibres [186], hollow nanospheres [187], and carbon monoliths [188, 189] have been designed to enhance lithiation capability and cycling stability because the transportation length of lithium ions will be shorten through porous structures and electrode/electrolyte interface for the charge-transfer reaction would also be improved. Doping non-carbon elements such as nitrogen in the porous carbon structures is another method to enhance the electrochemical performance. Nitrogen-doped porous carbons with high surface area (2381 m^2/g) and high nitrogen doping (about 16%) were produced through KOH activation of poly-pyrrole nano-fibre webs. The material showed excellent capacity, rate capability and stable stability when utilized as anode material for Li^+ ion batteries. In addition, the specific capacity of this material could reach 943 mAh/g even at 2 A/g after 600 cycles because of the reduced transport length and facile transport channel of lithium ions and high nitrogen doping [190].

5.2.2 Lithium sulphur batteries

Lithium sulphur batteries consist of a lithium metal anode and a sulphur cathode. The cathode involves the reaction of $S_8 + 16\ Li^+ + 16\ e^- \leftrightarrow 8\ Li_2S$, which provides high theoretical capacity of 1672 mAh/g. However, some shortcomings still hindered the development of lithium sulphur batteries because of the poor electrical conductivity of sulphur, the dissolution and shuttling effects of lithium polysulfides in organic liquid electrolytes resulted in poor coulombic efficiency, low utilization and poor cycle stability [191, 192]. In order to solve these problems, porous carbon materials have also been investigated as host for sulphur because porous carbon materials can strongly absorb polysulfides and buffer the volume expansion. Recently, various carbon hosts such as microporous carbons [193], mesoporous carbons [194], hierarchical porous carbons [195], hollow carbon spheres [163], carbon fibres [196], nitrogen-doped carbons [197] have also been designed to overcome these drawbacks. Nazar et al. designed and synthesized polymer-modified mesoporous carbon-sulphur composites with reversible capacities up to 1320 mAh/g and good cycling ability. This superior performance was attributed to mesoporous structure of carbon hosts which provided more active sites between lithium ions and sulphur, constrained the growth of sulphur, generated more electrical contact area with sulphur and aided in trapping the formed polysulphides during reaction. The use of polymer on the carbon surface assisted the restriction of diffusion of large anions out of electrodes, leading to excellent electrochemical performance [194]. They also used mesoporous silica SBA-15 as polysulphide reservoir based on the design principle of drug delivery, which was incorporated in carbon/sulphur materials. These kind of materials provided superior adsorption and desorption ability of poly-sulphur ions, leading to the enhanced cycling ability and coulombic efficiency compared with CMK-3/sulphur composites [198].

5.2.3 Lithium-air batteries

Lithium-air batteries are also one of the most significant energy storage systems and their capacities are theoretically calculated to be 10 times higher than that of traditional lithium ion batteries [199]. Schematic cell configuration for one type of Li-air battery is shown in Fig. 1.8a. They usually involved a porous oxygen-breathing electrode as the cathode and a lithium metal as the anode. During the discharge process, lithium ions transport from the anode to the cathode and react with oxygen to form Li_2O or Li_2O_2 inside the porous cathode material. At the same time, the electrons pass through the external circuit to the cathode. During change process, oxygen will be released through decomposition of Li_2O or Li_2O_2 through oxygen evolution reaction [200]. Large surface area and high pore volume are necessary for superior electrochemical performance of lithium-air batteries because large surface area would provide more active sites and high pore volume could be related to the more accumulated discharge products [201]. For example, bimodal porous carbons with inter pore size about 28 nm (as shown in Fig. 1.8b) have been synthesized by Xia et al. and exhibited about 40% higher discharge capacity than the commercial

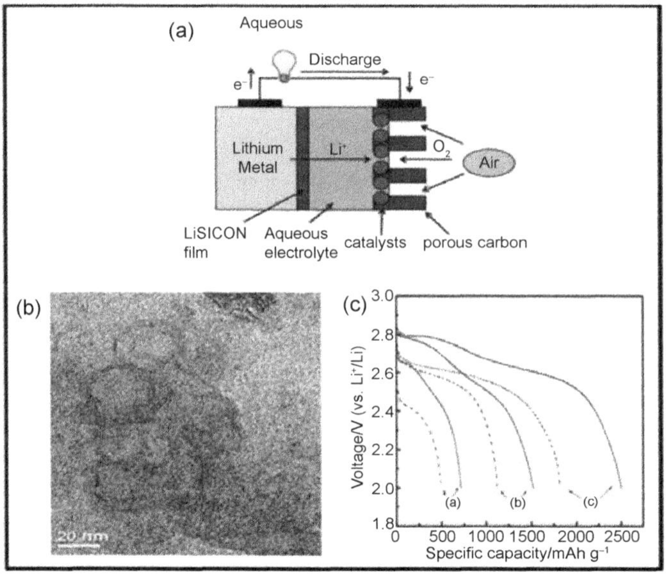

Figure 1.8. (a) Schematic cell configurations for the one type of Li-air battery [204]. (b) TEM image mesoporous carbons (MC) and (c) the discharge curve of MC (solid) and Super P carbon black (dash) at 0.5 mA/cm^2, 0.2 mA/cm^2, 0.1 mA/cm^2 [202].

carbons (as shown in Fig. 1.8c). The main factor was ascribed to large pore volume and ultra-large mesoporous structures [202]. In addition, the porous structure of carbon materials also affected the charge and discharge behaviour of lithium batteries. If the pore size is too small, the pores would be blocked by either electrolyte or lithium oxide. On the contrary, for large pores, electrochemical polarization would be caused by the growing lithium oxide because of poor conductivity, leading to the termination of discharge process [203]. Recently, nitrogen-enriched mesoporous carbon materials showed 1.73 times higher discharge ability (about 4500 mAh/g) than commercial carbons (BP 2000). More solid products were held in the large mesopores with suitable size. The interconnected macropore channels and low affinity to organic electrolytes facilitated oxygen diffusion, leading to high electrochemical performance of lithium-air batteries.

5.2.4 Na-ion batteries

Rechargeable sodium ion batteries have also attracted a lot of attention due to abundant supply and low cost for replacement to lithium ion batteries [205]. Due to larger radius than lithium ions, sodium ions have stronger coordination in the host lattices and are hard to be hosted in the interstitial space of most materials' crystallographic structure. Porous carbons have been investigated as anode materials because porous structure and large interlayer distance facilitated insertion and extraction of sodium ions [206]. The reversible sodium capacity of these carbons is usually between 100 and 300 mAh/g [207–210]. Tirado et al. reported that carbon microspheres which were prepared from resorcinol/formaldehyde mixture exhibited high reversible capability (about 285 mAh/g). This was attributed to the interaction between the porous carbon structure and reversible inserted sodium ions [210]. Recently, Liu et al. also reported large mesopore and micropore in the carbon spheres and the connectivity and conductivity of the mesoporous network were also responsible for good electrochemical performance for sodium batteries such as a high initial capacity (410 mAh/g) and a good retention (125 mAh/g after 100 cycles) [211].

5.2.5 Na-sulphur batteries

Sodium–sulphur (Na–S) batteries are one type of molten salt battery including the molten sodium and molten sulphur separated by beta-alumina solid electrolytes [212]. Due to reasonable power and energy densities, stability, abundance and low price, sodium sulphur batteries have become a sustainable and environmentally-friendly technology. But sodium sulphur batteries still met with difficulties as lithium sulphur batteries such as the shuttling effect of polysulphides and low electric conductivity of sulphur [194]. High operation temperatures (above 300°C) also restricted the application of lithium sulphur batteries. Choi et al. recently designed carbon–sulphur composite fibres which can be performed at room temperature with good sodiation and desodiation activity (about 500 cycles) and high rate performance. This was attributed to the stable atomic configuration of sulphur in the calcined polyacrylonitrile matrix overcoming the shortcomings of lithium sulphur batteries and the fibre structure facilitating the sodiation and desodiation process [213].

5.3 Supercapacitors

Supercapacitors have become one of the dominate power supplies with high energy and power densities and long cycling life and thus been considered as an excellent candidate for alternative batteries. A representation of an electrochemical double layer capacitor (in its charged state) is shown in Fig. 1.9a [214]. Electrochemical capacitors can accumulate electrical energy due to formation of electrical double layer between electrodes and electrolytes. They can be classified into two categories: Electrical Double-Layer Capacitors (EDLCs) with a non-Faradic process based on carbon materials such as activated carbon, graphene, carbon nanotubes and

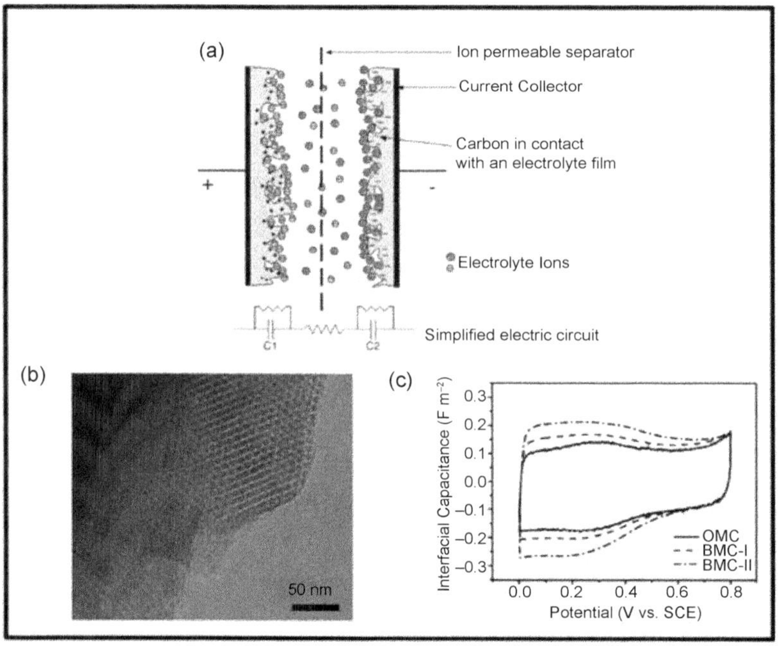

Figure 1.9. (a) Representation of an electrochemical double layer capacitor (in its charged state) [214]. (b) SEM image and (c) cyclic voltammogram of carbon spheres [79].

carbon nanofibres and pseudo-capacitors with a Faradic process using RuO_2, MnO_2, Co_3O_4 and conducting polymers [215]. To acquire superior electrochemical behaviour of super-capacitor, it is highly desirable to design carbon materials with optimal porous structure, high surface area, tuneable surface functionalities, appropriate wettability and excellent stability and electronic conductivity [48, 216, 217].

Micro-porous and macro-porous carbons were designed to enhance the capacitance through increasing surface area to facilitate ion transportation when carbon materials were used as electrode materials [218–221]. An ordered mesoporous carbon material, templated from Na–Y zeolite via nanocasting method, exhibited a narrow pore-size distribution within the micropore range. A large gravimetric capacitance (340 F/g) and better electrical performance than commercial carbon were shown by using this carbon material in aqueous solution. The adequate pore-size distribution and ordered structure make ions diffusion easy in the channels, indicating the enhanced electrochemical performance is not only derived from pore shape and pseudocapacitive effects but also from the pore shape and void tortuosity. In addition, the presence of nitrogen and oxygen functional groups on the surface of synthesized carbons led to a pseudocapacitive effect to facilitate electron transfer in acidic atmosphere [218].

The electrochemical capacitance of carbon electrodes could also be improved via surface modification of porous carbon materials with heteroatoms such as B [79], P [222], N [223, 224], and S [225]. The hetero atom such as nitrogen on the carbon surface could lead to the pseudocapacitance interaction which is usually related to a charge or mass transfer between the electrode material and the ions of the electrolyte [224]. More protons could be attracted by introducing electron donor and charge-density of the space-charge layer could be enhanced to strengthen the redox reactions [224]. For example, Cheng et al. prepared boron-doped ordered mesoporous carbons with a diameter of about 7 nm (Fig. 1.9b) by using sucrose as a carbon source and boric acid as a boron source through the hard template method (mesoporous SBA-15). They also found interfacial capacitance of the boron-doped carbon increased about 1.6 times higher than non-doped material in both acidic (as shown in Fig. 1.9c) and alkaline conditions, because low-level boron doping improved oxygen chemisorption on the surface and increased charge carrier concentration and Fermi level density of states [79].

Coating of the porous carbon surface with conducting polymer such as polyaniline (PANI) [226–228] could also affect the electrochemical performances. Three dimensionally ordered macroporous carbons with a thin layer PANI coating were fabricated to improve the electrical performance. An excellent capacitance of 1490 F/g, great rate performance and cycling ability were observed because the carbon matrix provided 3D interconnected microporous structure for fast ion transportation, good electrical conductivity, large surface area for deposition of PANI and a thin PANI coating layer endowed the pseudocapacitive contribution for the enhanced behaviour [227]. However, the decreased surface area and conductivity of carbons, the blockage of porous structure, strong swelling capability, weak surface adhesive ability and shortened cycling life could also occur when polymers were coated on the surface of carbon template [227, 229].

6. Conclusions and Outlook

Functional nano-porous carbons from template methods for energy storage and conversion were reviewed in this chapter. In the beginning, porous carbon materials were introduced according to pore size classification. Pore size is of great importance to electrochemical reactions for energy storage devices and electro-conductivity for fuel cells because micro-porous carbons could provide high surface area to enhance reaction rate, while macro-porous carbon materials could facilitate mass transfer and diffusion of reactants and products in the reaction. In view of material preparation, the hard template method, soft template method and hard-soft template method are the three main common strategies to prepare ordered porous carbon materials. Moreover, carbon precursors, porous structures, pore sizes, pore volumes and surface areas are summarized in tables. It is obvious that porous carbons with tuneable pore size, porous structure and morphologies can be achieved through template methods. With regard to the introduction of heteroatoms in the carbon matrix, monodoped, dual doped and multi-doped porous carbons from template methods for oxygen reduction reaction in fuel cells are reviewed. Recent investigations from this field reported that heteroatoms doped porous carbons showed superior electrochemical properties compared to undoped carbons and commercial Pt/C because of the enhanced porosity and the tailored electronic structure through the incorporation of heteroatoms into the carbon matrix. Therefore, these heteroatom-doped porous carbons can be considered as metal-free electro-catalysts to pave new ways to substitute the commercial Pt/C materials for cathode electrodes. Surface modification towards the functionalization of porous carbons is necessary because carbons materials are usually used as catalysts support to improve the metal dispersion and adsorption of metallic species on the carbon surface, leading to the enhanced electrical conductivity and application behaviour.

Tremendous developments have been made for the preparation of nanoporous carbons with uniform structure and morphology from template methods, but there are still a lot of promising opportunities for the advancement of modifying porous carbons to optimize their performance in energy-related applications. In terms of synthesis method, it is hard to avoid the hard-template etching by acid or base solution and reach the requirement of template recycling. In addition, the synthesis of microporous carbons by using soft template method is still a challenge. In order to find an environmentally-friendly way for the preparation of ordered porous carbons with tuneable pore size (especially for less than 2 nm), porous structure and morphology, new templates will be explored. Additionally, the relationship among pore size, heteroatom doped effect, graphitization degree, surface modification and morphology need to be further understood to fabricate superior electrochemical devices. To improve electrochemical performance and electrocapacitance, it is desirable to design hierarchical structural carbon materials with different levels of porosity into carbon framework due to the enhanced mass diffusion, electron transportation, surface area and pore volume. In addition to the materials preparation, the utilization of environmentally friendly, low cost and abundant raw carbon precursors is highly desirable for further advancement of rechargeable electrochemical energy storage and conversion devices. What is more, the prepared ordered porous carbon materials can not only be used in energy-related applications, but also can be operated in the area of gas separation, absorption and storage or supports for bio-molecular immobilization in the field of biosensing.

References

[1] Zhang, L.P. and Xia, Z.H. 2011. Mechanisms of oxygen reduction reaction on nitrogen-doped graphene for fuel cells. *J. Phys. Chem. C*, 115(22): 11170–11176.

[2] Schuster, J., He, G., Mandlmeier, B. et al. 2012. Spherical ordered mesoporous carbon nanoparticles with high porosity for lithium-sulfur batteries. *Angew. Chem. Int. Edit.*, 51(15): 3591–3595.

[3] Gong, K.P., Du, F., Xia, Z.H. et al. 2009. Nitrogen-doped carbon nanotube arrays with high electrocatalytic activity for oxygen reduction. *Science*, 323(5915): 760–764.

[4] Dai, L.M. 2013. Functionalization of graphene for efficient energy conversion and storage. *Acc. Chem. Res.*, 46(1): 31–42.

[5] Sun, X.M. and Li, Y.D. 2004. Colloidal carbon spheres and their core/shell structures with noble-metal nanoparticles. *Angew. Chem. Int. Edit.*, 43(5): 597–601.

[6] Liu, C., Li, F., Ma, L.-P. et al. 2010. Advanced materials for energy storage. *Adv. Mater.*, 22(8): E28–E62.

[7] Cleghorn, S.J.C., Ren, X., Springer, T.E. et al. 1997. PEM fuel cells for transportation and stationary power generation applications. *Int. J. Hydrogen Energ.*, 22(12): 1137–1144.

[8] Hou, P.X., Orikasa, H., Yamazaki, T. et al. 2005. Synthesis of nitrogen-containing microporous carbon with a highly ordered structure and effect of nitrogen doping on H2O adsorption. *Chem. Mater.*, 17(20): 5187–5193.

[9] Joo, S.H., Choi, S.J., Oh, I. et al. 2001. Ordered nanoporous arrays of carbon supporting high dispersions of platinum nanoparticles. *Nature*, 412(6843): 169–172.

[10] Kang, S., Yu, J.S., Kruk, M. et al. 2002. Synthesis of an ordered macroporous carbon with 62 nm spherical pores that exhibit unique gas adsorption properties. *Chem. Commun.*, 16: 1670–1671.

[11] Fang, B.Z., Kim, J.H., Kim, M. et al. 2009. Ordered hierarchical nanostructured carbon as a highly efficient cathode catalyst support in proton exchange membrane fuel cell. *Chem. Mater.*, 21(5): 789–796.

[12] Corma, A. 1997. From microporous to mesoporous molecular sieve materials and their use in catalysis. *Chem. Rev.*, 97(6): 2373–2419.

[13] Kyotani, T., Nagai, T., Inoue, S. et al. 1997. Formation of new type of porous carbon by carbonization in zeolite nanochannels. *Chem. Mater.*, 9(2): 609–615.

[14] Zhao, X.J., Zhao, H.Y., Zhang, T.T. et al. 2014. One-step synthesis of nitrogen-doped microporous carbon materials as metal-free electrocatalysts for oxygen reduction reaction. *J. Mater. Chem. A*, 2(30): 11666–11671.

[15] Zhao, Y.C., Zhao, L., Mao, L.J. et al. 2013. One-step solvothermal carbonization to microporous carbon materials derived from cyclodextrins. *J. Mater. Chem. A*, 1(33): 9456–9461.

[16] Lee, J., Han, S. and Hyeon, T. 2004. Synthesis of new nanoporous carbon materials using nanostructured silica materials as templates. *J. Mater. Chem.* 14(4): 478–486.

[17] Ryoo, R., Joo, S.H., Kruk, M. et al. 2001. Ordered mesoporous carbons. *Adv. Mater.*, 13(9): 677–681.

[18] Kyotani, T. 2000. Control of pore structure in carbon. *Carbon*, 38(2): 269–286.

[19] Schuth, F. 2003. Endo- and exotemplating to create high-surface-area inorganic materials. *Angew. Chem. Int. Edit.*, 42(31): 3604–3622.

[20] Yang, H.F. and Zhao, D.Y. 2005. Synthesis of replica mesostructures by the nanocasting strategy. *J. Mater. Chem.*, 15(12): 1217–1231.

[21] Ryoo, R., Joo, S.H. and Jun, S. 1999. Synthesis of highly ordered carbon molecular sieves via template-mediated structural transformation. *J. Phys. Chem. B*, 103(37): 7743–7746.

[22] Sang, L.C., Vinu, A. and Coppens, M.O. 2011. Ordered mesoporous carbon with tunable, unusually large pore size and well-controlled particle morphology. *J. Mater. Chem.*, 21(20): 7410–7417.

[23] Vinu, A., Miyahara, M., Sivamurugan, V. et al. 2005. Large pore cage type mesoporous carbon, carbon nanocage: a superior adsorbent for biomaterials. *J. Mater. Chem.*, 15(48): 5122–5127.
[24] Ariga, K., Vinu, A., Miyahara, M. et al. 2007. One-pot separation of tea components through selective adsorption on pore-engineered nanocarbon, carbon nanocage. *J. Am. Chem. Soc.*, 129(36): 11022–11023.
[25] Lee, J., Sohn, K. and Hyeon, T. 2001. Fabrication of novel mesocellular carbon foams with uniform ultralarge mesopores. *J. Am. Chem. Soc.*, 123(21): 5146–5147.
[26] Fuertes, A.B. and Centeno, T.A. 2005. Mesoporous carbons with graphitic structures fabricated by using porous silica materials as templates and iron-impregnated polypyrrole as precursor. *J. Mater. Chem.*, 15(10): 1079–1083.
[27] Kresge, C.T., Leonowicz, M.E., Roth, W.J. et al. 1992. Ordered mesoporous molecular-sieves synthesized by a liquid-crystal template mechanism. *Nature*, 359(6397): 710–712.
[28] Beck, J.S., Vartuli, J.C., Roth, W.J. et al. 1992. A new family of mesoporous molecular-sieves prepared with liquid-crystal templates. *J. Am. Chem. Soc.*, 114(27): 10834–10843.
[29] Zhao, D.Y., Feng, J.L., Huo, Q.S. et al. 1998. Triblock copolymer syntheses of mesoporous silica with periodic 50 to 300 angstrom pores. *Science*, 279(5350): 548–552.
[30] Zhao, D.Y., Huo, Q.S., Feng, J.L. et al. 1998. Nonionic triblock and star diblock copolymer and oligomeric surfactant syntheses of highly ordered, hydrothermally stable, mesoporous silica structures. *J. Am. Chem. Soc.*, 120(24): 6024–6036.
[31] Zhang, F.Q., Meng, Y., Gu, D. et al. 2005. A facile aqueous route to synthesize highly ordered mesoporous polymers and carbon frameworks with Ia(3)over-bard bicontinuous cubic structure. *J. Am. Chem. Soc.*, 127(39): 13508–13509.
[32] Yoon, S.B., Sohn, K., Kim, J.Y. et al. 2002. Fabrication of carbon capsules with hollow macroporous core/mesoporous shell structures. *Adv. Mater.*, 14(1): 19–21.
[33] Yu, J.S., Kang, S., Yoon, S.B. et al. 2002. Fabrication of ordered uniform porous carbon networks and their application to a catalyst supporter. *J. Am. Chem. Soc.*, 124(32): 9382–9383.
[34] Kyotani, T., Tsai, L.F. and Tomita, A. 1996. Preparation of ultrafine carbon tubes in nanochannels of an anodic aluminum oxide film. *Chem. Mater.*, 8(8): 2109–2113.
[35] Rodriguezreinoso, F. and Molinasabio, M. 1992. Activated carbons from lignocellulosic materials by chemical and or physical activation—an overview. *Carbon*, 30(7): 1111–1118.
[36] Kyotani, T. 2006. Synthesis of various types of nano carbons using the template technique. *B. Chem. Soc. Jpn.*, 79(9): 1322–1337.
[37] Woo, S.-W., Dokko, K., Nakano, H. et al. 2008. Preparation of three dimensionally ordered macroporous carbon with mesoporous walls for electric double-layer capacitors. *J. Mater. Chem.*, 18(14): 1674–1680.
[38] Meng, Y., Gu, D., Zhang, F. et al. 2005. Ordered mesoporous polymers and homologous carbon frameworks: amphiphilic surfactant templating and direct transformation. *Angew. Chem. Int. Edit.*, 44(43): 7053–7059.
[39] Wang, Z.Y., Kiesel, E.R. and Stein, A. 2008. Silica-free syntheses of hierarchically ordered macroporous polymer and carbon monoliths with controllable mesoporosity. *J. Mater. Chem.*, 18(19): 2194–2200.
[40] Lee, J.S., Joo, S.H. and Ryoo, R. 2002. Synthesis of mesoporous silicas of controlled pore wall thickness and their replication to ordered nanoporous carbons with various pore diameters. *J. Am. Chem. Soc.*, 124(7): 1156–1157.
[41] Lu, F., Wu, S.H., Hung, Y. et al. 2009. Size effect on cell uptake in well-suspended, uniform mesoporous silica nanoparticles. *Small*, 5(12): 1408–1413.
[42] Lee, H.I., Kim, J.H., You, D.J. et al. 2008. Rational synthesis pathway for ordered mesoporous carbon with controllable 30 to 100-angstrom pores. *Adv. Mater.*, 20(4): 757–762.
[43] Lei, Z.B., Xiao, Y., Dang, L.Q. et al. 2006. Fabrication of ultra-large mesoporous carbon with tunable pore size by monodisperse silica particles derived from seed growth process. *Micropor. Mesopor. Mat.*, 96(1-3): 127–134.

[44] Taguchi, A., Smatt, J.H. and Linden, M. 2003. Carbon monoliths possessing a hierarchical, fully interconnected porosity. *Adv. Mater.*, 15(14): 1209–1211.
[45] Su, B.L., Vantomme, A., Surahy, L. et al. 2007. Hierarchical multimodal mesoporous carbon materials with parallel macrochannels. *Chem. Mater.*, 19(13): 3325–3333.
[46] Wang, Z.Y., Li, F., Ergang, N.S. et al. 2006. Effects of hierarchical architecture on electronic and mechanical properties of nanocast monolithic porous carbons and carbon-carbon nanocomposites. *Chem. Mater.*, 18(23): 5543–5553.
[47] Morishita, T., Soneda, Y., Tsumura, T. et al. 2006. Preparation of porous carbons from thermoplastic precursors and their performance for electric double layer capacitors. *Carbon*, 44(12): 2360–2367.
[48] Wang, D.-W., Li, F., Liu, M. et al. 2008. 3D aperiodic hierarchical porous graphitic carbon material for high-rate electrochemical capacitive energy storage. *Angew. Chem. Int. Edit.*, 47(2): 373–376.
[49] Liang, J., Zheng, Y., Chen, J. et al. 2012. Facile oxygen reduction on a three-dimensionally ordered macroporous graphitic C3N4/Carbon composite electrocatalyst. *Angew. Chem. Int. Edit.*, 51(16): 3892–3896.
[50] Liang, C.D., Hong, K.L., Guiochon, G.A. et al. 2004. Synthesis of a large-scale highly ordered porous carbon film by self-assembly of block copolymers. *Angew. Chem. Int. Edit.*, 43(43): 5785–5789.
[51] Tanaka, S., Nishiyama, N., Egashira, Y. et al. 2005. Synthesis of ordered mesoporous carbons with channel structure from an organic-organic nanocomposite. *Chem. Commun.*, 16: 2125–2127.
[52] Liang, C.D. and Dai, S. 2006. Synthesis of mesoporous carbon materials via enhanced hydrogen-bonding interaction. *J. Am. Chem. Soc.*, 128(16): 5316–5317.
[53] Zhang, F.Q., Gu, D., Yu, T. et al. 2007. Mesoporous carbon single-crystals from organic-organic self-assembly. *J. Am. Chem. Soc.*, 129(25): 7746-7747.
[54] Liu, L., Wang, F.Y., Shao, G.S. et al. 2010. A low-temperature autoclaving route to synthesize monolithic carbon materials with an ordered mesostructure. *Carbon*, 48(7): 2089–2099.
[55] Huang, Y., Cai, H.Q., Feng, D. et al. 2008. One-step hydrothermal synthesis of ordered mesostructured carbonaceous monoliths with hierarchical porosities. *Chem. Commun.*, 23: 2641–2643.
[56] Liang, C.D., Li, Z.J. and Dai, S. 2008. Mesoporous carbon materials: Synthesis and modification. *Angew. Chem. Int. Edit.*, 47(20): 3696–3717.
[57] Zhang, F., Meng, Y., Gu, D. et al. 2005. A facile aqueous route to synthesize highly ordered mesoporous polymers and carbon frameworks with Ia3d bicontinuous cubic structure. *J. Am. Chem. Soc.*, 127(39): 13508–13509.
[58] Zhang, J.Y., Deng, Y.H., Wei, J. et al. 2009. Design of amphiphilic ABC triblock copolymer for templating synthesis of large-pore ordered mesoporous carbons with tunable Pore wall thickness. *Chem. Mater.*, 21(17): 3996–4005.
[59] Deng, Y.H., Cai, Y., Sun, Z.K. et al. 2010. Controlled synthesis and functionalization of ordered large-pore mesoporous carbons. *Adv. Funct. Mater.*, 20(21): 3658–3665.
[60] Deng, Y.H., Liu, J., Liu, C. et al. 2008. Ultra-large-pore mesoporous carbons templated from poly(ethylene oxide)-b-polystyrene diblock copolymer by adding polystyrene homopolymer as a pore expander. *Chem. Mater.*, 20(23): 7281–7286.
[61] Fang, Y., Gu, D., Zou, Y. et al. 2010. A low-concentration hydrothermal synthesis of biocompatible ordered mesoporous carbon nanospheres with tunable and uniform size. *Angew. Chem. Int. Edit.*, 49(43): 7987–7991.
[62] Yang, T., Liu, J., Zhou, R. et al. 2014. N-doped mesoporous carbon spheres as the oxygen reduction reaction catalysts. *J. Mater. Chem. A*, 2(42): 18139–18146.
[63] Liu, J., Yang, T., Wang, D.-W. et al. 2013. A facile soft-template synthesis of mesoporous polymeric and carbonaceous nanospheres. *Nat. Commun.*, 4.

[64] Holland, B.T., Blanford, C.F. and Stein, A. 1998. Synthesis of macroporous minerals with highly ordered three-dimensional arrays of spheroidal voids. *Science*, 281(5376): 538–540.
[65] Zakhidov, A.A., Baughman, R.H., Iqbal, Z. et al. 1998. Carbon structures with three-dimensional periodicity at optical wavelengths. *Science*, 282(5390): 897–901.
[66] Deng, Y.H., Liu, C., Yu, T. et al. 2007. Facile synthesis of hierarchically porous carbons from dual colloidal crystal/block copolymer template approach. *Chem. Mater.*, 19(13): 3271–3277.
[67] Li, N.W., Zheng, M.B., Feng, S.Q. et al. 2013. Fabrication of hierarchical macroporous/mesoporous carbons via the dual-template method, and the restriction effect of hard template on shrinkage of mesoporous polymers. *J. Phys. Chem. C*, 117(17): 8784–8792.
[68] Lei, Z., Zhang, Y., Wang, H. et al. 2001. Fabrication of well-ordered macroporous active carbon with a microporous framework. *J. Mater. Chem.*, 11(8): 1975–1977.
[69] Bo, X.J. and Guo, L.P. 2013. Ordered mesoporous boron-doped carbons as metal-free electrocatalysts for the oxygen reduction reaction in alkaline solution. *Phys. Chem. Chem. Phys.*, 15(7): 2459–2465.
[70] Wang, H. and Kong, A. 2014. Mesoporous fluorine-doped carbon as efficient cathode material for oxygen reduction reaction. *Mater. Lett.*, 136(0): 384–387.
[71] Yang, D.S., Bhattacharjya, D., Inamdar, S. et al. 2012. Phosphorus-doped ordered mesoporous carbons with different lengths as efficient metal-free electrocatalysts for oxygen reduction reaction in alkaline media. *J. Am. Chem. Soc.*, 134(39): 16127–16130.
[72] Wang, H., Bo, X.J., Zhang, Y.F. et al. 2013. Sulfur-doped ordered mesoporous carbon with high electrocatalytic activity for oxygen reduction. *Electrochim. Acta*, 108: 404–411.
[73] Wu, X.X. and Radovic, L.R. 2004. Ab initio molecular orbital study on the electronic structures and reactivity of boron-substituted carbon. *J. Phys. Chem. A*, 108(42): 9180–9187.
[74] Lowell, C.E. 1967. Solid solution of boron in graphite. *J. Am. Ceram. Soc.*, 50(3): 142–144.
[75] Zhong, D.H., Sano, H., Uchiyama, Y. et al. 2000. Effect of low-level boron doping on oxidation behavior of polyimide-derived carbon films. *Carbon*, 38(8): 1199–1206.
[76] Radovic, L.R., Karra, M., Skokova, K. et al. 1998. The role of substitutional boron in carbon oxidation. *Carbon*, 36(12): 1841–1854.
[77] Zhao, X.C., Wang, A.Q., Yan, J.W. et al. 2010. Synthesis and electrochemical performance of heteroatom-incorporated ordered mesoporous carbons. *Chem. Mater.*, 22(19): 5463–5473.
[78] Zhao, X.C., Zhang, Q., Zhang, B.S. et al. 2012. Dual-heteroatom-modified ordered mesoporous carbon: Hydrothermal functionalization, structure, and its electrochemical performance. *J. Mater. Chem.*, 22(11): 4963–4969.
[79] Wang, D.W., Li, F., Chen, Z.G. et al. 2008. Synthesis and electrochemical property of boron-doped mesoporous carbon in supercapacitor. *Chem. Mater.*, 20(22): 7195–7200.
[80] Wang, T., Zhang, C., Sun, X. et al. 2012. Synthesis of ordered mesoporous boron-containing carbon films and their corrosion behavior in simulated proton exchange membrane fuel cells environment. *J. Power Sources*, 212(0): 1–12.
[81] Xia, Y.D. and Mokaya, R. 2004. Synthesis of ordered mesoporous carbon and nitrogen-doped carbon materials with graphitic pore walls via a simple chemical vapor deposition method. *Adv. Mater.*, 16(17): 1553–1558.
[82] Xia, Y.D. and Mokaya, R. 2005. Generalized and facile synthesis approach to N-doped highly graphitic mesoporous carbon materials. *Chem. Mater.*, 17(6): 1553–1560.
[83] Mane, G.P., Talapaneni, S.N., Anand, C. et al. 2012. Preparation of highly ordered nitrogen-containing mesoporous carbon from a gelatin biomolecule and its excellent sensing of acetic acid. *Adv. Funct. Mater.*, 22(17): 3596–3604.
[84] Liu, R.L., Wu, D.Q., Feng, X.L. et al. 2010. Nitrogen-doped ordered mesoporous graphitic arrays with high electrocatalytic activity for oxygen reduction. *Angew. Chem. Int. Edit.*, 49(14): 2565–2569.

[85] Yang, C.-M., Weidenthaler, C., Spliethoff, B. et al. 2005. Facile template synthesis of ordered mesoporous carbon with polypyrrole as carbon precursor. *Chem. Mater.*, 17(2): 355–358.

[86] Guo, Y.X., He, J.P., Wang, T. et al. 2011. Enhanced electrocatalytic activity of platinum supported on nitrogen modified ordered mesoporous carbon. *J. Power Sources*, 196(22): 9299–9307.

[87] Silva, R., Voiry, D., Chhowalla, M. et al. 2013. Efficient metal-free electrocatalysts for oxygen reduction: polyaniline-derived N- and O-doped mesoporous carbons. *J. Am. Chem. Soc.*, 135(21): 7823–7826.

[88] Liu, L., Deng, Q.F., Ma, T.Y. et al. 2011. Ordered mesoporous carbons: citric acid-catalyzed synthesis, nitrogen doping and CO2 capture. *J. Mater. Chem.*, 21(40): 16001–16009.

[89] Li, Z.J., Del Cul, G.D., Yan, W.F. et al. 2004. Fluorinated carbon with ordered mesoporous structure. *J. Am. Chem. Soc.*, 126(40): 12782–12783.

[90] Chamssedine, F., Dubois, M., Guerin, K. et al. 2007. Reactivity of carbon nanofibers with fluorine gas. *Chem. Mater.*, 19(2): 161–172.

[91] Wan, Y., Qian, X., Jia, N.Q. et al. 2008. Direct triblock-copolymer-templating synthesis of highly ordered fluorinated mesoporous carbon. *Chem. Mater.*, 20(3): 1012–1018.

[92] Fulvio, P.F., Brown, S.S., Adcock, J. et al. 2011. Low-temperature fluorination of soft-templated mesoporous carbons for a high-power lithium/carbon fluoride battery. *Chem. Mater.*, 23(20): 4420–4427.

[93] Wang, H.L., Gao, Q.M. and Hu, J. 2010. Preparation of porous doped carbons and the high performance in electrochemical capacitors. *Micropor. Mesopor. Mat.*, 131(1-3): 89–96.

[94] Shin, Y.S., Fryxell, G., Um, W.Y. et al. 2007. Sulfur-functionalized mesoporous carbon. *Adv. Funct. Mater.*, 17(15): 2897–2901.

[95] Valle-Vigon, P., Sevilla, M. and Fuertes, A.B. 2013. Functionalization of mesostructured silica-carbon composites. *Mater. Chem. Phys.*, 139(1): 281–289.

[96] Sevilla, M., Fuertes, A.B. and Mokaya, R. 2011. Preparation and hydrogen storage capacity of highly porous activated carbon materials derived from polythiophene. *Int. J. Hydrogen Energ.*, 36(24): 15658–15663.

[97] Sevilla, M. and Fuertes, A.B. 2012. Highly porous S-doped carbons. *Micropor. Mesopor. Mat.*, 158: 318–323.

[98] Kwon, K., Jin, S.A., Pak, C. et al. 2011. Enhancement of electrochemical stability and catalytic activity of Pt nanoparticles via strong metal-support interaction with sulfur-containing ordered mesoporous carbon. *Catal. Today*, 164(1): 186–189.

[99] Zhao, X.C., Zhang, Q., Chen, C.M. et al. 2012. Aromatic sulfide, sulfoxide, and sulfone mediated mesoporous carbon monolith for use in supercapacitor. *Nano Energy*, 1(4): 624–630.

[100] Guo, J., Luo, Y., Lua, A.C. et al. 2007. Adsorption of hydrogen sulphide (H2S) by activated carbons derived from oil-palm shell. *Carbon*, 45(2): 330–336.

[101] Morris, E.A., Kirk, D.W., Jia, C.Q. et al. 2012. Roles of sulfuric acid in elemental mercury removal by activated carbon and sulfur-impregnated activated carbon. *Environ. Sci. Technol.*, 46(14): 7905–7912.

[102] Guo, H.L. and Gao, Q.M. 2009. Boron and nitrogen co-doped porous carbon and its enhanced properties as supercapacitor. *J. Power Sources*, 186(2): 551–556.

[103] Jiang, H.L., Zhu, Y.H., Feng, Q. et al. 2014. Nitrogen and phosphorus dual-doped hierarchical porous carbon foams as efficient metal-free electrocatalysts for oxygen reduction reactions. *Chem. Eur. J.*, 20(11): 3106–3112.

[104] Liu, Z., Nie, H.G., Yang, Z. et al. 2013. Sulfur-nitrogen co-doped three-dimensional carbon foams with hierarchical pore structures as efficient metal-free electrocatalysts for oxygen reduction reactions. *Nanoscale*, 5(8): 3283–3288.

[105] Liu, J., Qiao, S.Z., Liu, H. et al. 2011. Extension of the stober method to the preparation of monodisperse resorcinol-formaldehyde resin polymer and carbon spheres. *Angew. Chem. Int. Edit.*, 50(26): 5947–5951.
[106] Li, H.F., Xi, H.A., Zhu, S.M. et al. 2006. Preparation, structural characterization, and electrochemical properties of chemically modified mesoporous carbon. *Microporous and Mesoporous Materials*, 96(1-3): 357–362.
[107] Huwe, H. and Froba, M. 2003. Iron(III) oxide nanoparticles within the pore system of mesoporous carbon CMK-1: intra-pore synthesis and characterization. *Micropor. Mesopor. Mat.*, 60(1-3): 151–158.
[108] Minchev, C., Huwe, H., Tsoncheva, T. et al. 2005. Iron oxide modified mesoporous carbons: Physicochemical and catalytic study. *Micropor. Mesopor. Mat.*, 81(1-3): 333–341.
[109] Li, H.F., Xi, H.A., Zhu, S.M. et al. 2006. Nickel oxide nanocrystallites embedded within the wall of ordered mesoporous carbon. *Mater. Lett.*, 60(7): 943–946.
[110] Li, H.F., Zhu, S.M., Xi, H.A. et al. 2006. Nickel oxide nanocrystallites within the wall of ordered mesoporous carbon CMK-3: Synthesis and characterization. *Micropor. Mesopor. Mat.*, 89(1-3): 196–203.
[111] Huwe, H. and Froba, M. 2007. Synthesis and characterization of transition metal and metal oxide nanoparticles inside mesoporous carbon CMK-3. *Carbon*, 45(2): 304–314.
[112] Fan, J., Wang, T., Yu, C. et al. 2004. Ordered, nanostructured tin-based oxides/carbon composite as the negative-electrode material for lithium-ion batteries. *Adv. Mater.*, 16(16): 1432–1436.
[113] Fulvio, P.F., Liang, C.D., Dai, S. et al. 2009. Mesoporous carbon materials with ultra-thin pore walls and highly dispersed nickel nanoparticles. *Eur. J. Inorg. Chem.*, 5: 605–612.
[114] Sun, Z.K., Sun, B., Qiao, M.H. et al. 2012. A general chelate-assisted co-assembly to metallic nanoparticles-incorporated ordered mesoporous carbon catalysts for fischer-tropsch synthesis. *J. Am. Chem. Soc.*, 134(42): 17653–17660.
[115] Tang, J., Liu, J., Torad, N.L. et al. 2014. Tailored design of functional nanoporous carbon materials toward fuel cell applications. *Nano Today*, 9(3): 305–323.
[116] Cheng, P.Z. and Teng, H.S. 2003. Electrochemical responses from surface oxides present on HNO3-treated carbons. *Carbon*, 41(11): 2057–2063.
[117] Otake, Y. and Jenkins, R.G. 1993. Characterization of oxygen-containing surface complexes created on a microporous carbon by air and nitric-acid treatment. *Carbon*, 31(1): 109–121.
[118] Eto, M. and Growcock, F.B. 1983. Effect of oxidizing environment on the strength of H451, Pgx and Ig-11 graphites. *Carbon*, 21(2): 135–147.
[119] Guha, A., Lu, W.J., Zawodzinski, T.A. et al. 2007. Surface-modified carbons as platinum catalyst support for PEM fuel cells. *Carbon*, 45(7): 1506–1517.
[120] Jun, S., Choi, M., Ryu, S. et al. 2003. Ordered mesoporous carbon molecular sieves with functionalized surfaces, in Studies in Surface Science and Catalysis, R.R.W.-S.A.C.W.L. Sang-Eon Park and C. Jong-San, Editors. Elsevier. p. 37–40.
[121] Yu, D., Wang, Z., Ergang, N.S. et al. 2007. Surface functionalization of templated porous carbon materials, in Studies in Surface Science and Catalysis, S.Q.Y.T. Dongyuan Zhao and Y. Chengzhong, Editors. Elsevier. p. 365–368.
[122] Tamai, H., Shiraki, K., Shiono, T. et al. 2006. Surface functionalization of mesoporous and microporous activated carbons by immobilization of diamine. *J. Colloid Interface Sci.*, 295(1): 299–302.
[123] Pittman, C.U., Wu, Z., Jiang, W. et al. 1997. Reactivities of amine functions grafted to carbon fiber surfaces by tetraethylenepentamine. Designing interfacial bonding. *Carbon*, 35(7): 929–943.
[124] Delamar, M., Hitmi, R., Pinson, J. et al. 1992. Covalent modification of carbon surfaces by grafting of functionalized aryl radicals produced from electrochemical reduction of diazonium salts. *J. Am. Chem. Soc.*, 114(14): 5883–5884.

[125] Li, Z., Yan, W. and Dai, S. 2005. Surface functionalization of ordered mesoporous carbons a comparative study. *Langmuir*, 21(25): 11999–12006.
[126] Li, Z. and Dai, S. 2005. Surface functionalization and pore size manipulation for carbons of ordered structure. *Chem. Mater.*, 17(7): 1717–1721.
[127] Pandurangappa, M., Lawrence, N.S., Compton, R.G. 2002. Homogeneous chemical derivatisation of carbon particles: a novel method for funtionalising carbon surfaces. *Analyst*, 127(12): 1568–1571.
[128] Wildgoose, G.G., Leventis, H.C., Davies, I.J. et al. 2005. Graphite powder derivatised with poly-l-cysteine using "building-block" chemistry-a novel material for the extraction of heavy metal ions. *J. Mater. Chem.*, 15(24): 2375–2382.
[129] Choi, M. and Ryoo, R. 2007. Mesoporous carbons with KOH activated framework and their hydrogen adsorption. *J. Mater. Chem.*, 17(39): 4204–4209.
[130] Wang, X.Q., Liu, R., Waje, M.M. et al. 2007. Sulfonated ordered mesoporous carbon as a stable and highly active protonic acid catalyst. *Chem. Mater.*, 19(10): 2395–2397.
[131] Lee, J.B., Park, Y.K., Yang, O.B. et al. 2006. Synthesis of porous carbons having surface functional groups and their application to direct-methanol fuel cells. *J. Power Sources*, 158(2): 1251–1255.
[132] Wang, L.F., Zhao, Y., Lin, K.F. et al. 2006. Super-hydrophobic ordered mesoporous carbon monolith. *Carbon*, 44(7): 1336–1339.
[133] Chen, S.G., Wei, Z.D., Qi, X.Q. et al. 2012. Nanostructured polyaniline-decorated Pt/C@PANI core-shell catalyst with enhanced durability and activity. *J. Am. Chem. Soc.*, 134(32): 13252–13255.
[134] Avouris, P., Chen, Z. and Perebeinos, V. 2007. Carbon-based electronics. *Nat. Nano*, 2(10): 605–615.
[135] Kim, M., Hwang, S. and Yu, J.-S. 2007. Novel ordered nanoporous graphitic C3N4 as a support for Pt-Ru anode catalyst in direct methanol fuel cell. *J. Mater. Chem.*, 17(17): 1656–1659.
[136] Yuan, D., Yuan, X., Zou, W. et al. 2012. Synthesis of graphitic mesoporous carbon from sucrose as a catalyst support for ethanol electro-oxidation. *J. Mater. Chem.*, 22(34): 17820–17826.
[137] Ishikawa, M., Sugimoto, T., Kikuta, M. et al. 2006. Pure ionic liquid electrolytes compatible with a graphitized carbon negative electrode in rechargeable lithium-ion batteries. *J. Power Sources*, 162(1): 658–662.
[138] Gupta, G., Slanac, D.A., Kumar, P. et al. 2010. Highly stable pt/ordered graphitic mesoporous carbon electrocatalysts for oxygen reduction. *J. Phys. Chem. C*, 114(24): 10796–10805.
[139] Gierszal, K.P., Jaroniec, M., Kim, T.-W. et al. 2008. High temperature treatment of ordered mesoporous carbons prepared by using various carbon precursors and ordered mesoporous silica templates. *New J. Chem.*, 32(6): 981–993.
[140] Yang, H., Yan, Y., Liu, Y. et al. 2004. A simple melt impregnation method to synthesize ordered mesoporous carbon and carbon nanofiber bundles with graphitized structure from pitches. *J. Phys. Chem. B*, 108(45): 17320–17328.
[141] Kim, T.-W., Park, I.-S. and Ryoo, R. 2003. A synthetic route to ordered mesoporous carbon materials with graphitic pore walls. *Angew. Chem. Int. Edit.*, 42(36): 4375–4379.
[142] Zhang , F., Yan, Y., Yang, H. et al. 2005. Understanding effect of wall structure on the hydrothermal stability of mesostructured silica SBA-15. *J. Phys. Chem. B*, 109(18): 8723–8732.
[143] Ehrburger-Dolle, F., Morfin, I., Geissler, E. et al. 2003. Small-angle X-ray scattering and electron microscopy investigation of silica and carbon replicas with ordered porosity. *Langmuir*, 19(10): 4303–4308.
[144] Yang, C.-M., Weidenthaler, C., Spliethoff, B. et al. 2004. Facile template synthesis of ordered mesoporous carbon with polypyrrole as carbon precursor. *Chem. Mater.*, 17(2): 355–358.

[145] Ji, X., Herle, P.S., Rho, Y. et al. 2007. Carbon/MoO2 composite based on porous semi-graphitized nanorod assemblies from *in situ* reaction of tri-block polymers. *Chem. Mater.*, 19(3): 374–383.
[146] Zhu, W., Ren, J., Gu, X. et al. 2011. Synthesis of hermetically-sealed graphite-encapsulated metallic cobalt (alloy) core/shell nanostructures. *Carbon*, 49(4): 1462–1472.
[147] Gao, W., Wan, Y., Dou, Y. et al. 2011. Synthesis of partially graphitic ordered mesoporous carbons with high surface areas. *Adv. Energy Mater.*, 1(1): 115–123.
[148] Yuan, J., Giordano, C. and Antonietti, M. 2010. Ionic liquid monomers and polymers as precursors of highly conductive, mesoporous, graphitic carbon nanostructures. *Chem. Mater.*, 22(17): 5003–5012.
[149] Schaefer, Z.L., Gross, M.L., Hickner, M.A. et al. 2010. Uniform hollow carbon shells: nanostructured graphitic supports for improved oxygen-reduction catalysis. *Angew. Chem. Int. Edit.*, 122(39): 7199–7202.
[150] Fuertes, A.B. and Alvarez, S. 2004. Graphitic mesoporous carbons synthesised through mesostructured silica templates. *Carbon*, 42(15): 3049–3055.
[151] Lee, K.T., Ji, X.L., Rault, M. et al. 2009. Simple synthesis of graphitic ordered mesoporous carbon materials by a solid-state method using metal phthalocyanines. *Angew. Chem. Int. Edit.*, 48(31): 5661–5665.
[152] Su, F., Zeng, J., Bao, X. et al. 2005. Preparation and characterization of highly ordered graphitic mesoporous carbon as a Pt catalyst support for direct methanol fuel cells. *Chem. Mater.*, 17(15): 3960–3967.
[153] Lei, Z., Bai, S., Xiao, Y. et al. 2008. CMK-5 mesoporous carbon synthesized via chemical vapor deposition of ferrocene as catalyst support for methanol oxidation. *J. Phys. Chem. C*, 112(3): 722–731.
[154] Zhang, W.H., Liang, C., Sun, H. et al. 2002. Synthesis of ordered mesoporous carbons composed of nanotubes via catalytic chemical vapor deposition. *Adv. Mater.*, 14(23): 1776–1778.
[155] Sun, X., He, J.P. Tang, J. et al. 2012. Structural and electrochemical characterization of ordered mesoporous carbon-reduced graphene oxide nanocomposites. *J. Mater. Chem.*, 22(21): 10900–10910.
[156] Su, F., Zhao, X.S., Wang, Y. et al. 2007. Bridging mesoporous carbon particles with carbon nanotubes. *Micropor. Mesopor. Mat.*, 98(1-3): 323–329.
[157] Fulvio, P.F., Mayes, R.T., Wang, X.Q. et al. 2011. "Brick-and-mortar" self-assembly approach to graphitic mesoporous carbon nanocomposites. *Adv. Funct. Mater.*, 21(12): 2208–2215.
[158] Liu, R., Mahurin, S.M., Li, C. et al. 2011. Dopamine as a carbon source: The controlled synthesis of hollow carbon spheres and yolk-structured carbon nanocomposites. *Angew. Chem. Int. Edit.*, 50(30): 6799–6802.
[159] Yang, T., Zhou, R., Wang, D.W. et al. 2015. Hierarchical mesoporous yolk-shell structured carbonaceous nanospheres for high performance electrochemical capacitive energy storage. *Chem. Commun. (Camb.)*, 51(13): 2518–21.
[160] Chen, L.F., Huang, Z.H., Liang, H.W. et al. 2014. Three-dimensional heteroatom-doped carbon nanofiber networks derived from bacterial cellulose for supercapacitors. *Adv. Funct. Mater.*, 24(32): 5104–5111.
[161] Jin, Z.Y., Lu, A.H., Xu, Y.Y. et al. 2014. Ionic liquid-assisted synthesis of microporous carbon nanosheets for use in high rate and long cycle life supercapacitors. *Adv. Mater.*, 26(22): 3700–3705.
[162] Hao, G.P., Li, W.C., Qian, D. et al. 2011. Structurally designed synthesis of mechanically stable poly(benzoxazine-co-resol)-based porous carbon monoliths and their application as high-performance CO2 capture sorbents. *J. Am. Chem. Soc.*, 133(29): 11378–11388.
[163] Zhang, C.F., Wu, H.B., Yuan, C.Z. et al. 2012. Confining sulfur in double-shelled hollow carbon spheres for lithium-sulfur batteries. *Angew. Chem. Int. Edit.*, 51(38): 9592–9595.

[164] Su, F.B., Zhao, X.S., Wang, Y. et al. 2006. Hollow carbon spheres with a controllable shell structure. *J. Mater. Chem.*, 16(45): 4413–4419.

[165] Liu, J., Qiao, S.Z., Budi Hartono, S. et al. 2010. Monodisperse yolk–shell nanoparticles with a hierarchical porous structure for delivery vehicles and nanoreactors. *Angew. Chem. Int. Edit.*, 49(29): 4981–4985.

[166] Liu, J., Yang, H.Q., Kleitz, F. et al. 2012. Yolk–shell hybrid materials with a periodic mesoporous organosilica shell: ideal nanoreactors for selective alcohol oxidation. *Adv. Funct. Mater.*, 22(3): 591–599.

[167] Liu, J., Qiao, S.Z., Chen, J.S. et al. 2011. Yolk/shell nanoparticles: new platforms for nanoreactors, drug delivery and lithium-ion batteries. *Chem. Commun.*, 47(47): 12578–12591.

[168] Yang, T., Liu, J., Zheng, Y. et al. 2013. Facile fabrication of core–shell-structured Ag@ carbon and mesoporous yolk–shell-structured Ag@carbon@silica by an extended stöber method. *Chem. Eur. J.*, 19(22): 6942–6945.

[169] Suzuki, N., Liu, J. and Yamauchi, Y. 2014. Recent progress on the tailored synthesis of various mesoporous fibers toward practical applications. *New J. Chem.*, 38(8): 3330–3335.

[170] Steinhart, M., Liang, C.D., Lynn, G.W. et al. 2007. Direct synthesis of mesoporous carbon microwires and nanowires. *Chem. Mater.*, 19(10): 2383–2385.

[171] Rodriguez, A.T., Chen, M., Chen, Z. et al. 2006. Nanoporous carbon nanotubes synthesized through confined hydrogen-bonding self-assembly. *J. Am. Chem. Soc.*, 128(29): 9276–9277.

[172] Jun, S., Joo, S.H., Ryoo, R. et al. 2000. Synthesis of new, nanoporous carbon with hexagonally ordered mesostructure. *J. Am. Chem. Soc.*, 122(43): 10712–10713.

[173] Zhuang, X.D., Zhang, F., Wu, D.Q. et al. 2014. Graphene coupled schiff-base porous polymers: Towards nitrogen-enriched porous carbon nanosheets with ultrahigh electrochemical capacity. *Adv. Mater.*, 26(19): 3081–3086.

[174] Hao, G.P., Lu, A.H., Dong, W. et al. 2013. Sandwich-type microporous carbon nanosheets for enhanced supercapacitor performance. *Adv. Energy Mater.*, 3(11): 1421–1427.

[175] Qian, D., Lei, C., Hao, G.P. et al. 2012. Synthesis of hierarchical porous carbon monoliths with incorporated metal-organic frameworks for enhancing volumetric based CO2 capture capability. *Acs Appl. Mater. Inter.*, 4(11): 6125–6132.

[176] Hao, G.P., Li, W.C., Qian, D. et al. 2010. Rapid synthesis of nitrogen-doped porous carbon monolith for CO2 capture. *Adv. Mater.*, 22(7): 853–857.

[177] Hao, G.P., Li, W.C., Wang, S.A. et al. 2011. Lysine-assisted rapid synthesis of crack-free hierarchical carbon monoliths with a hexagonal array of mesopores. *Carbon*, 49(12): 3762–3772.

[178] Liang, H.W., Wei, W., Wu, Z.S. et al. 2013. Mesoporous metal-nitrogen-doped carbon electrocatalysts for highly efficient oxygen reduction reaction. *J. Am. Chem. Soc.*, 135(43): 16002–16005.

[179] Service, R.F. 2002. Fuel Cells: Shrinking fuel cells promise power in your pocket. *Science*, 296(5571): 1222–1224.

[180] Shao, Y.Y., Sui, J.H., Yin, G.P. et al. 2008. Nitrogen-doped carbon nanostructures and their composites as catalytic materials for proton exchange membrane fuel cell. *Appl. Catal. B-Environ.*, 79(1-2): 89–99.

[181] Wang, D.W. and Su, D.S. 2014. Heterogeneous nanocarbon materials for oxygen reduction reaction. *Energy Environ. Sci.*, 7(2): 576–591.

[182] Shen, M., Ruan, C., Chen, Y. et al. 2015. Covalent entrapment of cobalt–iron sulfides in N-doped mesoporous carbon: Extraordinary bifunctional electrocatalysts for oxygen reduction and evolution reactions. *Acs Appl. Mater. Inter.*, 7(2): 1207–1218.

[183] Tao, G., Zhang, L., Chen, L. et al. 2015. N-doped hierarchically macro/mesoporous carbon with excellent electrocatalytic activity and durability for oxygen reduction reaction. *Carbon*, 86(0): 108–117.

[184] Yan, J., Meng, H., Xie, F.Y. et al. 2014. Metal free nitrogen doped hollow mesoporous graphene-analogous spheres as effective electrocatalyst for oxygen reduction reaction. *J. Power Sources*, 245: 772–778.
[185] Cheng, F., Tao, Z., Liang, J. et al. 2008. Template-directed materials for rechargeable lithium-ion batteries. *Chem. Mater.*, 20(3): 667–681.
[186] Li, C.C., Yin, X.M., Chen, L.B. et al. 2009. Porous carbon nanofibers derived from conducting polymer: Synthesis and application in lithium-ion batteries with high-rate capability. *J. Phys. Chem. C*, 113(30): 13438–13442.
[187] Han, F.D., Bai, Y.J., Liu, R. et al. 2011. Template-free synthesis of interconnected hollow carbon nanospheres for high-performance anode material in lithium-ion batteries. *Adv. Energy Mater.*, 1(5): 798–801.
[188] Hu, Y.S., Adelhelm, P., Smarsly, B.M. et al. 2007. Synthesis of hierarchically porous carbon monoliths with highly ordered microstructure and their application in rechargeable lithium batteries with high-rate capability. *Adv. Funct. Mater.*, 17(12): 1873–1878.
[189] Lee, K.T., Lytle, J.C., Ergang, N.S. et al. 2005. Synthesis and rate performance of monolithic macroporous carbon electrodes for lithium-ion secondary batteries. *Adv. Funct. Mater.*, 15(4): 547–556.
[190] Qie, L., Chen, W.-M., Wang, Z.-H. et al. 2012. Nitrogen-doped porous carbon nanofiber webs as anodes for lithium ion batteries with a superhigh capacity and rate capability. *Adv. Mater.*, 24(15): 2047–2050.
[191] Yang, Y., Yu, G., Cha, J.J. et al. 2011. Improving the performance of lithium–sulfur batteries by conductive polymer coating. *ACS Nano*, 5(11): 9187–9193.
[192] Tao, X., Chen, X., Xia, Y. et al. 2013. Highly mesoporous carbon foams synthesized by a facile, cost-effective and template-free Pechini method for advanced lithium-sulfur batteries. *J. Mater. Chem. A*, 1(10): 3295–3301.
[193] Zhang, B., Qin, X., Li, G.R. et al. 2010. Enhancement of long stability of sulfur cathode by encapsulating sulfur into micropores of carbon spheres. *Energy Environ. Sci.*, 3(10): 1531–1537.
[194] Ji, X., Lee, K.T. and Nazar, L.F. 2009. A highly ordered nanostructured carbon-sulphur cathode for lithium-sulphur batteries. *Nat. Mater.*, 8(6): 500–506.
[195] Ding, B., Yuan, C., Shen, L. et al. 2013. Encapsulating sulfur into hierarchically ordered porous carbon as a high-performance cathode for lithium–sulfur batteries. *Chem. Eur. J.*, 19(3): 1013–1019.
[196] Zheng, G., Yang, Y., Cha, J.J. et al. 2011. Hollow carbon nanofiber-encapsulated sulfur cathodes for high specific capacity rechargeable lithium batteries. *Nano Lett.*, 11(10): 4462–4467.
[197] Sun, X.-G., Wang, X., Mayes, R.T. et al. 2012. Lithium–sulfur batteries based on nitrogen-doped carbon and an ionic-liquid electrolyte. *ChemSusChem.*, 5(10): 2079–2085.
[198] Ji, X.L., Evers, S., Black, R. et al. 2011. Stabilizing lithium-sulphur cathodes using polysulphide reservoirs. *Nat. Commun.*, 2.
[199] Lu, Y.C., Gasteiger, H.A., Parent, M.C. et al. 2010. The influence of catalysts on discharge and charge voltages of rechargeable Li-oxygen batteries. *Electrochem. Solid-State Lett.*, 13(6): A69–A72.
[200] Zhang, S.S., Foster, D. and Read, J. 2010. Discharge characteristic of a non-aqueous electrolyte Li/O2 battery. *J. Power Sources*, 195(4): 1235–1240.
[201] Xiao, J., Mei, D.H., Li, X.L. et al. 2011. Hierarchically porous graphene as a lithium-air battery electrode. *Nano Lett.*, 11(11): 5071–5078.
[202] Yang, X.-h., He, P. and Xia, Y.-y. 2009. Preparation of mesocellular carbon foam and its application for lithium/oxygen battery. *Electrochem. Commun.*, 11(6): 1127–1130.
[203] Zhang, Y.N., Zhang, H.M., Li, J. et al. 2013. The use of mixed carbon materials with improved oxygen transport in a lithium-air battery. *J. Power Sources*, 240: 390–396.

[204] Lee, J.S., Kim, S.T., Cao, R. et al. 2011. Metal-air batteries with high energy density: Li-Air versus Zn-Air. *Adv. Energy Mater.*, 1(1): 34–50.

[205] Slater, M.D., Kim, D., Lee, E. et al. 2013. Sodium-Ion batteries. *Adv. Funct. Mater.*, 23(8): 947–958.

[206] Sun, Y.K., Chen, Z.H., Noh, H.J. et al. 2012. Nanostructured high-energy cathode materials for advanced lithium batteries. *Nat. Mater.*, 11(11): 942–947.

[207] Tang, K., Fu, L.J., White, R.J. et al. 2012. Hollow carbon nanospheres with superior rate capability for sodium-based batteries. *Adv. Energy Mater.*, 2(7): 873–877.

[208] Wenzel, S., Hara, T., Janek, J. et al. 2011. Room-temperature sodium-ion batteries: Improving the rate capability of carbon anode materials by templating strategies. *Energy Environ. Sci.*, 4(9): 3342–3345.

[209] Cao, Y.L., Xiao, L.F., Sushko, M.L. et al. 2012. Sodium ion insertion in hollow carbon nanowires for battery applications. *Nano Lett.*, 12(7): 3783–3787.

[210] Alcantara, R., Lavela, P., Ortiz, G.F. et al. 2005. Carbon microspheres obtained from resorcinol-formaldehyde as high-capacity electrodes for sodium-ion batteries. *Electrochem. Solid-State Lett.*, 8(4): A222–A225.

[211] Liu, J., Liu, H., Yang, T.Y. et al. 2014. Mesoporous carbon with large pores as anode for Na-ion batteries. *Chinese Sci. Bull.*, 59(18): 2186–2190.

[212] Hueso, K.B., Armand, M. and Rojo, T. 2013. High temperature sodium batteries: status, challenges and future trends. *Energy Environ. Sci.*, 6(3): 734–749.

[213] Hwang, T.H., Jung, D.S., Kim, J.-S. et al. 2013. One-dimensional carbon–sulfur composite fibers for Na–S rechargeable batteries operating at room temperature. *Nano Lett.*, 13(9): 4532–4538.

[214] Pandolfo, A.G. and Hollenkamp, A.F. 2006. Carbon properties and their role in supercapacitors. *J. Power Sources*, 157(1): 11–27.

[215] An, G.-H. and Ahn, H.-J. 2013. Activated porous carbon nanofibers using Sn segregation for high-performance electrochemical capacitors. *Carbon*, 65(0): 87–96.

[216] Simon, P. and Gogotsi, Y. 2008. Materials for electrochemical capacitors. *Nat. Mater.*, 7(11): 845–854.

[217] Huang, J., Sumpter, B.G. and Meunier, V. 2008. Theoretical model for nanoporous carbon supercapacitors. *Angew. Chem. Int. Edit.*, 47(3): 520–524.

[218] Ania, C.O., Khomenko, V., Raymundo-Piñero, E. et al. 2007. The large electrochemical capacitance of microporous doped carbon obtained by using a zeolite template. *Adv. Funct. Mater.*, 17(11): 1828–1836.

[219] Largeot, C., Portet, C., Chmiola, J. et al. 2008. Relation between the ion size and pore size for an electric double-layer capacitor. *J. Am. Chem. Soc.*, 130(9): 2730–2731.

[220] Chmiola, J., Largeot, C., Taberna, P.-L. et al. 2008. Desolvation of ions in subnanometer pores and its effect on capacitance and double-layer theory. *Angew. Chem. Int. Edit.*, 47(18): 3392–3395.

[221] Chmiola, J., Yushin, G., Gogotsi, Y. et al. 2006. Anomalous increase in carbon capacitance at pore sizes less than 1 nanometer. *Science*, 313(5794): 1760–1763.

[222] Hulicova-Jurcakova, D., Puziy, A.M., Poddubnaya, O.I. et al. 2009. Highly stable performance of supercapacitors from phosphorus-enriched carbons. *J. Am. Chem. Soc.*, 131(14): 5026–5027.

[223] Hulicova, D., Yamashita, J., Soneda, Y. et al. 2005. Supercapacitors prepared from melamine-based carbon. *Chem. Mater.*, 17(5): 1241–1247.

[224] Hulicova-Jurcakova, D., Kodama, M., Shiraishi, S. et al. 2009. Nitrogen-enriched nonporous carbon electrodes with extraordinary supercapacitance. *Adv. Funct. Mater.*, 19(11): 1800–1809.

[225] Jeong, H.-K., Jin, M., Ra, E.J. et al. 2010. Enhanced electric double layer capacitance of graphite oxide intercalated by poly(sodium 4-styrensulfonate) with high cycle stability. *ACS Nano*, 4(2): 1162–1166.

[226] Fan, L.Z., Hu, Y.S., Maier, J. et al. 2007. High electroactivity of polyaniline in supercapacitors by using a hierarchically porous carbon monolith as a support. *Adv. Funct. Mater.*, 17(16): 3083–3087.
[227] Zhang, L.L., Li, S., Zhang, J. et al. 2009. Enhancement of electrochemical performance of macroporous carbon by surface coating of polyaniline†. *Chem. Mater.*, 22(3): 1195–1202.
[228] Sivakkumar, S.R., Kim, W.J., Choi, J.-A. et al. 2007. Electrochemical performance of polyaniline nanofibres and polyaniline/multi-walled carbon nanotube composite as an electrode material for aqueous redox supercapacitors. *J. Power Sources*, 171(2): 1062–1068.
[229] Fusalba, F., Ho, H.A., Breau, L. et al. 2000. Poly(Cyano-Substituted Diheteroareneethylene) as active electrode material for electrochemical supercapacitors. *Chem. Mater.*, 12(9): 2581–2589.

CHAPTER 2

Nanoporous Carbon for Capacitive Energy Storage

Clement Bommier and *Xiulei Ji**

ABSTRACT

As an extension of traditional "parallel-plate" capacitors, Electrochemical Capacitors (ECs) hold a transformative role in the area of energy storage. Whereas traditional capacitors operate by storing charges on two polarized plates, separated by a dielectric material, ECs operate on the formation of Electrical Double Layer (EDL) at each electrode surface, connected in series. The formation of EDLs allows for much greater amounts of energy to be stored. Of course, such improvements are only made possible through the use of the suitable electrode materials, which must display a high electrical conductivity, along with a high microporosity and the right amount of surface functional groups. Moreover, these materials must also be cost effective—which explains why this chapter focuses exclusively on carbon-based materials. Herein, we will first focus on theoretical principles by discussing the fundamental physical properties of capacitor materials, along with structure-property relationships of carbon materials and their electrochemical performance in ECs. The chapter will then introduce several synthetic techniques as well as the use of graphene in ECs. Lastly, the chapter will conclude with a discussion on pseudocapacitance and the effect of heteroatom doping in carbon in electrochemical performance. This work is intended to give interested audiences an introduction and a starting point towards the synthesis and use of carbon materials for use in ECs.

Keywords: Electrochemical capacitors, Carbon materials, Porous carbon, Carbon synthesis, Pseudocapacitance, Doped carbon

---

Department of Chemistry, Oregon State University, Corvallis, Oregon, United States.
* Corresponding author: david.ji@oregonstate.edu

1. Early History of Capacitors

The earliest capacitors were co-invented in the mid 1700s by Ewald Georg von Kleist, a German scientist and Pieter Van Musschenbroek, a professor of math and physics at the University of Leyden in Holland. Unfortunately for von Kleist, these were eventually named as Leyden jars, which became the foundation of the modern day capacitors. Later on, the English chemist Faraday developed the early physical concepts of capacitor technology, where the unit of capacitance is named after him as Farad.

The early types of capacitors were mostly of the "parallel-plate" model that involved two charged plates of opposite charges separated by a vacuum or dielectrics. In the 1950s, a new category of capacitor was invented. Instead of relying on metal plates to store the charges and vacuum to keep them separated, this new capacitor relies on porous carbon materials for charge storage and an electrolyte solution to separate them. The capacitors utilizing electrical double layers were originally invented in 1954 by Becker and Ferry from General Electric. However, it was 1978 when NEC first commercialized such a device by licensing the technology from Standard Oil Company of Ohio [1], and gave it a brand name of SuperCapacitor. These have since then adopted other designations, such as electrochemical capacitors (ECs), electrical double layer capacitors (EDLCs), supercapacitors and ultracapacitors.

Since the early inception of the ECs in the 1950s, the applications of these capacitors have continued to increase. They are now being sought for use in electronic devices, transportation and cranking. Though there has been much progress, there may still be knowledge gaps to fill before such devices can be used more widely with desired intent—and porous carbon materials are going to play a large role in this.

2. Electrochemical Principles

2.1 *The parallel plate capacitor*

Going back to general physics principles, a capacitor electrode is a conductor that can store electrons balanced by net charges. If connected to the ground, the current will flow until the net charges dissipate. This is unlike a battery as a capacitor does not store energy through chemical reactions but with polarization.

To further elaborate, we can introduce the concept of a parallel plate capacitor. In the fully discharged state, the plates are at the same potential and separated by a distance *d*, which is "filled" with a vacuum or a dielectric material. During charging, polarized charges, i.e., electrons and electron holes, are stored on two opposite plates (Fig. 2.1).

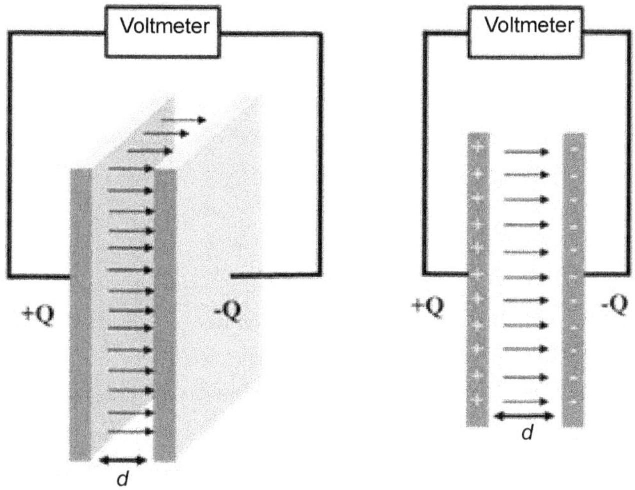

Figure 2.1. A parallel plate capacitor in the charged state with polarized plates and a net electric field as indicated by the arrows.

From these basic principles we can define the term capacitance that alludes to how many charges a capacitor electrode can hold per unit voltage difference between the two plate electrodes. This can be described by the equation relating charge, voltage and capacitance as

$$C = Q/V \tag{1}$$

where C is the capacitance of one electrode, Q is quantity of charges and V is the potential difference between two electrodes.

In the case of a parallel plate capacitor, we can further describe the electric field between the positive and negative plates as

$$E = \frac{Q}{\varepsilon_0 A} \tag{2}$$

where E is the electric field, Q is the stored charge of the capacitor, ε_0 is the permittivity of the phase between electrodes, and A is the surface area of one electrode plate. Since the electric field between the two plates is uniform, we can then write

$$E = \frac{V}{d} \tag{3}$$

where d is the distance between the two plates. Combining the equations 2 and 3, we can now re-write capacitance as

$$C = \frac{\varepsilon_0 A}{d} \tag{4}$$

which relates capacitance to the area of one of the charged plates. From the equation above, we can see that it is in our best interest to maximize the A and to minimize the d in order to enhance the capacitance.

Having described capacitance, we now need to connect it to something more tangible: energy. Charging the capacitor requires inputting work to polarize the plates. By considering the work equation

$$W = Fd \tag{5}$$

and the equation of force in an electric field

$$F = qE \tag{6}$$

we can write

$$W = d \cdot qE \tag{7}$$

If we consider our charge carriers to be electrons and consider moving them at infinitesimal increments, and that the distance multiplied by the electric field can be rewritten as the voltage, we can write

$$\Delta W = V\Delta q \tag{8}$$

Through integration, the total work needs to move the entirety of the charge to one side of the capacitor is equal to

$$W = \frac{Q^2}{2C} \tag{9}$$

Since this is the work done to charge the capacitor, the potential energy in the charged state, assuming that there is no self-discharge, is then equal to the work. By substituting for Q using the first equation, we can then write the more well-known equation of the energy stored in a capacitor as

$$U = \frac{1}{2}CV^2 \tag{10}$$

While we stated earlier that maximizing the area of the plate and minimizing the distance between the two polarized plates are necessary to optimize the capacitance; we can see from the equation above that those are not the only two factors that determine the energy stored in a capacitor. It is also necessary to maximize the potential difference between the two polarized sides. In fact, one could claim that to enlarge the device voltage is even more important, as doubling the capacitance only doubles the energy, while doubling the potential difference increases the energy by a factor of four.

For any energy storage devices, it is important to consider the specific energy in units of Watt·hour/kg and energy density in Watt·hour/liter. While an ideal capacitor should be very light, very small and store a large amount of energy, this is not always possible as some characteristics must usually be forgone to make others possible. The importance of specific energy vs. energy density depends on the applications of the ECs. Lastly, two more metrics that capacitors are usually judged by are specific power and power density. Power refers to how much energy can be dispensed per second, and can be calculated through the equation

$$P_{max} = I_{max} V_{max} \tag{11}$$

where P_{max} is the highest power in Watts, I_{max} is the highest current in Amperes, and V_{max} is initial discharge voltage of the cellin volts. This can then be rewritten as

$$P_{max} = \frac{V_{max}^{\ 2}}{R_s} \tag{12}$$

where R_s is the Equivalent Series Resistance (ESR). The specific power and power density are expressed in units of Watt/kg and Watt/liter, respectively.

While the description of capacitors thus far has focused on parallel plate capacitors and relevant ways of characterizing their performances, electrochemical capacitors (ECs) abide very similar physics principles. Please note that in this chapter ECs do not include pseudocapacitors that are based on Faradaic reactions. Much like the parallel plate capacitors, ECs operate through storage of charges on polarized surfaces, which are often composed of carbon materials that have a much more complex morphology as they are porous materials with very high surface area (Fig. 2.2).

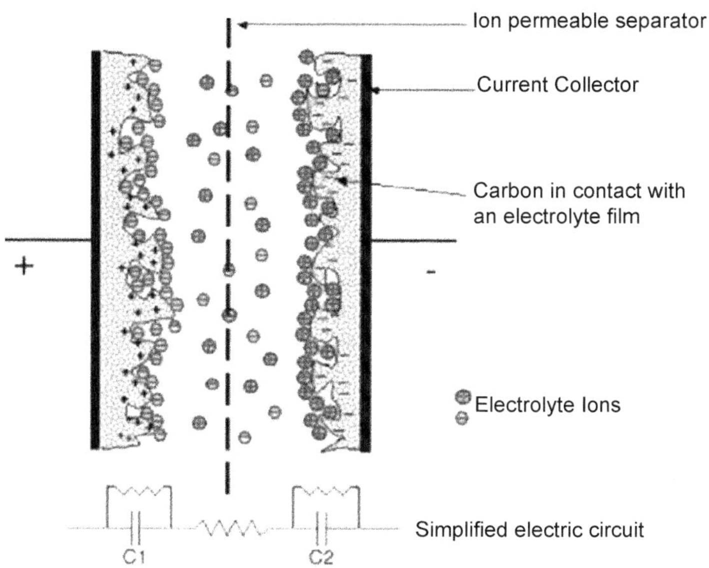

Figure 2.2. Schematic for a symmetric electrochemical capacitor with a simplified circuit diagram at the bottom [2]. Reprinted with permission Ref. 2 © 2006 Elsevier.

This leads to the formation of an electrical double layer (EDL) at the electrode/electrolyte interface as the charges on the inner surface of the electrode materials are balanced by ionic charges from the electrolyte at the interface. The formation of such a double layer has also earned ECs another name: Electrical Double Layer Capacitors (ELDCs).

2.2 The electrical double layer

The theories governing the formation of the EDL are a bit more involved than they were for parallel plate capacitors, as we have to account for the behavior of solvated ionic charges in solution. Though the descriptions of the theories provided below are not meant to provide an in-depth treatise on EDL theory, they are necessary to gain a fundamental understanding of the EDL formation in order to design optimal materials and electrolytes for use in ECs. A more in-depth explanation of EDL formation can be obtained elsewhere [3–5].

The first theory was proposed by Helmholtz in the 1800s. This model proposed that the surface of a terminal should be treated as a capacitor, with charges of opposite signs on either side. On the electrolyte side of the electrode-electrolyte interface, there is a thin uniform layer of charged species, referred to as the Helmholtz plane. Beyond the Helmholtz plane, there is the bulk solution, which for all intents and purposes, is not affected by the presence of the double layer. According to the Helmholtz model, the potential drops linearly from a finite value at the electrode-electrolyte interface to 0 V at the junction of the Helmholtz plane and the bulk solution. Though this model is representative of the EDL, it does have some noticeable shortcomings. First, it does not consider the mobility of ions in solution. Additionally, it does not consider the presence of charges in solution, as the charges are confined to the Helmholtz plane (Fig. 2.3).

For these reasons, the EDL model was revised by Gouy and Chapman in the early 1900s. Gouy-Chapman introduced the concept that the solvated ionic charges were not rigidly distributed, but free to move around. By introducing thermal motion to the system, they showed that the electrostatically attracted ionic species have a tendency to move away from the charged surface—as entropy would dictate, thus giving the provisions for a diffuse layer. However, the presence of electrostatic forces keeps the double-layer largely intact. As further developed compared to the Helmholtz model, which has the charged species solely on the surface, the Gouy-Chapman model stated that the distribution of ions, and therefore the potential, decreases exponentially through Boltzmann statistics in relation to the distance from the electrode surface (Fig. 2.4).

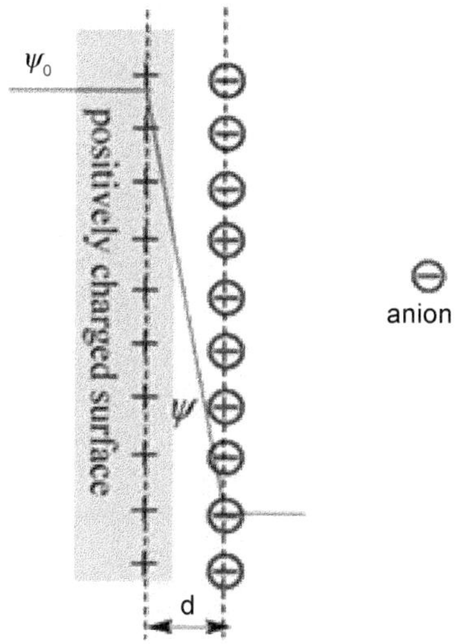

Figure 2.3. Graphical representation of the Helmholtz model for EDL charge storage with a slope representing the voltage potential as a function of distance [6]. Reprinted with permission from Ref. 6 © 2009 Royal Society of Chemistry.

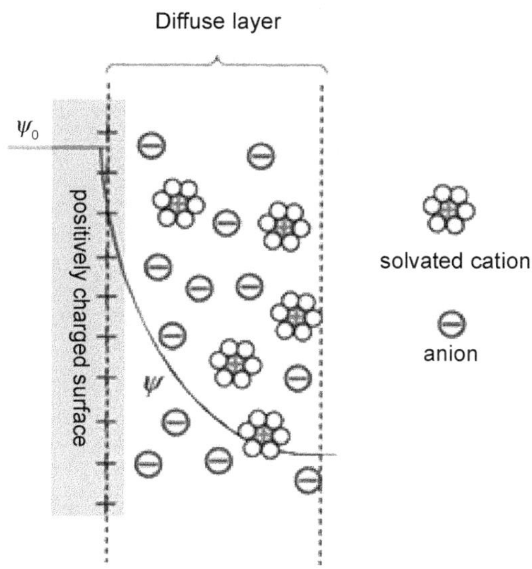

Figure 2.4. Graphical representation of the Gouy-Chapman model for EDL charge storage, along with the slope of the potential as a function of distance [6]. Reprinted with permission from Ref. 6 © 2009 Royal Society of Chemistry.

Though improved from the original Helmholtz model, the Gouy-Chapman model still had some shortcomings. One of its main drawbacks was that the charged species were considered to be point charges. Thus, since the ions do not have measurable sizes in this model, they can, in theory, approach the polarized electrode surface without a limit, thus causing the capacitance to rise without a limit. However, such is not the case in actuality: ions are not point charges and capacitance at the electrode/electrolyte interface does not rise to infinity. As a result, Stern provided a third model in 1924. Stern's model accounted for the size of the ions, and in a way is a combination of the models from Helmholtz and Gouy-Chapman. According to Stern, the immediate electrode/electrolyte interface is composed of a rigid charge layer, specifically adsorbed, referred to as the Stern layer, which accounts for the size of the ions, and a diffuse layer, adsorbed electrostatically, when moving away from the Stern plane. The specifically adsorbed ions in the Stern plane and the electrostatically attracted nonspecifically adsorbed nearest solvated ions are referred to as the Inner Helmholtz Plane (IHP) and the Outer Helmholtz Plane (OHP), respectively (Fig. 2.5). Beyond OHP, nonspecifically adsorbed ions are distributed in a three dimensional region, referred as diffuse layer.

The descriptions of these three theories—especially the latter two, give a critical insight to the roles of the pores and their sizes of electrodes in ECs, as different pore curvatures (i.e., different pore sizes) will affect the formation of charge layers at the electrode surface [7] while influencing

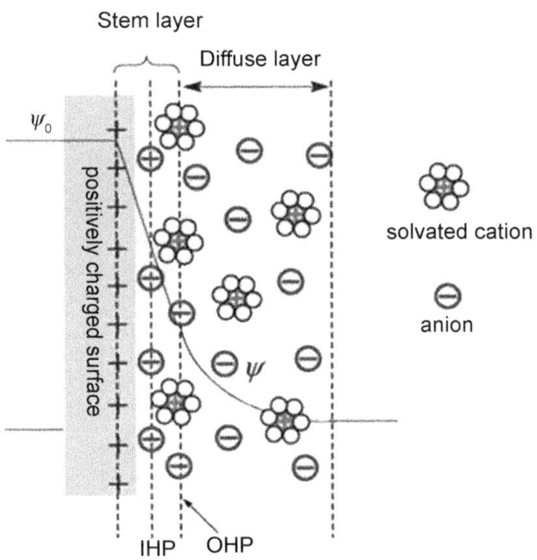

Figure 2.5. Graphical representation of the Stern model for EDL charge storage, along with the slope of the potential as a function of distance [6]. Reprinted with permission from Ref. 6 © 2009 Royal Society of Chemistry.

diffusion, electrosorption and desolvation of the ions in the electrolyte solution [8, 9]. Furthermore, the formation and presence of the EDL, as well as the ion concentration give rise to such phenomenon as overscreening [10] and overfilling [11], which ultimately affect both the overall capacitance and the power performance of the ECs.

3. Carbon Electrode Materials

Carbon is an ideal material for use in ECs. Owing to its many allotropes and morphological configurations, ranging from graphite, graphene, nanotubes, fullerene to diamond, it can be highly conductive, possesses a high surface area and has a tunable pore structure, thus giving us a good control of the final material properties. Furthermore, it also has chemical inertness, good temperature stability, low cost and low toxicity, which makes it an ideal electrode material for practical ECs.

3.1 Carbon structures

To understand the many different types of carbon structures, it is first necessary to understand its intrinsic bonding properties. Carbon can form three different types of covalent bonds with itself. There is the 1D sp^1 carbyne bond, the 2D sp^2 planar graphene bond and the tetrahedral 3D sp^3 diamond bond. All forms of carbon materials that exist have bonds falling into one of these categories. It must be said that there are some special forms of carbon, such as carbon nanotubes (CNTs) and fullerenes that are sometimes referred to as "buckyballs"; however, even such forms of carbon are made through the deformed sp^2 bonds.

On the other hand, the overwhelming majority of carbon materials used in ECs can be broadly ascribed by the label of "amorphous" or "disordered", which is a carbon without a high level of crystallinity. This is not to say that amorphous carbons are all the same: quite the contrary there are different types of amorphous carbons. Ultimately, the name of the carbon matters little when designing materials for supercapacitors. What more important is the physical characteristics of the carbon: the electrical conductivity, the surface area and the porosity.

3.2 Physical properties of carbon materials in ECs

As discussed earlier, carbon materials used in ECs mostly fall under the category of disordered carbons. However, the types of carbon bonds and the atomic structures have distinctive impacts on the performance of carbon as an EC material. One of the first issues to consider is the electrical conductivity of the carbon, which is decided by the type of bonds in the carbon as well as their connectivity. A graphene based carbon with a connected sp^2 network, which is akin to a conjugated π-bond network with delocalized electrons [12], is expected to have higher electrical conductivity than a less-graphenic carbon, which is composed of a mixture of sp^2 and sp^3 carbon bonds. The electrical conductivity of the EC material plays a direct factor in the specific power of the EC, as power is inversely proportional to the series resistance.

3.2.1 Surface area

When evaluating the morphological properties of the carbon materials, two specifically stand out: the Specific Surface Area (SSA) and the porosity. Looking at the governing equations for the properties of capacitor electrodes, we can see that capacitance is directly related to the surface area of the charged plates. Empirically, this is logical as we can argue that a larger surface area allows for more charged species to be stored, thus causing an increase in the capacitance. The general relationship between surface area and capacitance had been demonstrated in some studies, which showed a clear trend between increases in surface area associated with increases in capacitance [13, 14, 15]. However, other studies suggested that the correlation is not as irrefutable as the elementary equations suggest [16, 17]. Kotz et al. reported that although there exists a correlation between capacitance and surface area, increases in capacitance plateaus out at roughly 1200 m^2 g^{-1} (Fig. 2.6) [17]. At surface areas beyond this, the net capacitance does not rise accordingly. The authors further showed that such

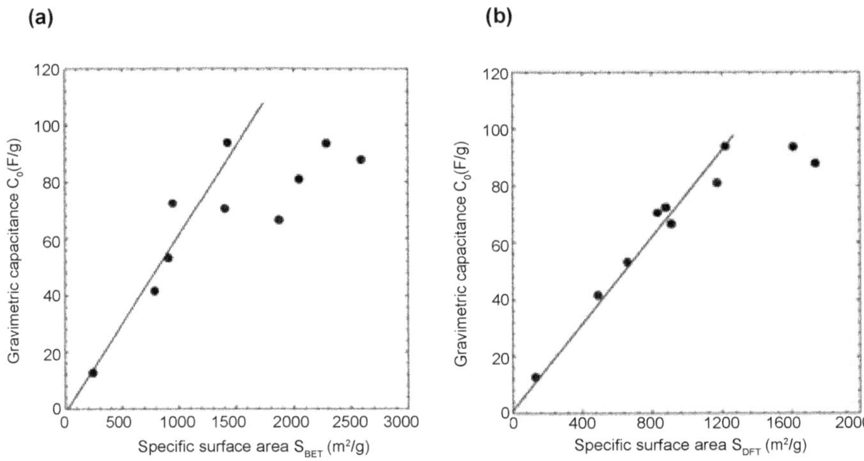

Figure 2.6. Relationship between (a) Gravimetric capacitance and BET surface area. (b) Gravimetric capacitance and DFT surface area [17]. Reprinted with permission from Ref. 17 © 2005 Elsevier.

results were not obtained due to errors in measuring surface area. In the past it had been noted that physiosorption models, such as the Brunauer-Emmett-Teller (BET), Barrett-Joyner-Halenda (BJH) models and the newer Density Functional Theory (DFT) model led to surface areas that were up to 20% different on the same material—especially when high surface areas are involved [16]. However, even accounting for these effects led to the same results. Thus the authors of this paper concluded that while increases in surface area were associated with increases in capacitance, as the pores grow large, the thickness of the pore walls becomes thin enough that there begins to be some negative inter-pore electrostatic interactions, caused by like charges repelling each other.

3.2.2 Porosity

These findings about surface area showed that instead of solely considering surface area of the carbon materials, it was equally as important—if not more—to consider the pore structure.

The pore structure is usually defined through pore size distribution. The aforementioned "pores" are empty spaces within the carbon structure that are bounded by carbon features, such as curved graphene sheets. The pores are not to be confused with the empty interlayer spacing that exists between individual graphene sheets in graphite and graphitic nanodomains—although a large enough dilation of the interlayer spacing could be construed at a "pore" (Fig. 2.7).

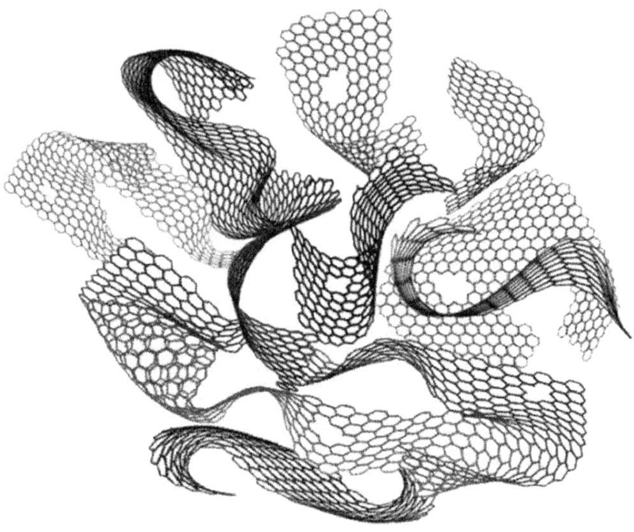

Figure 2.7. An example of a porous carbon material, as solved by a computer simulation [18]. Reprinted with permission from Ref. 18 © 2013 Nature Publishing Group.

The pores that separate the various domains of the structure can fall into three broad categories: micropores that are of a size of 2 nm or less, mesopores that are between 2 and 50 nm and macropores that are the pores greater than 50 nm. All contribute to the overall porosity of the materials; however, their contribution and role in the electrochemical charge storage vary greatly, in terms of both capacitance and power.

The pore size distribution plays a large role in determining capacitance. In terms of capacitance it was long thought that micropores were not beneficial to capacitance as solvated ions would not be able to access them for fault of being too large and not being able to shed their solvation shell. As a result, such micropores would not contribute charge storage, and as such would not be beneficial in increasing the capacitance. It was later revealed that ions in the electrolyte could in fact shed their solvation shell, and effectively be stored in pores smaller than 1 nm [19]. The energy density would be optimized if the size of the micropores matched that of the ion in the electrolyte—essentially suggesting a one-ion-per-pore storage to be the most ideal [20, 21].

These results showed why the method of increasing surface area, as suggested by the traditional model, eventually fails. The traditional area model assumes that the interaction at the EDL is between an ion and a flat surface. This works well for describing charge storage at macropores, as the difference between the radius of the ion and the curvature of the pore is large enough to be modeled as a flat surface. However, such a model fails to accurately represent what occurs in micropores and mesopores. For a while, researchers made the incorrect assumption that micropores sized less than 1 nm could not contribute to charge storage, as the desolvation energy was too great to be overcome.

This led to the proposal of a new model for charge storage by Huang et al. who proposed that charge storage in micropores could be modeled through desolvated ions in a hollow cylinder while mesopores could be modeled by solvated ions in a hollow cylinder [22, 23]. With these assumptions, micropores modeled from the electric-wire in cylinder (EWCC) model makes the equation for the capacitance equivalent to

$$C = \frac{\varepsilon_r \varepsilon_0 A}{b\ \mathrm{In}(b/a_0)} \tag{13}$$

where a_0 is the radius of the ion present in the cylinder, and b is the outer radius of the cylinder. Meanwhile, the equation for mesopores that is from the Electric-Double Cylinder Capacitors (EDCC) model, capacitance can be obtained through the equation

$$C = \frac{2\pi \varepsilon_r \varepsilon_0 A}{b\ \mathrm{In}[b/b-d]} \tag{14}$$

where b is the radius of the outer cylinder, which is the distance from the center of the pore to the pore wall, and d is the distance between the inner-cylinder and the outer-cylinder, which is equivalent to the thickness of the EDL. It is important not to confuse the d in eq. 14 with the d in the other equations, which represents the distance between two oppositely charge plates.

We can now combine these two previous equations with the more general equation for macropores to create a model for the total capacitance as a function of porosity

$$C = \sum_i \frac{\varepsilon_{r,micro}\,\varepsilon_0 A_{i,micro}}{b_i \ln[b_i/a_0]} + \sum_j \frac{\varepsilon_{r,meso}\varepsilon_0 A_{j,meso}}{b_j \ln[b_j/(b_j - d)]} + \sum_k \frac{\varepsilon_{r,macro}\varepsilon_0 A_{k,macro}}{d} \tag{15}$$

As seen in eq. 15, different types of pores have different contributions to the capacitance. This helps to explain why surface area and capacitance are not perfectly correlated, especially when the surface area is increased through the enlargement of micropores into mesopores and macropores. Different pores have different storage mechanisms, and this is clearly reflected in the overall capacitance (Fig. 2.8).

The explanation for the optimization of capacitance with a highly microporous material was rationalized with the aid of Molecular Dynamics (MD) by Merlet et al. [24]. The study suggested that micropores are best positioned to mitigate the overscreening effect in the electrolyte. What is the overscreening effect? Using the positive electrode as an example, the overscreening effect arises from the fact that there are more negative

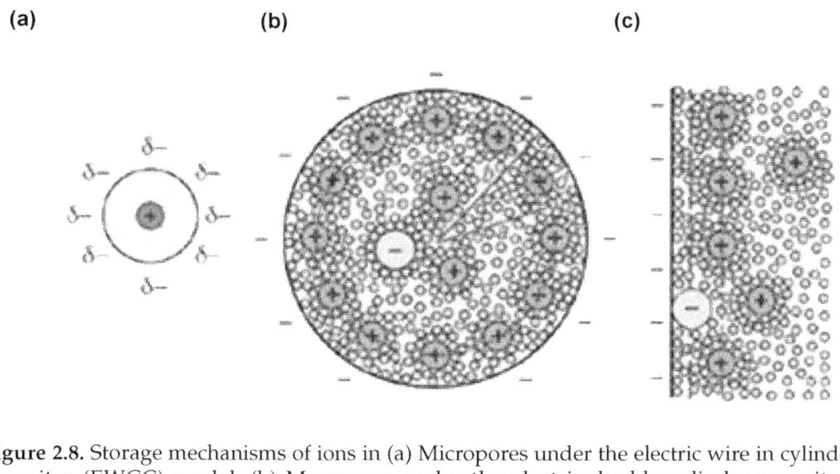

Figure 2.8. Storage mechanisms of ions in (a) Micropores under the electric wire in cylinder capacitor (EWCC) model. (b) Mesopores under the electric-double cylinder capacitors (EDCC) model. (c) Macropores under the traditional parallel plate model [23]. Reprinted with permission from Ref. 23 © 2008 John Wiley & Sons.

charges present at the electrode surface than are needed to balance out the positive charge on the electrode. As a consequence, a 2nd layer of positive charges is formed behind the 1st layer of negative charges to balance the extra charges. This process repeats itself until the point that the oscillations in the net charges of the layers fade away, and the concentration of the electrolyte matches that of the bulk electrolyte. At planar electrode surfaces, those typically found in macropores, the overscreening effect is much more significant than at the curved electrode surface associated with micropores.

Thus, this overscreening effect prevents the close adsorption of ions to the electrode surface, as they are also drawn towards the electrolyte side by the charges in the 2nd layer. However in the case of micropores, there is a much smaller overscreening effect, making it possible for the charges to be adsorbed more closely to the electrode surface, thereby greatly improving its capacitance (Fig. 2.9). Optimization of capacitance with a high concentration of micropores is useful, but we also have to take into account the rate capability of ECs when evaluating their electrochemical properties. The rate capability is highly dependent on the ionic mobility in the electrolyte and diffusion lengths as both factors are critical in determining the overall diffusivity. This in turn determines the power and rate performance: the faster the charged species can diffuse through the carbon material, the faster they can be stored on the EDL and vice versa. Additional ways to improve

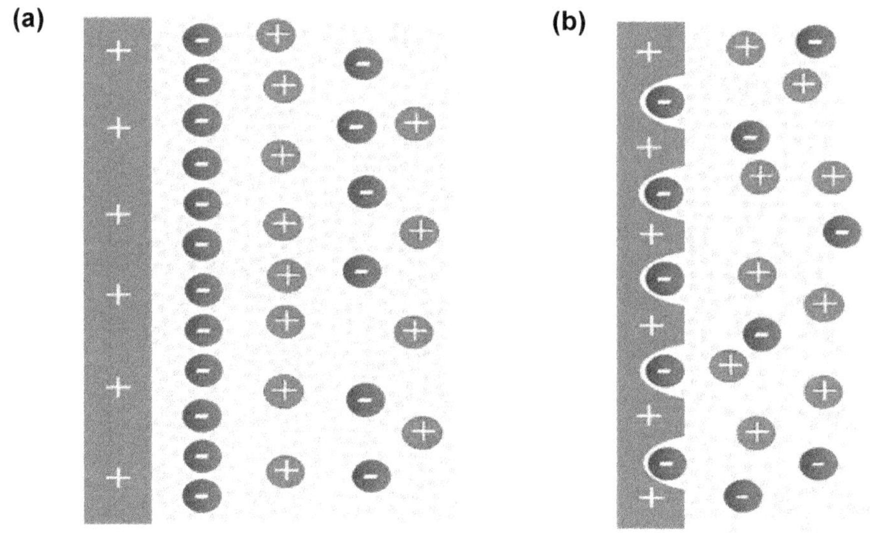

Figure 2.9. Overscreening effect in (a) Macropores. (b) Micropores. We can see from the figure that the conformal fit of the charged species inside the micropores allows for a smaller distance between the electrode surface and ions, leading to a higher efficiency. This is not seen with the macropores.

the power performance of EC are to tune the ionophobicity/ionophilicity of the micropores with ionophobic pores being able to provide a better power performance as they can be charged nearly an order of magnitude faster than ionophilic pores due to a difference in diffusion mechanisms [25]. In fact, it has even been suggested that ionphobic pores not only increase the rate performance, but can also lead to higher energy densities, as the higher voltages are needed to fill ionophobic pores [26].

Cycle life of any energy storage device is critical for practical applications. Some concerns arise as micropores are prone to electroactuation, a process whereby the dimensions of the pores change due to the injection of charge which changes the C-C bond length, with negative charges causing an expansion of the bond length, while positive charges cause a contraction [27]. This effect causes a stress in the carbon structure, with smaller pores experiencing a higher strain [28, 29]. As a result, exclusively microporous materials are prone to degrade faster than materials with a larger proportion of mesopores—though such effects may be mitigated by a proper choice of electrolyte.

Thus, in the context of power and cycle life, an exclusively microporous material has some drawbacks. The diffusion length increases from the bulk electrolyte to the inner pores, restriction of the ionic mobility, the desolvation step at the pore opening and the stress caused by the expansion all impede performance. Thus, including mesoporosity/macroporosity into the structure may prove to be useful.

Despite the potential benefits of more porous materials, simply increasing the porosity is not a sure way of improving the power performance. Unchecked increases in the porosity can lead to unintended consequences. First, it could lead to a drop in overall capacitance. The material may have a good rate performance, but that counts for little when there are no charges to store. The second is that increases in porosity could result in the destruction of conductive carbon features, such as the sp^2 bonds, thereby leading to a drop in electrical conductivity. This would also lead to a decrease in power performance.

3.2.3 Conclusion on physical properties of carbon materials

There are many ways of optimizing the performance of carbon materials in an EC depending on its intended use. For an optimal capacitance, a highly microporous carbon material is needed. For high power, a mesoporous material would be better. The real challenge lies in making a material that can have both a high capacitance and a high power. This is very difficult, as some features that enable one preclude the other—however, this challenge maybe overcome through the materials design and tailoring of surface properties.

4. Carbon Synthesis

Having discussed the operational characteristics of ECs as well as the physical properties of carbon materials needed for these devices, we now turn our attention to their synthesis.

Carbon materials for use in EC can come from a variety of sources. It can be obtained from natural graphite or from hydrocarbon derivative sources, such as coal pitch, petroleum coke and pitch coke. These sources are usually referred to as "soft-carbons" for their ability to graphitize when annealed at high temperatures. Carbon materials can also be synthesized through pyrolysis of biomass products, often leading to "hard-carbon", labeled as such for its inability to graphitize thermally (Fig. 2.10). Lastly, specialty carbons such as graphene, CNTs, carbon onions, and nanodiamonds can also be synthesized through more involved—relative to the soft and hard carbons—synthesis methods.

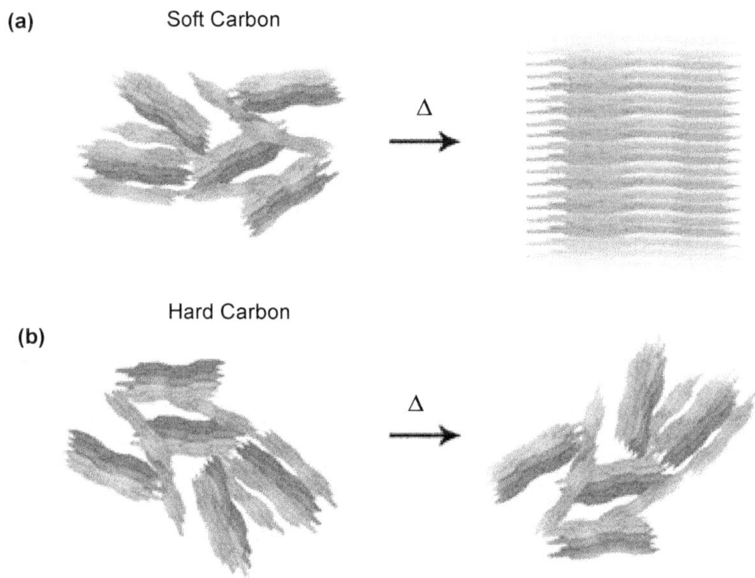

Figure 2.10. Graphical representations of the consequences of the thermal process for the two amorphous carbon categories (a) Soft carbon (b) Hard carbon.

4.1 Soft carbon

The non-natural graphite carbons are usually annealed to increase the electrical conductivity, as it will extend the conjugated sp^2 π-network, while driving out minor amounts of impurities, such as oxygen and sulfur. These products, which include polymers derived from aromatic hydrocarbons, are usually rich in sp^2 bonds, and thus tend to graphitize when annealed at temperatures higher than 2500°C. Due to their ability to re-arrange their structures into one analogous with graphite, these carbons can also be labeled as "soft-carbons". Though graphite has a high electrical conductivity, which is ideal, it also has a low overall porosity, which does not bode well for increased charge storage, power and capacitance.

4.2 Hard carbon

This carbon is popular for use in ECs as it can be formed from the pyrolysis of many organic precursors, which can range from agricultural waste [30, 31], wood products [32–36] to food waste [37, 38]. This makes the cost of the raw material extremely cheap. In fact the only costs associated with obtaining the raw material are those of collecting and transporting it to the milling farm. In addition to the organic precursors, a selected number of polymers that do not fall into the "soft-carbon" category can also be pyrolyzed to form hard carbon. The pyrolysis serves to remove the oxygen and hydrogen atoms from the structure, leaving behind a carbon backbone.

This can be done under either a vacuum or inert gases such as argon or nitrogen, since the failure to remove the evolved gaseous species, such as H_2O and H_2, can end up consuming the carbon material. Additionally, the resulting carbon typically needs to be treated in an acid bath to remove residual metals atoms, such as Na, K, Fe and other remaining artifacts from the organic material.

Unlike the soft-carbons, the precursors have a higher concentration of sp^3 bonds, and as a consequence the final carbon structure does not re-arrange into a graphitic one by a thermal treatment. Under the Franklin's card-house model of nongraphitizable carbon, this material consists of few layered graphene nanodomains that are connected by thermally stable sp^3 bonds and separated by empty voids [39].

Considering the plethora of hard-carbon precursors, as well as their vastly different starting structures, the nature of the precursors as well as the pyrolysis process have very influential effects on the final structural properties of the hard carbon material, including porosity, average graphite nanodomain size, sp^2/sp^3 ratio and electrical conductivity. Naturally, these different structural properties lead some precursors to having better electrochemical properties than others.

Though the concept of the pyrolysis is simple the process is much more complicated. In addition to water, there are many other compounds that are released during pyrolysis, all of which affect the formation of the final carbon structure. There have been a few studies that focus on the effects of pyrolysis on the properties of the final carbon materials. For example,

Wornat et al. studied the pyrolysis process of pine wood and switchgrass, where they identified three processes, including devolatilization that causes release of oxygen and hydrogen rich gases, vaporization of alkali metals, such as K and Na, and short-range order formation of carbon structures [38].

4.3 Specialty carbon

Lastly, specialty carbon products for ECs, such as graphene [40, 41], CNTs [42], carbon onions [43] and nanodiamonds [44] can be obtained through synthetic processes that range from relatively simple procedures such as the Hummer's method, to more involved ones such as Chemical Vapor Deposition (CVD), electric arc discharge and plasma etching.

One advantage to these specialty carbon materials is that there is a greater control over their structural property during the synthesis processes, and undesirable characteristics are easier to be avoided. This is unlike synthesis of soft or hard carbon, which is a lot more difficult to control. Additionally these carbons, especially in the case of the sp^2 family of carbons, usually have greater electrical conductivity than soft and hard carbons, which improves their rate performance as EC electrode materials. However, specialty carbon materials in most cases are expensive to synthesize. CVD, arc-discharge and plasma etching usually require specialized equipment. More commonplace methods, such as the Hummer's method for synthesizing graphene oxide and eventually graphene, are time consuming and require the use of corrosive oxidizing chemicals. Their wide applicability is thus dependent on the discovery of a simple and more cost-effective synthesis.

5. Synthesizing Porosity

The development of pores within the carbon material is of paramount importance to increase the capacitance and power performance of the ECs. One of the most common and cheapest ways of developing porosity and increasing surface area is achieved through activation: differential gasification of carbon.

Activation relies on etching the carbon structure, through either "physical" or chemical methods as to increase the surface area/porosity of the final material. While increases in surface area and porosity are typically related, the notion of increasing porosity deserves more consideration. Porosities are not created equal: there is microporosity, mesoporosity and macroporosity and the choice of the activation method can determine which type of porosity will be enhanced.

Additionally, porous materials can also be formed through tailor made structures such as the pyrolysis Metal-Organic-Framework (MOF) [45–47], polyaromatic-framework (PAF) materials [48, 49], special polymer materials [50], or templating methods (Fig. 2.11) [51–53]. These methods offer excellent control over the porosity of the final carbon material: the structure of the final carbon material will be close to the original structure, or in the case of the template, the negative images. Unfortunately, these methods have the drawback of being both labor and cost intensive.

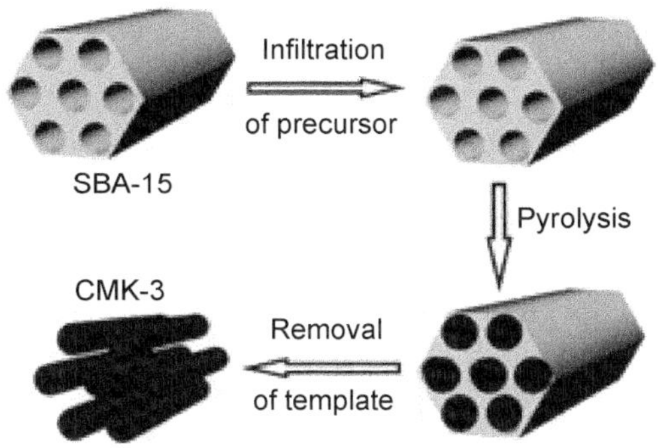

Figure 2.11. Synthesis of porous carbon materials using a template [54]. Reprinted with permission from Ref. 54 © 2014 Royal Society of Chemistry.

6. Activation Methods

Activation methods fall under two different categories: "physical activation" and chemical activation. Though similar in principle, since the goal is to increase the porosity, the two methods operate with different mechanisms, each having their own advantages and drawbacks.

6.1 Physical activation

Physical activation relies on the modification of a pre-existing carbon structure. Such activation relies on gas-solid reactions, which utilizes chemical reactions. Some of the most common physical activation methods involve CO_2 activation and steam activation, though other gases, such as NH_3, have also shown to work well as activation agents.

6.1.1 CO_2 activation

The CO_2 activation involves the Boudouard reaction described below:

$$C(s) + CO_2(g) \xrightarrow{\Delta} 2CO(g)$$

this type of activation is done by flowing CO_2 gas in a furnace at elevated temperatures with a carbon material inside (Fig. 2.12).

Tuning the temperature of the tube furnace, the flow rate of the CO_2 gas, as well as the duration time of the activation reaction allows for control of the degree to which the carbon material is activated. Through simple ΔG calculations, it can be shown that the CO_2 activation reaction will become spontaneous at temperatures above 700°C. This leads to a tradeoff when considering the activation parameters. Achieving a high degree of porosity at temperatures close to 700°C can be a lengthy process. This can be sped up if a higher activation temperature is used, as it will favor the formation of CO, thus leading to more activation in a shorter amount of time. However it was demonstrated that use of higher temperatures also favors the formation of larger pore sizes, which is counter-productive if one is aiming to increase solely the microporosity.

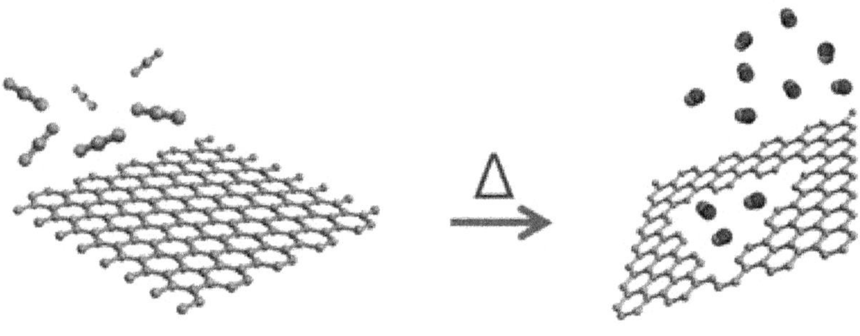

Figure 2.12. Schematic illustration representing the concept of CO_2 activation of carbon. Please note that this is only a representation, and it is not yet clear if sp^2 bonds are the first carbon atoms to be removed.

6.1.2 Steam (H_2O) activation

Another widely used form of carbon activation is done through steam activation. Steam activation involves mixing H_2O in the gas flow to react with solid carbon at elevated temperatures. This will lead to the following two reactions [55].

$$C(s) + H_2O(g) \xrightarrow{\Delta} CO(g) + H_2(g)$$

$$C(s) + 2H_2O(g) \xrightarrow{\Delta} CO_2(g) + 2H_2(g)$$

It can be noticed that in the following two reactions, we obtain CO, CO_2 and H_2 as products, which can then be re-used in the following reactions causing further activation to occur.

$$C(s) + 2H_2(g) \xrightarrow{\Delta} CH_4(g)$$

$$C(s) + CO_2(g) \xrightarrow{\Delta} 2CO(g)$$

Furthermore, the CO, H_2 and secondarily produced CH_4 can engage in feedback loops leading to the further formation of activation agents through the reactions

$$CO(g) + H_2O(g) \xrightarrow{\Delta} CO_2(g) + H_2(g)$$

$$CH_4(g) + H_2O(g) \xrightarrow{\Delta} CO(g) + 3H_2(g)$$

As with the CO_2 activation reaction, some parameters that can be controlled to influence this reaction include the furnace temperature of the reaction, the reaction duration time and the flow rate of the steam used in the reaction. As with the CO_2 activation method, the thermodynamics of steam activation necessitate a temperature greater than 700°C, with higher temperatures and greater flow rates leading to greater amounts of activation in a shorter time. However, there is still the tradeoff between micropores formed in a longer time at low temperatures and larger pores created in a shorter time at high temperatures. Mi et al. found that higher steam flow rates increased the overall volume of the mesopores in their material from 25 to 75% at an activation temperature of 800°C when the flow rate was increased from 0.02 mL/min to 0.12 mL/min [56].

6.1.3 NH_3 activation

The use of NH_3 as an activation agent was also found to increase the porosity and surface area of carbon materials. Systematic studies by Luo et al. showed that annealing cellulose or cellulose-derived N-doped carbon under a NH_3 flow led to activation through the following reaction:

$$4NH_3(g) + 3C(s) \xrightarrow{\Delta} 3CH_4(g) + 2N_2(g)$$

Such a reaction actually has a dual effect on the resulting carbon materials, as it was found to be an effective way to both increase the porosity as well as dope nitrogen atoms into the carbon structure [57, 58]. When tested for capacitance, such NH_3 activated material had a specific capacitance of 120 F/g in an aqueous electrolyte. However, special precautions must be taken before employing such a method, as the use of ammonia gas at high temperatures has a risk of explosion.

6.1.4 Conclusion on physical activation

The physical activation methods: CO_2 activation and H_2O activation are both viable options, whose basic principles are well understood. NH_3 is also an option, but safety concerns make it less ideal, especially if used for large-scale production. It is worth noting that steam activation generates a plethora of side reactions and feedback loops, which can make the process harder to control. As a comparison, CO_2 activation has very little chance of generating side reactions, especially if the starting carbon material is devoid of oxygen and hydrogen impurities. However, as we will soon see, both H_2O and CO_2 activation are much simpler mechanisms than those involving chemical activation, mainly because the physical activation does not interfere with the pyrolysis process, which is highly complex.

6.2 Chemical activation

Chemical activation is the other major type of activation mechanism. Unlike physical activation, chemical activation is a "one-step" process, where the pyrolysis of the biomass or polymer materials, along with the activation process is done simultaneously. In the process, a variety of chemical activation agents are added to the carbon precursor prior to pyrolysis. During pyrolysis, the chemical products will alter the decomposition of the carbon precursor and subsequently change the morphology of the final structure. The chemical agents that have been used for chemical activation can range from alkali based compounds, such as KOH, NaOH and K_2CO_3, to acids such as H_3PO_4 and metal salt products such as $ZnCl_2$ and $FeCl_3$.

The activation mechanisms for these chemical activations are more complex than those of the physical activations, as there are a lot more possibilities for side reactions.

6.2.1 KOH activation

In the case of KOH activation, the reaction mechanism is broken down in the following manner. First, the alkali hydroxide reacts with the carbon to form an alkali metal, hydrogen gas and a carbonate [58].

$$6KOH(s) + C(s) \xrightarrow{\Delta} 2K(s) + 3H_2(g) + 2K_2CO_3(s)$$

From thermodynamic principles, this first reaction can occur at temperatures as low as 570°C. If the temperature of the reaction rises to 700°C and above, the H_2 gas will go on to activate the forming carbon material through the same reactions that had been described with physical activation. Following the initial breakdown of the alkali hydroxide, the metal product along with the carbonate begins to impact the activation of the carbon material. The presence of the carbonate and the alkali metal were confirmed by XRD measurements at an intermediate stage during activation. Potassium in this process plays a very unique role as K atoms can intercalate into the gallery of as-formed carbon structures under heat, forming K-carbon intercalation compound. This is directly related to the activation effect of KOH. Lastly, at temperatures greater than 760°C, the carbonate may begin to decompose, leading to the formation of the CO_2 activation agent as well as a further activation reaction [59].

$$K_2CO_3(s) \xrightarrow{\Delta} K_2O(s) + CO_2(g)$$

$$K_2CO_3(s) + C(s) \xrightarrow{\Delta} K_2O(s) + 2CO(g)$$

Additionally, more feedback loops can become involved in the reaction when we consider the presence of the pure metal along with the breakdown of the metal oxide. The pure metal, in this case K, can combine with CO_2 as such

$$2K(s) + CO_2(g) \xrightarrow{\Delta} K_2O(s) + CO(g)$$

the metal oxide can then also become involved in the activation through the reaction

$$K_2O(s) + C(s) \xrightarrow{\Delta} 2K(s) + CO(g)$$

From the reactions above, the feedback loops can lead to extremely large degrees of activation. However, these feedback loops as well as the contribution of the K_2CO_3 in activation can largely be avoided if the reaction temperatures are kept below 750°C. In fact, much literature published on the use of KOH as an activation material typically concluded that an activation temperature of 700°C is the most conducive to forming an optimal microporosity. This may be due to the fact that lower reaction temperatures will not cause pore widening or pore collapse, as would be at higher temperatures. The activation mechanism with NaOH also relies on the formation of a sodium carbonate for the activation mechanism [60]. However, Na metal cannot be intercalated into carbon structure by thermal treatment.

Furthermore, it was suggested by Wang et al. that a KOH activation used in conjunction with a hydrothermal treatment resulted in exfoliation of graphene sheets while activating them, leading to high surface area porous graphene nanosheets (Fig. 2.13) [61]. The porous exfoliated graphene sheets had a reported capacitance of 113 F g^{-1} at a current rate of 100 A g^{-1}, with an equivalent energy density of 19 Wh kg^{-1} using an organic electrolyte. An even higher capacitance and rate performance from using KOH activation was reported by Kang et al. who used KOH activation on a seaweed-derived carbon, with a capacitance of 280 F g^{-1} at a current rate of

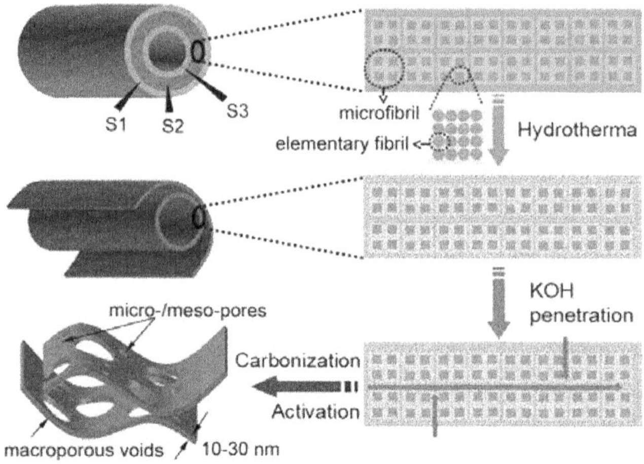

Figure 2.13. Schematic illustrating the pyrolysis and activation of hemp fiber into porous graphene nanosheets [61]. Reprinted with permission from Ref. 61 © 2013 American Chemical Society.

100 A g^{-1} in an aqueous electrolyte [62]. This corresponds to an energy density of 42 Wh kg^{-1}. The high electrochemical performance of the seaweed was attributed to the intrinsic presence of the porosity due to the structure of the precursor material.

6.2.2 H_3PO_4 activation

In the case of activation using acidic chemical agents, such as H_3PO_4, the theoretical mechanisms for the activation reactions are not as developed as the ones for the alkali salts: most of the theories accounting for the activation have come from empirical observations such as physisorption measurements of final structures. Undoubtedly, the introduction of the acids changes the pyrolytic process, a process whose mechanism still eludes us, without even considering the addition of acidic additives. Thus, the field has yet to develop a comprehensive carbon activation mechanism for acids though a few ideas have been proposed.

In the case of activation with acidic agents, it is hypothesized that the acid content first causes a breakdown in the organic precursor at very low temperatures, leading to a re-arrangement of the precursor [63]. Additionally, the addition of acids protonates the oxygens of the glycosylic "linker" bonds in the cellulose chains, which affects the hydrolysis of such bonds during pyrolysis [64]. As the pyrolysis temperature increases, the acids begin to breakdown, forming oxides on the surface of the carbon material. These products eventually along with the re-arranged structure lead to different properties obtained in the final materials.

Despite tenuous understanding of the acidic activation process, numerous studies have revealed methods of controlling the degree of activation, as well as the optimal temperatures needed for the reaction. When controlling activation, it was demonstrated that only small amounts of H_3PO_4 are needed to be added to the precursor to develop the microporosity, and that larger amounts generally cause an increase in larger pore sizes [65]. Most of the literature published on H_3PO_4 activation has suggested that a reaction temperature of 450°C to 500°C was best suited for the development of microporosity [66, 67]. The low temperatures used are evidence that this activation method works by affecting the early stages of pyrolytic process. Whereas other activation methods depend on the consumption of carbon by an etchant at higher temperatures, the use of an acidic agent bypasses this, and instead forms an intrinsic porosity. This is extremely useful, as the lower temperatures promote both a higher microporosity as well as a higher yield of the final carbon material.

On the other hand, the use of acidic materials as activation agents also results in different surface properties of the carbon material. In a study that compared activation via H_3PO_4 activation versus CO_2 activation, SEM images of the final carbon products revealed a much rougher surface percolated with cavities in the case of the H_3PO_4 activation, compared to a much smoother surface in the carbon activated via CO_2. The authors of that study concluded that activation using acidic materials had its greatest impact on the surface of the carbon, as opposed to CO_2 activation that had a greater impact on the inside structure [68].

One should note that not all acids can be used as activation agents. H_3PO_4 is a highly capable activation agent, creating high microporosity during pyrolysis. This may have to do with its capability of dehydration and its lack of oxidizing power. In contrast, mixing H_2SO_4 into carbon precursors do not increase the surface area at all but decreases the surface area of the resulting carbon.

6.2.3 $ZnCl_2$ activation

It is well known that addition of $ZnCl_2$ into the carbon precursors changes the dynamics of the pyrolysis process, leading to the formation of pores. $ZnCl_2$ liquefies at as low as 290°C and boils at 732°C, leaving a porous carbon material. Though the boiling point is at 732°C, one must remember that this is in the case of a pure material, and that the boiling point of an impure material is expected to be slightly lower. Hu et al. reported a significant weight loss occurred between 500°C and 750°C, with a peak coming around 690°C through thermogravimetric analysis (TGA) and Differential Scanning Calorimetry (DSC) measurements. This weight loss was attributed to the vaporization of the $ZnCl_2$, which may eventually cause the porosity. The

TGA and DSC measurements were also supported by visual observation of white smoke that was released from the carbon char at these temperatures during the pyrolysis process [69].

Although the exact mechanism of the metal salt activation process is unknown, there has been a substantial amount of research done to understand the relationship between the activation process involving metal compounds and the properties of the final materials. These have shown that using higher amounts of activation agents usually leads to the formation of larger pore sizes, while smaller amounts of activation material make for a more microporous carbon. This had been attributed to a "pore-widening" effect of the metal particles. Additionally, it has also been shown that annealing at higher temperatures also causes a drop in microporosity, which is accompanied by an increase in mesoporosity and macroporosity [70, 71]. Such an increase is explained by the collapse or merging of the micropores at higher temperatures. In fact it was shown that the optimal porosity using $ZnCl_2$ activation was at an annealing temperature of 700°C, with temperatures of 800°C or higher leading to an overall larger pore size distribution along with a lower surface area [72]. Lastly, it was also demonstrated through both Scanning Electron Microscopy (SEM) and Atomic Force Microscopy (AFM) that the surface properties of the activated carbon were very impacted by the process (Fig. 2.14) [73].

Using a silk precursor material, along with $ZnCl_2$ activation, Hou et al. synthesized a carbon material with a surface area close to 2500 m^2/g and a specific capacitance of 155 F/g at a current rate of 10 A/g [74]. Similarly, $ZnCl_2$ activation of glucose yielded a porous carbon with a surface area of 2560 m^2/g and a capacitance of 196 F/g at a current rate of 0.5 A/g [75].

Figure 2.14. Polyanaline (PANI) synthesized (a) With a $ZnCl_2$ activation. (b) Without a $ZnCl_2$ activation [73]. Reproduced with permission from Ref. 73 © 2011 Springer.

6.2.4 Conclusion of chemical activation

In summary, it can be said that chemical activation has the advantage of being able to form highly microporous materials. Though the exact science of the micropore formation is still unknown, it has been experimentally demonstrated that proper use of chemical activation agents, coupled with the right pyrolysis conditions will lead to highly microporous materials. This is partially due to the fact that the chemical activation process can occur at temperatures as low as 400°C, with temperatures of 700°C shown to be optimal. The effective formation of porosity at lower temperatures is due to the fact that such pores form simultaneously with the pyrolysis process, which mainly occurs in the 400°C to 700°C range. The lower temperatures are beneficial, as higher temperatures have been shown to increase the phenomenon of pore collapse and pore widening, both of which lower the microporosity while increasing the macroporosity. By comparison, the lowest temperatures which CO_2 and steam activation can be effective are at 700°C, with higher temperatures being more beneficial to drive the activation reactions. However, these higher temperatures also favor the formation of larger pores.

While the low temperature reactions and high micropore formation are beneficial, chemical activation methods also have a few drawbacks. First, there is the use of chemical agents, many of which are highly corrosive. Additionally, these activation methods require extensive washing of the final carbon materials with water, and in the case of metal compound activation, acids, both of which bring about higher cost, especially if disposal and water treatment costs are considered.

7. Synthesis and Activation of Nanoporous Graphene

While great progress has been made on improving the performance of activated carbon materials, the unique structure of specialty carbons, such as graphene and nanotubes gives those carbons an electrical conductivity that is extremely difficult to replicate in carbon materials derived from "natural" sources. Furthermore, these carbon materials have very high theoretical surface areas, with single-walled nanotubes having a theoretical surface area of 1300 m^2/g and single layer graphene doubling that at 2630 m^2/g. However when defects are considered, higher surface areas are possible. In fact, computational simulations of a porous graphene material have estimated surface areas as high as 3600 m^2/g [18].

Despite the high theoretical surface areas, and the possibility of even higher actual surface area, for a while, researchers were unable to maximize the surface area of graphene to its full theoretical extent. The main challenge to synthesizing high surface area/high porosity graphene is to prevent the re-arranging of the individual graphene sheets into few layered domains due to the favorable electrostatic interactions. This failure of developing significant surface area was a problem for both the "top-down" and "bottom-up" synthesis approaches.

7.1 Top-down synthesis

The "top-down" approach involves the formation of Graphite Oxide (GO) followed by its exfoliation and reduction leading to graphene in the form of reduced Graphene Oxide (rGO). This is one of the most common synthesis procedures, as it leads to a high yield while being relatively simple and inexpensive and is well suited for the bulk production of graphene. Unfortunately, this approach is especially prone to the restacking of individual sheets during the exfoliation process. This is due to water molecules that facilitate the re-arranging and restacking of graphene sheets via hydrogen bonds [76]. Additionally, the absence of the oxygen functional groups on the graphene surface also leads to the restacking of the sheets, as the electrostatic interactions between neighboring sheets become favorable again [77].

One solution to the restacking issue was the addition of an activation step following the graphene exfoliation. Most of the activation processes that have been described earlier are applicable for increasing the surface area of graphene sheets obtained synthetically with some resulting in greater activation than others. For example, Zhu et al. reported a drastic increase in surface area to 3100 m^2/g by performing KOH activation on a microwave-exfoliated/reduced GO, where this material exhibits a specific capacitance of 166 F/g at a current rate of 5 A/g, and an energy density of 70 Wh/kg using an organic electrolyte [78].

Other ways of activating the graphene materials included adding $KMnO_4$ to the GO before the reduction and acid treatment, and hydrothermally reducing GO in the presence of H_2O_2. The $KMnO_4$ treatment employed by Fan et al. increased the surface area of the graphene nanosheets (GNS) from 267 m^2/g to 1320 m^2/g, resulting in a drastic capacitance increase [79]. The hydrothermal H_2O_2 treatment reported by Xu et al. showed an increase in surface area from 260 m^2/g to 830 m^2/g [80]. Acidic treatments with H_3PO_4 [81], HNO_3 [82] have also been demonstrated to increase the porosity of graphene sheets formed from GO, though the increase in porosity was not as high as the KOH activation treatments.

Addition of the activation agents helps to increase the surface area by both creating more defects, as well as preventing the restacking of the graphene layers. Prevention of the graphene restacking can also be done through a chemical method, whereby molecules or polymers are inserted between the GO sheets. One such method involved inserting functionalized carbon spheres between the GO sheets, as demonstrated by Wang et al. [77]. The functionalized carbon spheres act as "spacers" between the GO sheets, thereby preventing the restacking during the reduction. This effectively increased the surface area of the final graphene product from 77 m^2/g to 1250 m^2/g. Aside from the insertion of carbon nanospheres, the use of polymers was also effective.

Lastly, another example of reducing GO to yield a high surface area graphene structure for use in supercapacitors was via laser-induced reduction [83]. This method developed by El-Kady et al. led to the formation of a self-standing graphene electrode with a specific surface area of 1520 m^2/g and a specific capacitance of 265 F/g, retaining almost half of that capacitance at current rates of up to 1000 A/g.

7.2 Bottom-up synthesis

The bottom-up approach focuses on the formation of a highly porous graphene through a monomeric assembly from a carbon precursor, typically around templates or particles that can be removed later. This approach has some advantages over the top-down approach. The use of templates in the monomeric formation of the graphene structures makes it possible to form a controllable 3D hierarchical structure. Such a structure can be both highly microporous, which is advantageous for capacitance, and having a well-defined interconnected pore network, which is good for the power performance of ECs.

One example of such a synthesis was demonstrated by Li et al. who pyrolyzed a nickel containing ion exchange resin with KOH into a graphene structure. The presence of the nickel ions in the resin led to the formation of a three-dimensional porous structure, as the resin monomers formed a graphene structure surrounding the metal ions/atoms before being washed out. The use of this method led to an interconnected carbon structure with a surface area of 1810 m^2/g, which is substantially higher than what would have been obtained through pyrolysis of the resin on its own [84]. The hierarchically porous carbon structure showed a capacitance of 156 F/g at a current rate of 32 A/g and a total energy density of 38 Wh/kg while using an organic electrolyte.

Other "bottom-up" syntheses processes include chemical vapor deposition (CVD) onto a template that is then etched away, and pyrolysis of ordered co-polymers, which leaves behind a porous carbon structure. The

CVD process is advantageous in preparing edge-oriented graphene [85]. Aligning the graphene planes to be perpendicular to the current collector allows for a full exploitation of the graphene surface area, as opposed to graphene sheets which are stacked parallel to the current collector (Fig. 2.15). Such "bottom-up" approaches to graphene synthesis have been extensively covered in other studies [86, 87] where interconnected graphene materials are formed for energy storage and catalytic applications.

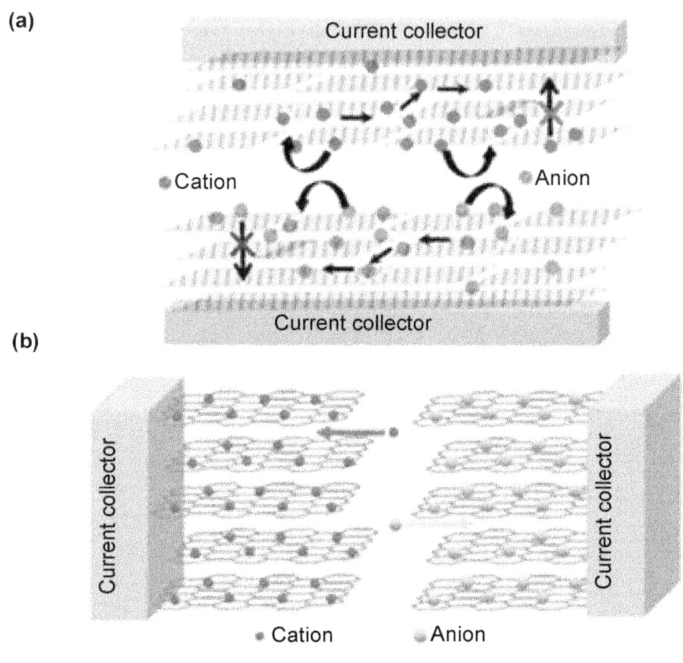

Figure 2.15. Schematic representation showing the difference between graphene aligned with a current collector on (a) The basal plane. (b) The edge plane, which is also referred to as vertically aligned graphite [85]. Reprinted with permission from Ref. 85 © 2011 American Chemical Society.

7.2.1 Metallothermic synthesis

In contrast to the typical bottom-up synthesis approaches, one recent method of graphene synthesis managed to use the same principles while bypassing the high-cost of templates and precursors to form an anoporous graphene with high porosity and high surface area. It was found in 2011 by Chakrabarti et al. that burning magnesium in the presence of dry ice yielded few-layered graphene from the reaction as follows:

$$2Mg(s) + CO_2(s) \xrightarrow{\Delta} 2MgO(s) + C(s)$$

Hot magnesium reduces CO_2, leading to a molecule-by-molecule synthesis of graphene at the nanoscale [88]. Though no specific surface area was mentioned for the products in the paper, it can be assumed that the presence of the few-layered graphene imbued a surface area much lower than the theoretical one of graphene. Improving on this method, Xing et al. synthesized a nanoporous graphene from a similar fashion; only the CO_2 source was gaseous, but not in dry ice (Fig. 2.16) [89]. However, as with the previous graphene synthesis process, one of the main challenges is to increase the surface area. The authors of this study were able to overcome such a challenge with the addition of metallic zinc to magnesium as an activation agent. The surface area of the original graphene structure was 890 m^2/g; however, the presence of zinc led to a highly nanoporous graphene with a final surface area to 1900 m^2/g, which is much closer to the theoretical surface area of graphene. The activated porous graphene displayed a capacitance of 170 F/g at a high scan rate of 2000 mV/s in cyclic voltammograms in an aqueous electrolyte.

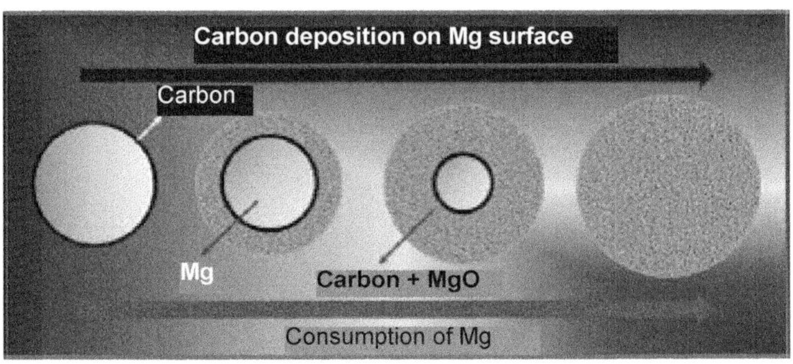

Figure 2.16. Schematic representation of the formation of graphene from the flow of CO_2 gas over magnesium metal [89]. Reproduced with permission from Ref. 89 © 2015 Elsevier.

7.3 Conclusion of porous graphene synthesis

With its high conductivity, high theoretical surface area and ability to form porous structures, porous graphene is a promising carbon material for use in ECs if certain properties such as the porosity and density can be maximized. The main challenge is to optimize the electrochemical performance with a minimal cost. While many graphene materials have been shown to have excellent properties as EC electrode materials, the cost of synthesis is still high when considering its commercial applications. This conundrum can be addressed if cheaper top-down or bottom-up methods, such as metallothermic methods, can be efficient enough to make high quality graphene materials with desirable properties for ECs.

8. Heteroatoms in Carbon Structures

Aside from porosity, the type of surface of the EC electrodes is critical towards its electrochemical performance. The presence of heteroatoms on the surface of an EC electrode will greatly affect the performance. This is a widely explored subject as heteroatoms in the structure bring about many interesting changes to the physical and electrochemical properties of the resulting material—especially in the case of graphene [81, 90]. Aside from the intentional doping of atoms in the carbon structures, we must also consider the fact that many of the carbon synthesis/activation processes will inevitably leave behind doped heteroatoms; despite attempts to remove them [91].

First and foremost, doped atoms can change the electronic structure of the host materials—much like the case of n-doped and p-doped semiconductors. In addition to changing the bulk properties of the structures, heteroatoms also change the surface and charge storage properties of carbon materials.

Some of the immediate properties that are affected by the presence of surface groups are the Potential of Zero Charge (PZC), electrical conductivity, and wettability of carbon materials due to different surface interactions. When the surface is coated with oxygen functionalities, the PZC is found to shift to higher potentials, whereas hydrogen functional groups typically lower the PZC. Typically, electrical conductivity tends to be lower in carbon materials that contain a lot of oxide functional groups, such as graphite oxide or graphene oxides, as they prevent conductive π-π interaction [92, 93]. On the other hand, the increased presence of surface oxides improves the wettability of the electrode surface, especially if the EC is using an aqueous based electrolyte. The improved wettability of the electrode surface is helpful, as there can be better contacts between the electrode surface and the electrolyte, which allows for smaller diffusion length for the ions to go from the bulk electrolyte solution to the carbon surface and vice versa. This allows for improved power performance (Fig. 2.17).

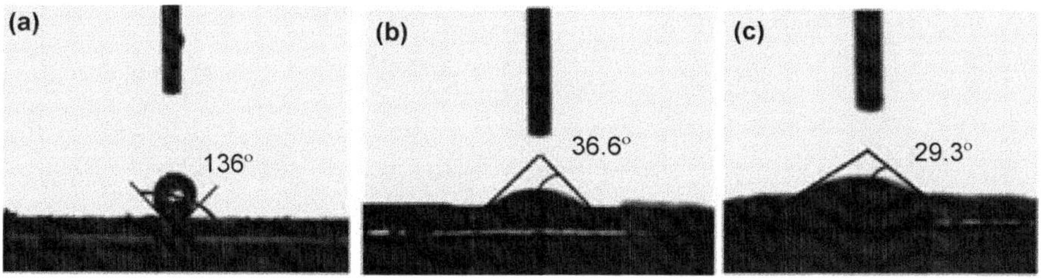

Figure 2.17. Contact angle measurements of the electrolyte on a series of carbon surface that are progressively more functionalized by nitrogen groups [94]. Reprinted with permission from Ref. 94 © Elsevier 2013.

9. Pseudocapacitance from Heteroatoms

Dopant atoms have a large impact on the electrochemical properties as they can also engage in pseudo capacitive storage. Whereas Electrical Double Layer (EDL) capacitive energy storage only involves the storage of charges at electrode surface through the formation of a double layer, pseudo capacitive storage involves fast solid-phase diffusion of ions accompanied with electron transfer—a process sometimes referred to as electrosorption [95]. In short, EDL capacitive charge storage does not involve a Faradaic reaction, whereas a pseudo capacitive one does.

Charge storage on carbon-based materials can be via different mechanisms. The presence of heterogeneous surface groups can change the very nature of how an EC operates; as such surface groups can store ions via pseudo capacitance. Due to the different kinetics of charge storage in an electrical double layer and electrosorption, the current contributions from the separate mechanisms can be distinguished through Cyclic Voltammetry (CV) measurements through the equation

$$I(V) = k_1 s^{1/2} + k_2 s \tag{16}$$

where $I(V)$ is the current as function of the scan voltage, k_1 is a constant for current contributions that are diffusion limited, and k_2 is a constant for "capacitive" current contributions, including EDL capacitance and pseudo capacitance; lastly the s term is the voltage scan rate [96]. Conducting multiple CV scans at different sweep rates can solve for both k_1 and k_2.

Detailed discussion on the subject of pseudo capacitor materials is beyond the scope of this chapter. There exists a host of viable candidates, especially when considering transition metal oxides such as RuO_2 and MnO_2. However, this subject cannot be ignored in the context of carbon materials, as dopant atoms in the carbon structure/on the carbon surface typically induce pseudo capacitive contributions.

10. Heteroatom Doping

As mentioned earlier in this chapter, a vast number of heteroatoms can be doped in the carbon structures, each bringing an additional layer of complexity to the electrochemical properties of the carbon materials, especially due to potential contributions to the total energy storage via pseudo capacitance. Though there are many types of heteroatoms that can be doped or anchored onto the carbon host, we will only focus on some representative non-metal elements in this chapter.

10.1 Oxygen

Hsieh et al. showed that a simple O_2 treatment of a carbon material at 250°C significantly increased the presence of the oxide-containing groups on carbon surface, such as carbonyl and quinone bonds. The oxygen-rich carbon material was then demonstrated to have a capacitance 25% greater than the carbon that had not undergone to O_2 treatment. Furthermore, it was also demonstrated that the O_2 treatment had very little impact on the carbon structure in terms of porosity and surface area, implying that the gains in capacitance were most likely a result of pseudo capacitive from the oxygen groups [97]. The increasing in capacitance stemming from an increase of surface oxygen groups was also corroborated in many other studies [98–100]. In addition to increasing capacitance, oxygen doping was shown to increase ionic conductivity, as revealed through impedance studies [101].

Though most studies focused on different methods of synthesizing oxygen-rich carbon surfaces, Hulicova-Jurcakova et al. [102] were able to show that certain functional groups contributed more than others in increasing the capacitance. In a systematic study evaluating surface functional groups through XPS measurements, they showed a well-defined linear relationship between the concentration of oxygen present in micropores and the capacitance. Interestingly, they found that including phenol and ether groups into their model caused a decrease in the linearity, thereby suggesting that such groups do not actively contribute to the pseudo capacitance, thus showing that not all oxygen functional groups contribute equally.

10.2 Nitrogen

Besides oxygen, nitrogen is the other most common type of heteroatom used to dope carbon structures used in ECs [103]. Doped nitrogen atoms alter the charge/donor properties of the host structures. The change in electronic properties of carbon is expected, as the extra electron of the nitrogen results in an n-type doping, which could raise the Fermi level and increase the electronic conductivity [104]. Furthermore, N-doping also leads to the pseudo capacitance. As a result, the increase in overall capacitance has been shown closely related to the amount of nitrogen doped into the carbon structure [105].

More specifically it was hypothesized by Lota et al. that the presence of nitrogen at "top" sites of a six membered carbon ring and the quaternary nitrogen were responsible for the pseudo capacitive reactions [106]. It was reasoned that the nitrogen present at a top site was more electrochemically active because of its strong electron donating characteristics. Contributions from quaternary nitrogens were attributed to their impact on the overall band structure as the n-doped effects of the nitrogen shrink the gap between the HOMO and LUMO, thus increasing the conductivity. It was also suggested by Jin et al.—although the context of their work was related to catalytic reactions as opposed to capacitors—that quaternary nitrogens are the most stable and as a consequence are able to facilitate redox reactions [107].

The effects positive effects of quaternary nitrogen in the carbon structure were further corroborated by Hulicova-Jurcakova et al. [102] who demonstrated that increases in quaternary nitrogen groups as well as bipolar nitrogen oxide groups resulted in improvements in capacitance and electrochemical performance at high current rates. The improvements at high current rates are most likely due to the better electronic conductivity brought on by these functional nitrogen groups.

Scouring the literature, there are several different approaches to synthesizing nitrogen doped carbon materials. One of them involves the pyrolysis an organic precursor containing nitrogen, such as proteins [108–110], coffee beans [37, 111] (containing nitrogen from caffeine) and even hair [112, 113]. The final content of the nitrogen in the carbon structure obtained by these methods is a function of both the starting nitrogen content, as well as the annealing temperature used in pyrolysis, with higher annealing temperatures leading to a smaller doping percentage of nitrogen. Aside from pyrolysis of organic polymers, nitrogen doped compound can also be obtained via NH_3 induced nitridation [57, 114, 115], and plasma-enhanced chemical vapor deposition [116–118] and solvothermal synthesis [119].

10.3 Boron

Moving on from oxygen and nitrogen doping, which have received the most attention in the literature, we continue to discuss lesser-explored—but still very useful—dopants such as boron [120, 121], sulfur [122, 123], phosphorous [122, 124, 125] and the halogens [126, 127].

Boron dopant atoms also make a good candidate for the doping of carbon, especially in the case of graphene. Considering that Boron usually eschews the octet rule in favor of forming three bonds, it is ideal to replace a sp^2 carbon in graphene. Replacing a sp^2 carbon with a boron atom yields a positively doped structure, where the boron atom can act as a Lewis acid—an electron pair acceptor—thus leading to pseudo capacitance.

B-doping of mesoporous carbon was studied by Wang et al. [128]. Boron was doped via the mixing of the boric acid with a carbon precursor—with the boron content being of less than 1%. In the case of this B-doped material, the increase in capacitance was attributed to both the surface oxygen groups introduced in the acid doping and p-type holes stemming from the electron deficiencies of boron embedded in the carbon matrix. Han et al. also synthesized boron-doped graphene nanoplatelets through reflux of a GO solution refluxed with a borane-tetrahydrofuran (THF) adduct [129]. Even though the surface area of the boron doped rGO was of only 466 m^2/g, the capacitance was still found to be very high, most likely from pseudo capacitive contributions from the boron and oxygen functional groups. In fact, when the capacitance was normalized for surface area, the boron-doped material was found to have a much higher areal capacitance than the undoped material.

Aside from pure boron doping, boron nitrogen co-doping of carbon material has also been investigated for the synthesis of supercapacitor materials. Such boron nitride doped structures include boron-nitride-carbon CNTs as investigated by Iyyamperumal et al. who used a CVD method to grow vertically aligned B/N/C CNTs [130]. This approach combines many advantages: the pseudo capacitance from the heteroatoms, the high electronic conductivity from the CNTs and the favorable ionic conductivity stemming from the vertically aligned CNTs—as had been discussed before. These structures demonstrated a capacitance of 312 F/g, nearly twice that of the non-aligned B/N/C CNTs, and three times that of undoped MWCNTs. Additionally, capacity retention was also found much improved in the vertically aligned B/N/C structure over that of the regular CNTs. While this method yielded very good results, its scalability is not ideal.

Fortunately there are other, and simpler, methods of co-doping boron and nitrogen. Guo et al. who synthesized them through the pyrolysis of a boron and nitrogen containing gel [131]. The gel was also infused with metal particles which were then etched out to increase porosity. Electrochemically, it was found that the doped carbon material had a capacitance nearly twice that of the dopant free material—despite having comparable surface areas. Other methods of co-doping carbon structures include plasma treatments of GO [132], pyrolysis of melamine diborate [133], zeoliticimidazolate frameworks (ZIFs) [134] and a laser induction process [135].

10.4 Sulfur

Sulfur doped carbon, as synthesized through the simultaneous pyrolysis and activation of a S-containing polymer by Gu et al. The authors suggested that the presence of sulfur in the polymer helped to resist the pore shrinkage that occurs during activation by the formation of S-bridges in the newly formed carbon structure. This was able to result in a high surface area carbon with a diminished amount of pore "bottle-necks" which allowed for improved power performance, as the capacitance stayed almost constant when the sweep rate increased from 1 mV sec^{-1} to 50 mV sec^{-1} [136].

Similar work by Zhang et al. showed that mesoporous carbon formed by the pyrolysis of a sulfonated pyrrole polymer was able to obtain a capacitance 40% greater than the regular carbon materials. The authors attributed this to the improved surface properties enabled by the sulfur and nitrogen functional groups on the carbon surface. Additionally, they were able to show that the doped carbon material possessed better capacity retention at higher current rates than the regular carbon, thereby also showing the effects of the surface groups of the electrochemical performance [137]. Similar to the other doped carbon materials, these carbons typically do not have as high of a surface area as activated carbons, but the presence of functional groups such as the sulfoxides and sulfones make up for the difference, leading these carbon materials to having a higher areal capacitance [138]. It is likely the presence of the functional groups improve the wettability of the carbon surface, while doping the carbon with more electrons, as nitrogen and sulfur are both n-type dopants.

Unlike co-doping of boron with nitrogen into carbon for ECs, there is not a lot of present literature that has explored co-doping of sulfur and nitrogen for ECs. These materials are more typically used for oxygen reduction electrodes. However a few examples do exist, such as co-doping from the pyrolysis of an organic material [139] as well as the preparation of graphene doped with carbon-nitrogen-sulfur quantum dots [140] were proven to have favorable outcomes.

10.5 Phosphorous

Phosphorous doping is largely studied in carbon EC materials even if most of the times it is not the original intent: any activation of carbon material that relies on using a phosphoric acid activation will forcibly have some leftover phosphorous groups on the surface, especially if low temperatures are used. This could help to explain why using phosphoric acid as an activation agent: not only helps increase microporosity, but it also induces pseudo capacitive effects, thereby increasing the overall capacitance in two ways.

This dual improvement, along with the relative ease of the synthesis, makes phosphorous doping of carbon materials highly attractive for EC materials. It has been shown by Karthika et al. that a simple activation of GO with phosphoric acid was able to yield a carbon material with a capacitance of 367 F g^{-1}, which was significantly higher than the graphene not activated [141]. Whether the improvement came from the changes in morphology brought on by the activation, or the presence of the phosphorous groups can be debated, as the two properties were changed at once.

In a similar fashion to the other dopant atoms, co-doping of phosphorous and nitrogen has also been investigated for use in ECs. Xu et al. were able to synthesize some phosphorous-doped carbon through the combination and subsequent pyrolysis of phytic acid and polyaniline (PANI) fibers [142]. The performance of the phosphorous-nitrogen co-doped carbon was superior to that of the solely nitrogen doped carbon, obtained from the pyrolysis of PANI, in terms of power performance, capacitance and capacity retention during long cycling. A similar carbon nanofiber approach was reported to synthesize co-doped material through electrospinning of polyacrolynitrile mixed with phosphoric acid [143]. The electrospun co-doped carbon fibers were then pyrolyzed at 800°C and tested for electrochemical performance. The results showed an interesting revelation: the fibers that had been doped with the highest phosphorous content had the lowest surface area at only 10 m^2/g and the highest phosphorous content. However, it is that material which had the highest capacitance of all the materials which were synthesized at 213 F g^{-1}, which was close to 100 F g^{-1} higher than the undoped carbon fibers that incidentally had a higher surface area. Similar results were also shown by Nasini et al. who also showed the improved performance of a phosphorous-nitrogen doped carbon synthesized from pyrolysis of tannin cross-linked to melamine in the presence of polyphosphoric acid which a capacitance of 271 F g^{-1} [144]. This shows that phosphorous doping of carbon materials may be a viable strategy towards improving performance of carbon materials in EC [145].

10.6 Halogens

Lastly, there have also been some studies conducted towards halogen-doped carbons and their performance as EC electrode materials. A study by Tanaike et al. doped both bromine and iodine into a variety of carbons by mixing the carbon materials in containers sealed with Br_2, I_2 and BrI gas [146]. It was found that the open circuit potentials of the doped carbon materials were more positive than the potentials of the undoped materials, especially in the case of the carbon synthesized in Br_2. This p-doping effect of the halogen atoms on the carbon material suggests that the incorporation of the dopant atoms is mostly on the surface of the carbon—as opposed to being embedded in the carbon matrix, as the replacement of a carbon atom in matrix by a halogen one would lead to the presence of extra electrons. When testing the electrochemical performance of electrode materials, it was found that the bromine doped material exhibited a greater capacitance that the undoped carbon, most likely due to pseudo capacitive contributions from the bromine atoms on the carbon surface.

A similar experimental approach was also tried by Bhattacharjya et al. which ball-milled graphite in the presence of I_2 and Br_2 [147]. The result of the synthesis was halogen edge-functionalized graphene sheets. The exfoliation of the graphite was confirmed through characterizations and the functionalizing of the graphene sheets was confirmed to be on the edges, likely due to the increased reactivity of the graphene edge bonds. When testing the electrochemical performance, it was found that both the I_2 and Br_2 functionalized materials displayed a higher capacitance than the H_2 control sample, with the I_2 functionalized carbon having the highest capacitance at 172 F g^{-1} compared to only 75 F g^{-1} for the H_2 control sample.

Acknowledgements

X. Ji thanks the financial support from the National Science Foundation of the United States, Award Number: 1507391.

References

[1] Becker, H.I. 1957. Low voltage electrolytic capacitor. *Google Patents*.

[2] Pandolfo, A.G. and Hollenkamp, A.F. 2006. Carbon properties and their role in supercapacitors. *J. Power Sources*, 157: 11–27.

[3] Conway, B.E. 1999. Electrochemical Supercapacitors: Scientific Fundamentals and Technological Applications. Plenum Pres, Springer Science+Business Media New York.

[4] Bard, A.J. and Faulkner, L.R. 2001. Electrochemical Methods: Fundamentals and Applications, Wiley.

[5] Berg, J.C. 2010. An Introduction to Interfaces & Colloids: The Bridge to Nanoscience. World Scientific, Singapore.

[6] Zhang, L.L. and Zhao, X.S. 2009. Carbon-based materials as supercapacitor electrodes. *Chem. Soc. Rev.*, 38: 2520.

[7] Feng, G., Jiang, D.-e. and Cummings, P.T. 2012. Curvature effect on the capacitance of electric double layers at ionic liquid/onion-like carbon interfaces. *J. Chem. Theory Comput.*, 8: 1058–1063.

[8] Pean, C., Merlet, C., Rotenberg, B. et al. 2014. On the dynamics of charging in nanoporous carbon-based supercapacitors. *ACS Nano.*, 8: 1576–1583.

[9] Pean, C., Daffos, B., Rotenberg, B. et al. 2015. Confinement, desolvation, and electrosorption effects on the diffusion of ions in nanoporous carbon electrodes. *J. Am. Chem. Soc.*, 137: 12627–12632.

[10] Kondrat, S., Georgi, N., Fedorov, M.V. et al. 2011. A superionic state in nano-porous double-layer capacitors: insights from Monte Carlo simulations. *Phys. Chem. Chem. Phys.*, 13: 11359.
[11] Lee, A., Kondrat, S., Oshanin, G. et al. 2014. Charging dynamics of supercapacitors with narrow cylindrical nanopores. *Nanotechnology*, 25: 315401.
[12] Popov, I.A., Bozhenko, K.V. and Boldyrev, A.I. 2011. Is graphene aromatic? *Nano Research*, 5: 117–123.
[13] Saliger, R., Fischer, U., Herta, C. et al. 1998. High surface area carbon aerogels for supercapacitors. *J. Non-Cryst. Solids*, 225: 81–85.
[14] Weng, T.-C. and Teng, H. 2001. Characterization of high porosity carbon electrodes derived from mesophase pitch for electric double-layer capacitors. *J. Electrochem. Soc.*, 148: A368.
[15] Lozano-Castelló, D., Cazorla-Amorós, D., Linares-Solano, A. et al. 2003. Influence of pore structure and surface chemistry on electric double layer capacitance in non-aqueous electrolyte. *Carbon*, 41: 1765–1775.
[16] Shi, H. 1996. Activated carbons and double layer capacitance. *Electrochim. Acta*, 41: 1633–1639.
[17] Barbieri, O., Hahn, M., Herzog, A. et al. 2005. Capacitance limits of high surface area activated carbons for double layer capacitors. *Carbon*, 43: 1303–1310.
[18] Zhang, L., Zhang, F., Yang, X. et al. 2013. Porous 3D graphene-based bulk materials with exceptional high surface area and excellent conductivity for supercapacitors. *Scientific Reports*, doi:10.1038/srep01408.
[19] Chmiola, J. 2006. Anomalous increase in carbon capacitance at pore sizes less than 1 nanometer. *Science*, 313: 1760–1763.
[20] Largeot, C., Portet, C., Chmiola, J. et al. 2008. Relation between the ion size and pore size for an electric double-layer capacitor. *J. Am. Chem. Soc.*, 130: 2730–2731.
[21] Largeot, C., Portet, C., Chmiola, J. et al. 2008. Relation between the ion size and pore size for an electric double-layer capacitor. *J. Am. Chem. Soc.*, 130: 2730–2731.
[22] Huang, J.S., Sumpter, B.G. and Meunier, V. 2008. Theoretical model for nanoporous carbon supercapacitors. *Angew. Chem. Int. Ed.*, 47: 520–524.
[23] Huang, J.S., Sumpter, B.G. and Meunier, V. 2008. A universal model for nanoporous carbon supercapacitors applicable to diverse pore regimes, carbon materials, and electrolytes. *Chem-Eur. J.*, 14: 6614–6626.
[24] Merlet, C., Rotenberg, B., Madden, P.A. et al. 2012. On the molecular origin of supercapacitance in nanoporous carbon electrodes. *Nat. Mater.*, 11: 306–310.
[25] Kondrat, S., Wu, P., Qiao, R. et al. 2014. Accelerating charging dynamics in subnanometre pores. *Nat. Mater.*, 13: 387–393.
[26] Kondrat, S. and Kornyshev, A.A. 2016. Pressing a spring: what does it take to maximize the energy storage in nanoporous supercapacitors? *Nanoscale Horiz.*
[27] Sun, G., Kürti, J., Kertesz, M. et al. 2002. Dimensional changes as a function of charge injection in single-walled carbon nanotubes. *J. Am. Chem. Soc.*, 124: 15076–15080.
[28] Hantel, M.M., Weingarth, D. and Kötz, R. 2014. Parameters determining dimensional changes of porous carbons during capacitive charging. *Carbon*, 69: 275–286.
[29] Rochester, C.C., Pruessner, G. and Kornyshev, A.A. 2015. Statistical mechanics of 'unwanted electroactuation' in nanoporous supercapacitors. *Electrochim. Acta*, 174: 978–984.
[30] Subramanian, V., Luo, C., Stephan, A.M. et al. 2007. Supercapacitors from activated carbon derived from banana fibers. *J. Phys. Chem. C*, 111: 7527–7531.
[31] Biswal, M., Banerjee, A., Deo, M. et al. 2013. From dead leaves to high energy density supercapacitors. *Energ. Environ. Sci.*, 6: 1249.
[32] Kalyani, P. and Anitha, A. 2013. Biomass carbon & its prospects in electrochemical energy systems. *Int. J. Hydrogen Energy*, 38: 4034–4045.
[33] Wei, L. and Yushin, G. 2012. Nanostructured activated carbons from natural precursors for electrical double layer capacitors. *Nano Energy*, 1: 552–565.

[34] Wang, D., Geng, Z., Li, B. et al. 2015. High performance electrode materials for electric double-layer capacitors based on biomass-derived activated carbons. *Electrochim. Acta*, 173: 377–384.
[35] Chen, H., Liu, D., Shen, Z. et al. 2015. Functional biomass carbons with hierarchical porous structure for supercapacitor electrode materials. *Electrochim. Acta*, 180: 241–251.
[36] Wang, H., Li, Z. and Mitlin, D. 2014. Tailoring biomass-derived carbon nanoarchitectures for high-performance supercapacitors. *ChemElectroChem.*, 1: 332–337.
[37] Rufford, T.E., Hulicova-Jurcakova, D., Zhu, Z. et al. 2008. Nanoporous carbon electrode from waste coffee beans for high performance supercapacitors. *Electrochem. Commun.*, 10: 1594–1597.
[38] Wornat, M.J., Hurt, R.H., Yang, N.Y.C. et al. 1995. Structural and compositional transformations of biomass chars during combustion. *Combust. Flame.*, 100: 131–143.
[39] Franklin, R.E. 1951. Crystallite growth in graphitizing and non-graphitizing carbons. *Proceedings of the Royal Society A: Mathematical, Physical and Engineering Sciences*, 209: 196–218.
[40] Chen, J., Li, C. and Shi, G. 2013. Graphene materials for electrochemical capacitors. *The Journal of Physical Chemistry Letters*, 4: 1244–1253.
[41] Huang, Y., Liang, J.J. and Chen, Y.S. 2012. An overview of the applications of graphene-based materials in supercapacitors. *Small*, 8: 1805–1834.
[42] An, K.H., Kim, W.S., Park, Y.S. et al. 2001. Electrochemical properties of high-power supercapacitors using single-walled carbon nanotube electrodes. *Adv. Funct. Mater.*, 11: 387–392.
[43] Pech, D., Brunet, M., Durou, H. et al. 2010. Ultrahigh-power micrometre-sized supercapacitors based on onion-like carbon. *Nat. Nanotechnol.*, 5: 651–654.
[44] Portet, C., Yushin, G. and Gogotsi, Y. 2007. Electrochemical performance of carbon onions, nanodiamonds, carbon black and multiwalled nanotubes in electrical double layer capacitors. *Carbon*, 45: 2511–2518.
[45] Liu, B., Shioyama, H., Akita, T. et al. 2008. Metal-organic framework as a template for porous carbon synthesis. *J. Am. Chem. Soc.*, 130: 5390–5391.
[46] Sun, J.-K. and Xu, Q. 2014. Functional materials derived from open framework templates/precursors: synthesis and applications. *Energ. Environ. Sci.*, 7: 2071.
[47] Yang, S.J., Kim, T., Im, J.H. et al. 2012. MOF-derived hierarchically porous carbon with exceptional porosity and hydrogen storage capacity. *Chem. Mater.*, 24: 464–470.
[48] Li, Y., Ben, T., Zhang, B. et al. 2013. Ultrahigh gas storage both at low and high pressures in KOH-activated carbonized porous aromatic frameworks. *Scientific Reports*, doi:10.1038/srep02420.
[49] Kuchta, B., Firlej, L., Mohammadhosseini, A. et al. 2012. Hypothetical high-surface-area carbons with exceptional hydrogen storage capacities: Open carbon frameworks. *J. Am. Chem. Soc.*, 134: 15130–15137.
[50] Dutta, S., Bhaumik, A. and Wu, K.C.W. 2014. Hierarchically porous carbon derived from polymers and biomass: effect of interconnected pores on energy applications. *Energy Environ. Sci.*, 7: 3574–3592.
[51] Velev, O.D. and Kaler, E.W. 2000. Structured porous materials via colloidal crystal templating: From inorganic oxides to metals. *Adv. Mater.*, 12: 531–534.
[52] Zhao, X.S., Su, F., Yan, Q. et al. 2006. Templating methods for preparation of porous structures. *J. Mater. Chem.*, 16: 637–648.
[53] Lee, J., Kim, J. and Hyeon, T. 2006. Recent progress in the synthesis of porous carbon materials. *Adv. Mater.*, 18: 2073–2094.
[54] Qu, Q., Yun, J., Wan, Z. et al. 2014. MOF-derived microporous carbon as a better choice for Na-ion batteries than mesoporous CMK-3. *RSC Adv.*, 4: 64692–64697.
[55] Hunt, B.E., Mori, S., Katz, S. et al. 1953. Reaction of carbon with steam at elevated temperatures. *Industrial & Engineering Chemistry*, 45: 677–680.

[56] Mi, J., Wang, X.-R., Fan, R.-J. et al. 2012. Coconut-shell-based porous carbons with a tunable micro/mesopore ratio for high-performance supercapacitors. *Energy Fuels*, 26: 5321–5329.
[57] Luo, W., Wang, B., Heron, C.G. et al. 2014. Pyrolysis of cellulose under ammonia leads to nitrogen-doped nanoporous carbon generated through methane formation. *Nano Lett.*, 14: 2225–2229.
[58] Raymundo-Piñero, E., Azaïs, P., Cacciaguerra, T. et al. 2005. KOH and NaOH activation mechanisms of multiwalled carbon nanotubes with different structural organisation. *Carbon*, 43: 786–795.
[59] Armandi, M., Bonelli, B., Geobaldo, F. et al. 2010. Nanoporous carbon materials obtained by sucrose carbonization in the presence of KOH. *Microporous Mesoporous Mater.*, 132: 414–420.
[60] Lillo-Ródenas, M.A., Cazorla-Amorós, D. and Linares-Solano, A. 2003. Understanding chemical reactions between carbons and NaOH and KOH. *Carbon*, 41: 267–275.
[61] Wang, H., Xu, Z., Kohandehghan, A. et al. 2013. Interconnected carbon nanosheets derived from hemp for ultrafast supercapacitors with high energy. *ACS Nano.*, 7: 5131–5141.
[62] Kang, D., Liu, Q., Gu, J. et al. 2015. "Egg-box"-assisted fabrication of porous carbon with small mesopores for high-rate electric double layer capacitors. *ACS Nano*.
[63] Jagtoyen, M. and Derbyshire, F. 1998. Activated carbons from yellow poplar and white oak by H3PO4 activation. *Carbon*, 36: 1085–1097.
[64] Dobele, G., Dizhbite, T., Rossinskaja, G. et al. 2003. Pre-treatment of biomass with phosphoric acid prior to fast pyrolysis. *J. Anal. Appl. Pyrolysis*, 68-69: 197–211.
[65] Molina-Sabio, M., RodRíguez-Reinoso, F., Caturla, F. et al. 1995. Porosity in granular carbons activated with phosphoric acid. *Carbon*, 33: 1105–1113.
[66] Molina-Sabio, M. and Rodríguez-Reinoso, F. 2004. Role of chemical activation in the development of carbon porosity. *Colloids and Surfaces A: Physicochemical and Engineering Aspects*, 241: 15–25.
[67] Al Bahri, M., Calvo, L., Gilarranz, M.A. et al. 2012. Activated carbon from grape seeds upon chemical activation with phosphoric acid: Application to the adsorption of diuron from water. *Chem. Eng. J.*, 203: 348–356.
[68] Teng, H., Yeh, T.-S. and Hsu, L.-Y. 1998. Preparation of activated carbon from bituminous coal with phosphoric acid activation. *Carbon*, 36: 1387–1395.
[69] Hu, Z., Srinivasan, M.P. and Ni, Y. 2001. Novel activation process for preparing highly microporous and mesoporous activated carbons. *Carbon*, 39: 877–886.
[70] Khalili, N.R., Campbell, M., Sandi, G. et al. 2000. Production of micro- and mesoporous activated carbon from paper mill sludge. *Carbon*, 38: 1905–1915.
[71] Olivares-Marín, M., Fernández-González, C., Macías-García, A. et al. 2006. Preparation of activated carbon from cherry stones by chemical activation with ZnCl2. *Appl. Surf. Sci.*, 252: 5967–5971.
[72] Kalderis, D., Bethanis, S., Paraskeva, P. et al. 2008. Production of activated carbon from bagasse and rice husk by a single-stage chemical activation method at low retention times. *Bioresour. Technol.*, 99: 6809–6816.
[73] Xiang, X., Liu, E., Huang, Z. et al. 2010. Preparation of activated carbon from polyaniline by zinc chloride activation as supercapacitor electrodes. *J. Solid State Electrochem.*, 15: 2667–2674.
[74] Hou, J., Cao, C., Idrees, F. et al. 2015. Hierarchical porous nitrogen-doped carbon nanosheets derived from silk for ultrahigh-capacity battery anodes and supercapacitors. *ACS Nano.*, 9: 2556–2564.
[75] Chang, B., Wang, Y., Pei, K. et al. 2014. ZnCl2-activated porous carbon spheres with high surface area and superior mesoporous structure as an efficient supercapacitor electrode. *RSC Adv.*, 4: 40546–40552.
[76] Acik, M., Mattevi, C., Gong, C. et al. 2010. The role of intercalated water in multilayered graphene oxide. *ACS Nano.*, 4: 5861–5868.

[77] Wang, M.-x., Liu, Q., Sun, H.-f. et al. 2012. Preparation of high-surface-area carbon nanoparticle/graphene composites. *Carbon*, 50: 3845–3853.
[78] Zhu, Y., Murali, S., Stoller, M.D. et al. 2011. Carbon-based supercapacitors produced by activation of graphene. *Science*, 332: 1537–1541.
[79] Fan, Z., Zhao, Q., Li, T. et al. 2012. Easy synthesis of porous graphene nanosheets and their use in supercapacitors. *Carbon*, 50: 1699–1703.
[80] Xu, Y., Lin, Z., Zhong, X. et al. 2014. Holey graphene frameworks for highly efficient capacitive energy storage. *Nat. Comm.*, doi:10.1038/ncomms5554.
[81] Wang, X., Sun, G., Routh, P. et al. 2014. Heteroatom-doped graphene materials: syntheses, properties and applications. *Chem. Soc. Rev.*, 43: 7067–7098.
[82] Xiao, N., Tan, H., Zhu, J. et al. 2013. High-performance supercapacitor electrodes based on graphene achieved by thermal treatment with the aid of nitric acid. *Acs Appl. Mater. Inter.*, 5: 9656–9662.
[83] El-Kady, M.F., Strong, V., Dubin, S. et al. 2012. Laser scribing of high-performance and flexible graphene-based electrochemical capacitors. *Science*, 335: 1326–1330.
[84] Li, Y., Li, Z. and Shen, P.K. 2013. Simultaneous formation of ultrahigh surface area and three-dimensional hierarchical porous graphene-like networks for fast and highly stable supercapacitors. *Adv. Mater.*, 25: 2474–2480.
[85] Yoo, J.J., Balakrishnan, K., Huang, J. et al. 2011. Ultrathin planar graphene supercapacitors. *Nano Lett.*, 11: 1423–1427.
[86] Jiang, L. and Fan, Z. 2014. Design of advanced porous graphene materials: from graphene nanomesh to 3D architectures. *Nanoscale*, 6: 1922–1945.
[87] Han, S., Wu, D., Li, S. et al. 2014. Porous graphene materials for advanced electrochemical energy storage and conversion devices. *Adv. Mater.*, 26: 849–864.
[88] Chakrabarti, A., Lu, J., Skrabutenas, J.C. et al. 2011. Conversion of carbon dioxide to few-layer graphene. *J. Mater. Chem.*, 21: 9491.
[89] Xing, Z., Wang, B., Gao, W. et al. 2015. Reducing CO2 to dense nanoporous graphene by Mg/Zn for high power electrochemical capacitors. *Nano Energy*, 11: 600–610.
[90] Pumera, M. 2014. Heteroatom modified graphenes: electronic and electrochemical applications. *J. Mater. Chem. C*, 2: 6454–6461.
[91] Chua, C.K., Ambrosi, A., Sofer, Z. et al. 2014. Chemical preparation of graphene materials results in extensive unintentional doping with heteroatoms and metals. *Chemistry - A European Journal*, 20: 15760–15767.
[92] Xu, B., Yue, S., Sui, Z. et al. 2011. What is the choice for supercapacitors: graphene or graphene oxide? *Energ. Environ. Sci.*, 4: 2826.
[93] Chen, Y., Zhang, X., Zhang, D. et al. 2011. High performance supercapacitors based on reduced graphene oxide in aqueous and ionic liquid electrolytes. *Carbon*, 49: 573–580.
[94] Fan, L.-Z., Qiao, S., Song, W. et al. 2013. Effects of the functional groups on the electrochemical properties of ordered porous carbon for supercapacitors. *Electrochim. Acta*, 105: 299–304.
[95] Conway, B.E., Birss, V. and Wojtowicz, J. 1997. The role and utilization of pseudocapacitance for energy storage by supercapacitors. *J. Power Sources*, 66: 1–14.
[96] Liu, T.C. 1998. Behavior of molybdenum nitrides as materials for electrochemical capacitors. *J. Electrochem. Soc.*, 145: 1882.
[97] Hsieh, C.-T. and Teng, H. 2002. Influence of oxygen treatment on electric double-layer capacitance of activated carbon fabrics. *Carbon*, 40: 667–674.
[98] Okajima, K., Ohta, K. and Sudoh, M. 2005. Capacitance behavior of activated carbon fibers with oxygen-plasma treatment. *Electrochim. Acta*, 50: 2227–2231.
[99] Bleda-Martínez, M.J., Maciá-Agulló, J.A., Lozano-Castelló, D. et al. 2005. Role of surface chemistry on electric double layer capacitance of carbon materials. *Carbon*, 43: 2677–2684.

[100] Oda, H., Yamashita, A., Minoura, S. et al. 2006. Modification of the oxygen-containing functional group on activated carbon fiber in electrodes of an electric double-layer capacitor. *J. Power Sources,* 158: 1510–1516.
[101] Liu, X., Wang, Y., Zhan, L. et al. 2010. Effect of oxygen-containing functional groups on the impedance behavior of activated carbon-based electric double-layer capacitors. *J. Solid State Electrochem.,* 15: 413–419.
[102] Hulicova-Jurcakova, D., Seredych, M., Lu, G.Q. et al. 2009. Combined effect of nitrogen- and oxygen-containing functional groups of microporous activated carbon on its electrochemical performance in supercapacitors. *Adv. Funct. Mater.,* 19: 438–447.
[103] Lu, Y., Huang, Y., Zhang, M. et al. 2014. Nitrogen-doped graphene materials for supercapacitor applications. *J. Nanosci. Nanotechnol.,* 14: 1134–1144.
[104] Hassan, F.M., Chabot, V., Li, J.D. et al. 2013. Pyrrolic-structure enriched nitrogen doped graphene for highly efficient next generation supercapacitors. *J. Mater. Chem. A,* 1: 2904–2912.
[105] Lota, G., Grzyb, B., Machnikowska, H. et al. 2005. Effect of nitrogen in carbon electrode on the supercapacitor performance. *Chem. Phys. Lett.,* 404: 53–58.
[106] Lota, G., Lota, K. and Frackowiak, E. 2007. Nanotubes based composites rich in nitrogen for supercapacitor application. *Electrochem. Commun.,* 9: 1828–1832.
[107] Jin, J., Fu, X., Liu, Q. et al. 2013. Identifying the active site in nitrogen-doped graphene for the VO2+/VO2+redox reaction. *ACS Nano.,* 7: 4764–4773.
[108] Li, Z., Xu, Z., Tan, X. et al. 2013. Mesoporous nitrogen-rich carbons derived from protein for ultra-high capacity battery anodes and supercapacitors. *Energ. Environ. Sci.,* 6: 871.
[109] Jin, H., Wang, X., Gu, Z. et al. 2015. A facile method for preparing nitrogen-doped graphene and its application in supercapacitors. *J. Power Sources,* 273: 1156–1162.
[110] Yun, Y.S., Cho, S.Y., Shim, J. et al. 2013. Microporous carbon nanoplates from regenerated silk proteins for supercapacitors. *Adv. Mater.,* 25: 1993–1998.
[111] Yun, Y.S., Park, M.H., Hong, S.J. et al. 2015. Hierarchically porous carbon nanosheets from waste coffee grounds for supercapacitors. *Acs Appl. Mater. Inter.,* 150203143836005.
[112] Si, W., Zhou, J., Zhang, S. et al. 2013. Tunable N-doped or dual N, S-doped activated hydrothermal carbons derived from human hair and glucose for supercapacitor applications. *Electrochim. Acta,* 107: 397–405.
[113] Qian, W., Sun, F., Xu, Y. et al. 2014. Human hair-derived carbon flakes for electrochemical supercapacitors. *Energy Environ. Sci.,* 7: 379–386.
[114] Maldonado, S., Morin, S. and Stevenson, K.J. 2006. Structure, composition, and chemical reactivity of carbon nanotubes by selective nitrogen doping. *Carbon,* 44: 1429–1437.
[115] Wang, X., Li, X., Zhang, L. et al. 2009. N-doping of graphene through electrothermal reactions with ammonia. *Science,* 324: 768–771.
[116] Jeong, H.M., Lee, J.W., Shin, W.H. et al. 2011. Nitrogen-doped graphene for high-performance ultracapacitors and the importance of nitrogen-doped sites at basal planes. *Nano Lett.,* 11: 2472–2477.
[117] Li, X., Wang, H., Robinson, J.T. et al. 2009. Simultaneous nitrogen doping and reduction of graphene oxide. *J. Am. Chem. Soc.,* 131: 15939–15944.
[118] Wang, H., Xie, M., Thia, L. et al. 2014. Strategies on the design of nitrogen-doped graphene. *The Journal of Physical Chemistry Letters,* 5: 119–125.
[119] Deng, D., Pan, X., Yu, L. et al. 2011. Toward N-doped graphene via solvothermal synthesis. *Chem. Mater.,* 23: 1188–1193.
[120] Paraknowitsch, J.P. and Thomas, A. 2013. Doping carbons beyond nitrogen: an overview of advanced heteroatom doped carbons with boron, sulphur and phosphorus for energy applications. *Energ. Environ. Sci.,* 6: 2839.
[121] Cermignani, W., Paulson, T.E., Onneby, C. et al. 1995. Synthesis and characterization of boron-doped carbons. *Carbon,* 33: 367–374.

[122] Garcia, A.G., Baltazar, S.E., Castro, A.H.R. et al. 2008. Influence of S and P doping in a graphene sheet. *J. Comput. Theor. Nanosci.*, 5: 2221–2229.
[123] Kiciński, W., Szala, M. and Bystrzejewski, M. 2014. Sulfur-doped porous carbons: Synthesis and applications. *Carbon*, 68: 1–32.
[124] Zhang, Y., Mori, T., Ye, J. et al. 2010. Phosphorus-doped carbon nitride solid: Enhanced electrical conductivity and photocurrent generation. *J. Am. Chem. Soc.*, 132: 6294–6295.
[125] Yang, D.-S., Bhattacharjya, D., Inamdar, S. et al. 2012. Phosphorus-doped ordered mesoporous carbons with different lengths as efficient metal-free electrocatalysts for oxygen reduction reaction in alkaline media. *J. Am. Chem. Soc.*, 134: 16127–16130.
[126] Singh, K.P., Song, M.Y. and Yu, J.-S. 2014. Iodine-treated heteroatom-doped carbon: conductivity driven electrocatalytic activity. *J. Mater. Chem. A*, 2: 18115–18124.
[127] Xu, J., Jeon, I.-Y., Seo, J.-M. et al. 2014. Edge-selectively halogenated graphene nanoplatelets (XGnPs, X = Cl, Br, or I) prepared by ball-milling and used as anode materials for mithium-ion batteries. *Adv. Mater.*, 26: 7317–7323.
[128] Wang, D.-W., Li, F., Chen, Z.-G. et al. 2008. Synthesis and electrochemical property of boron-doped mesoporous carbon in supercapacitor. *Chem. Mater.*, 20: 7195–7200.
[129] Han, J., Zhang, L.L., Lee, S. et al. 2013. Generation of B-doped graphene nanoplatelets using a solution process and their supercapacitor applications. *ACS Nano.*, 7: 19–26.
[130] Iyyamperumal, E., Wang, S. and Dai, L. 2012. Vertically aligned BCN nanotubes with high capacitance. *ACS Nano.*, 6: 5259–5265.
[131] Guo, H. and Gao, Q. 2009. Boron and nitrogen co-doped porous carbon and its enhanced properties as supercapacitor. *J. Power Sources*, 186: 551–556.
[132] Fujisawa, K., Cruz-Silva, R., Yang, K.-S. et al. 2014. Importance of open, heteroatom-decorated edges in chemically doped-graphene for supercapacitor applications. *J. Mater. Chem. A*, 2: 9532.
[133] Dou, S., Huang, X., Ma, Z. et al. 2015. A simple approach to the synthesis of BCN graphene with high capacitance. *Nanotechnology*, 26: 045402.
[134] Hao, F., Yao, Y., Li, Y. et al. 2015. Synthesis of high-concentration B and N co-doped porous carbon polyhedra and their supercapacitive properties. *RSC Adv.*, 5: 77527–77533.
[135] Peng, Z., Ye, R., Mann, J.A. et al. 2015. Flexible boron-doped laser-induced graphene microsupercapacitors. *ACS Nano.*, 9: 5868–5875.
[136] Gu, W., Sevilla, M., Magasinski, A. et al. 2013. Sulfur-containing activated carbons with greatly reduced content of bottle neck pores for double-layer capacitors: a case study for pseudocapacitance detection. *Energ. Environ. Sci.*, 6: 2465.
[137] Zhang, D., Hao, Y., Zheng, L. et al. 2013. Nitrogen and sulfur co-doped ordered mesoporous carbon with enhanced electrochemical capacitance performance. *J. Mater. Chem. A*, 1: 7584.
[138] Seredych, M. and Bandosz, T.J. 2013. S-doped micro/mesoporous carbon–graphene composites as efficient supercapacitors in alkaline media. *J. Mater. Chem. A*, 1: 11717.
[139] Xu, G., Han, J., Ding, B. et al. 2015. Biomass-derived porous carbon materials with sulfur and nitrogen dual-doping for energy storage. *Green Chem.*, 17: 1668–1674.
[140] Samantara, A.K., Chandra Sahu, S., Ghosh, A. et al. 2015. Sandwiched graphene with nitrogen, sulphur co-doped CQDs: an efficient metal-free material for energy storage and conversion applications. *J. Mater. Chem. A*, 3: 16961–16970.
[141] Karthika, P., Rajalakshmi, N. and Dhathathreyan, K.S. 2013. Phosphorus-doped exfoliated graphene for supercapacitor electrodes. *J. Nanosci. Nanotechnol.*, 13: 1746–1751.
[142] Xu, G., Ding, B., Pan, J. et al. 2015. Porous nitrogen and phosphorus co-doped carbon nanofiber networks for high performance electrical double layer capacitors. *J. Mater. Chem. A*.
[143] Yan, X., Liu, Y., Fan, X. et al. 2014. Nitrogen/phosphorus co-doped nonporous carbon nanofibers for high-performance supercapacitors. *J. Power Sources*, 248: 745–751.

[144] Nasini, U.B., Bairi, V.G., Ramasahayam, S.K. et al. 2014. Phosphorous and nitrogen dual heteroatom doped mesoporous carbon synthesized via microwave method for supercapacitor application. *J. Power Sources*, 250: 257–265.
[145] Huang, C., Sun, T. and Hulicova-Jurcakova, D. 2013. Wide electrochemical window of supercapacitors from coffee bean-derived phosphorus-rich carbons. *ChemSusChem.*, 6: 2330–2339.
[146] Tanaike, O., Yamada, Y., Yamada, K. et al. 2011. Electrochemical behavior of halogen-doped carbon materials as capacitor electrodes. 71–76.
[147] Bhattacharjya, D., Jeon, I.-Y., Park, H.-Y. et al. 2015. Graphene nanoplatelets with selectively functionalized edges as electrode material for electrochemical energy storage. *Langmuir*, 31: 5676–5683.

CHAPTER 3

Mesoporous Structured Materials as New Proton Exchange Membranes for Fuel Cells

Jinlin Lu,[1] *Jin Zhang*[2] and *San Ping Jiang*[2,*]

ABSTRACT

In this chapter, the working principle and structure of Proton Exchange Membrane Fuel Cells (PEMFCs) based on conventional membranes such as Nafion and PBI membranes is briefly introduced first. The emphasis of the chapter is on the development and status of various mesoporous structured PEMs, including polymeric and inorganic materials. The rationale for developing mesoporous structured PEMs is discussed, followed with the detailed review of the advances in mesoporous structured materials as PEMs including organic, inorganic and MOF and organic-inorganic membranes. Finally, the prospect of mesoporous structured materials as PEMs to operate at high temperature and low humidity conditions for next generation of PEMFCs is discussed.

Keywords: Fuel cells, Proton exchange membrane fuel cells, Proton Exchange Membrane, Nafion, Polybenzimidazole, Mesoporous structured materials, self-assembly, mesoporous Nafion, mesoporous TiO_2, sol-gel technique, Mesoporous alumina, ferroxane, nature of the surface groups, mesoporous silica, porous silica glass,

[1] School of Materials and Metallurgy, University of Science and Technology Liaoning, Anshan, China.

[2] Fuels and Energy Technology Institute & Department of Chemical Engineering, Curtin University, Perth, WA 6102, Australia.

* Corresponding author: s.jiang@curtin.edu.au

proton conduction, acidity of the material surface, sulfonic acid functionalization, phosphonic acid functionalized mesoporous-structured silica, composite silicate, functionalized mesoporous silica, interpenetrating polymer network, functionalized Hollow Mesoporous Silica, capillary condensation effect, proton diffusion mechanism, post-grafting method, co-condensation method, Periodic Mesoporous Organosilica, Pore structure, Acidity, Kelvin equation, capillary condensation effect, filler content, HPAs-functionalized meso-silica, post synthesis by impregnation, one-pot synthesis, HPW-meso-silica, HPW, button cells, stack cells

1. Introduction

The depletion of fossil fuel and environmental pollution have led to an urgent demand for alternative energy sources and more efficient energy conversion technologies. Fuel cells are electrochemical devices that can directly convert the chemical energy of fuels such as hydrogen, methanol, ethanol, methane and light hydrocarbons to electrical energy. The key advantages of the fuel cell technologies are their high energy utilization efficiency and very low pollutant emissions [1–3].

The efficiency of fuel cells is not limited by the Carnot cycle and in the case of H_2/O_2 fuel cells, the only product of the reaction is water and the reaction can be represented by the following two half-cell reactions:

Anode: $$H_2 \rightarrow 2H^+ + 2e^- \quad (1)$$

Cathode: $$\frac{1}{2}O_2 + 2H^+ + 2e^- \rightarrow H_2O \quad (2)$$

The overall reaction is:

$$H_2 + \frac{1}{2}O_2 = H_2O \quad (3)$$

Proton exchange membrane fuel cells (PEMFCs) use a Proton Exchange Membrane (PEM) as an electrolyte and Pt-based electrocatalysts as anode and cathode. In PEMFCs, hydrogen is split into protons and electrons on the anode. Protons are transported through the electrolyte to the cathode, and electrons move through an external circuit. On the cathode, protons, electrons and oxygen are combined to produce water and heat [4]. The flow of ionic charge through the electrolyte is balanced by the flow of electronic charge through the external circuit, producing electrical current or power. Figure 3.1 shows the working principle of a PEMFC.

PEMFCs operate at relatively low temperatures of around 80°C. Low temperature operation allows quick startup and shutdown

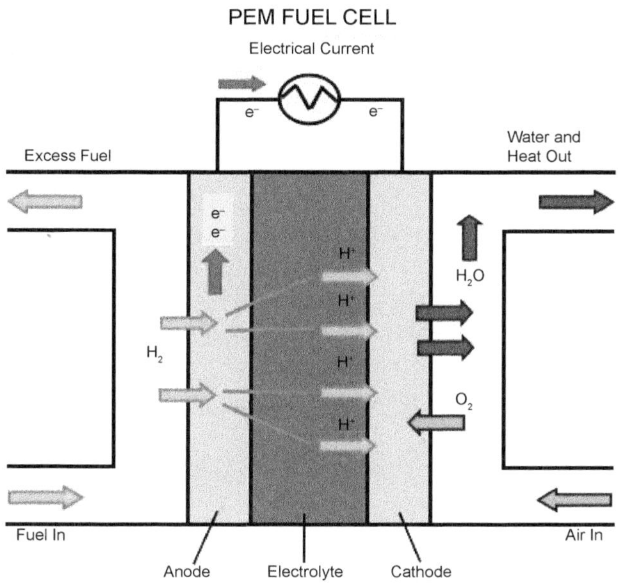

Figure 3.1. Principle operation diagram for a proton exchange membrane fuel cell.

and produces less wear on system components, resulting in better durability [3]. PEMFC is one of the most promising clean energy technologies under development in fuel cell families. The power output of a PEMFC depends critically on the performance and activity of its key components and the most important component in PEMFCs is the PEM. The development of PEMFCs technologies was initiated by the US General Electric Company (GE) in the 1950s. It was the introduction of Nafion membranes by DuPont that substantially accelerated the development and advancement of the PEMFC technologies. Nafion was initially developed and manufactured for membrane cells used in the production of chlorine and nowadays is the state-of-the-art and most common membrane electrolytes used in PEMFCs owing to their high proton conductivity and good thermal and chemical stability [5–7].

Figure 3.2 shows the chemical structure of Nafion. The perfluorinated backbone imparts a chemical and thermal stability, while the pendant sulfonic acid group imparts strong acidic characteristics for facile proton transport. The combination of extremely high hydrophobicity of the perfluorinated backbone with the extremely high hydrophilicity of the sulfonic acid functional groups gives rise to some hydrophobic/hydrophilic nano-separation. The sulfonic acid functional groups aggregate to form a hydrophilic domain. When this hydrophilic domain is hydrated, protonic charge carriers within inner space charge layers move by dissociation of

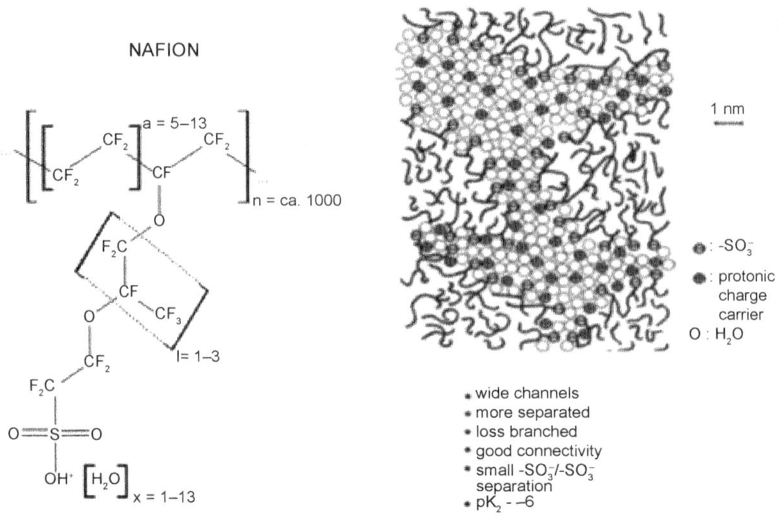

Figure 3.2. Schematic illustration of the microstructures of Nafion. (Reproduced from [12] with permission. Copyright (2001) Elsevier.)

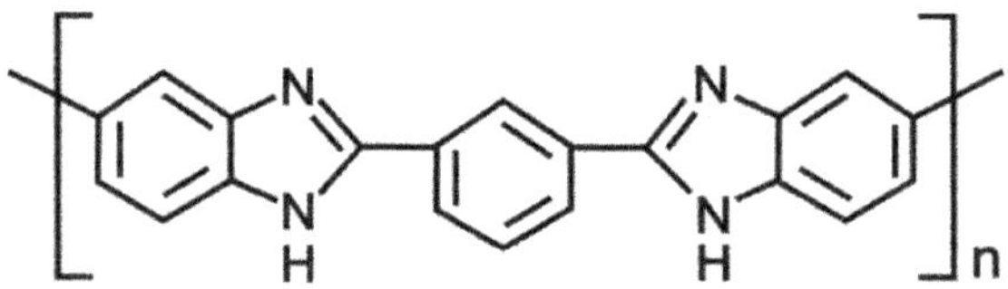

Figure 3.3. Structure of polybenzimidazole or PBI. (Reproduced from [15] with permission. Copyright (2004) John & Wiley Sons.)

the acidic functional groups. The occurrence of water dynamics assists proton conductance (see the right graph of Fig. 3.2). The costs of Nafion membranes remain high and hence many efforts are being made to develop cheaper, usually fluorine-free membrane materials. In addition, their proton conductivity is critically dependent on the humidity and water content in the membrane structure [8–11]. Thus development of an alternative PEM which can be operated at high temperature/low humidity remains a critical challenge.

Polybenzimidazole (PBI) is an aromatic heterocyclic polymer (Fig. 3.3) with a Tg of 425–436°C. It is a basic polymer (pKa = 5.5) which can be complex of strong acids or very strong bases [13]. PBI shows some tendency to take up water, thus explaining the low proton conductivity (~10^{-7} S cm^{-1}) as shown even by the non-modified polymer [13, 14]. PBI membrane doped with acids has received much attention mainly for use as PEM above 100°C [14–16]. Early mention of stabilized and plasticized PBI referred to PBI treated with sulfuric and Phosphoric Acid (PA), respectively. At high temperatures, it is difficult to keep Nafion hydrated, but the acid doped PBI material does not use water as a medium for proton conduction. It also exhibits better mechanical properties, higher strength and is cheaper than Nafion. However, acid leaching is a considerable issue and processing, mixing with catalyst to form ink, has proved problematic [17, 18].

A major issue of the state-of-the-art Nafion membranes is their high sensitivity to Relative Humidity (RH) primarily due to the fact that the Nafion membrane has a non-ordered nanostructure consisting of hydrophobic perfluorinated main chains surrounded by hydrophilic ionic domains that swell on hydration [19–21]. The growth of ionic domains during water sorption induces a phase-separation, where water-swollen ionic domains tend to interconnect with nanochannels and thus facilitate water and ion transport through the hydrophobic polymer matrix [22]. However, the water channels in the Nafion membrane are essentially random, which results in inherent limitations in the relatively low water retention ability of conventional Nafion membrane [10].

Mesoporous structured materials have attracted intensive attention because of the tunable mesoporous structure and the unique characteristic of capillary condensation, i.e., vapor condensation would occur favorably in meso- or nano-sized hydrophilic channels or pores below the saturation vapor pressure of the pure liquid due to an increased number of van der Waals interactions between the vapor phase molecules within the confined space of a capillary [23, 24]. Water condensation occurs at reduced humidity [25]. Thus, high specific surface area, ordered and interconnected mesopores, and high structural stability allow their potential applications as PEMs operating at elevated temperatures [26–36]. This chapter will start with the synthesis and characterization mesoporous structured Nafion membranes, followed by the detailed description of the development of functionalized inorganic mesoporous structured materials such as mesoporous silica as alternative high temperature PEMs for fuel cells.

2. Mesoporous Polymeric Proton Conducting Membranes

2.1 Synthesis of mesoporous Nafion

Mesoporous structures can be introduced into the Nafion membrane via a soft template method [37, 38]. In this method, a nonionic surfactant, F108 [poly(ethylene oxide)-poly(propylene oxide)-poly(ethylene oxide), PEO_{127}-PPO_{48}-PEO_{127}, Pluronic F108] was used to control the micelle size in the polymer mixture. The surfactant embedded in the synthesized micelle/Nafion precursor was removed by reflux with hot water, forming well-ordered mesoporous structures in Nafion (Fig. 3.4).

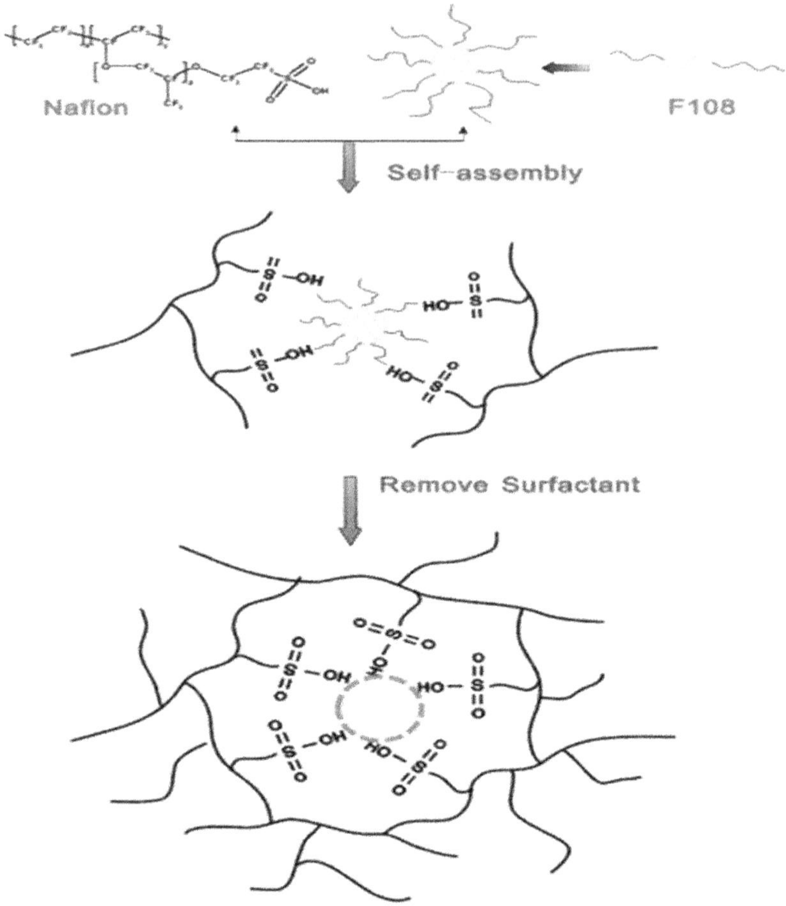

Figure 3.4. Synthesis principle of formation of mesoporous structures in Nafion via a surfactant-directed self-assembly process. The dotted circle refers to the ordered mesopores formed after the removal of the F108 surfactant. (Reproduced from [37] with permission. Copyright (2011) The Royal Society of Chemistry.)

However, the degree of mesoporous structure in Nafion via the soft template method is not high as indicated by the appearance of shoulder on the SAXS patterns of mesoporous Nafion [37]. Further studies showed that introduction of a colloidal silica mediated self-assembly method can produce highly ordered and periodic mesoporous Nafion membranes with controlled structural symmetries including 2D hexagonal (2D-H, *P6mm*), 3D face-centered (3D-FC, $Fm\bar{3}m$), 3D cubic-bicontinuous (3D-CB, $Ia\bar{3}d$) and 3D body-centered (3D-BC, $Im\bar{3}m$) structures [39, 40]. In this method, the colloidal silica was used as an interlink orbridge between the sulfonic chain

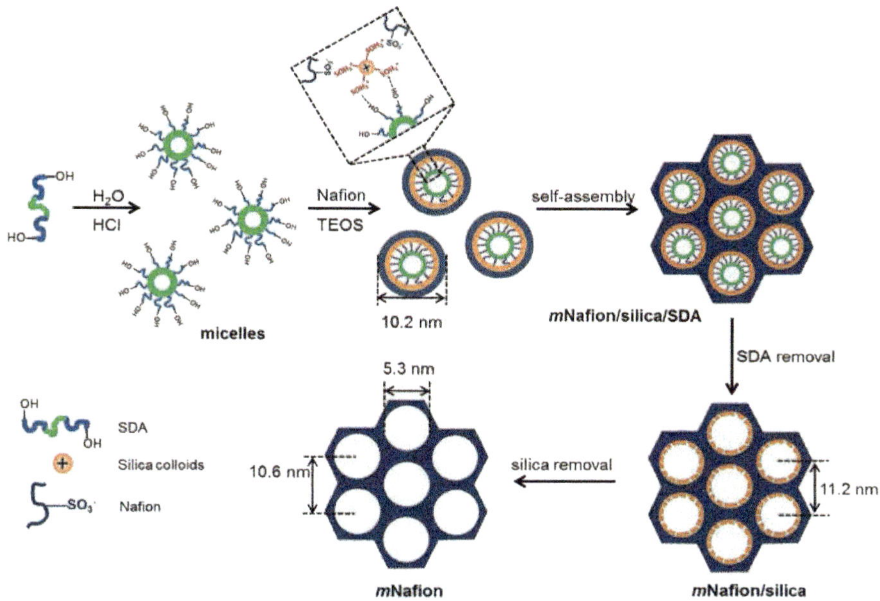

Figure 3.5. Schematic of a surfactant-directed and soft template method with the assistance of silica colloidal mediator for the formation of mesoporous Nafion (2D-H structured Nafion was selected as a representative). The transformation from random micelles to an ordered hexagonal structural phase in solution occurs at ~600 seconds. (Reproduced from [40] with permission. Copyright (2014) The Royal Society of Chemistry.)

of Nafion ionomers and the structure-directing block copolymer surfactants, creating ordered Nafion-silica-surfactant micelles via a cooperative self-assembly. An *in situ* time-resolved synchrotron Small Angle X-ray Scattering (SAXS) technique was used to study the micelles and mesoporous structure formation assisted with silica colloids [40]. The microstructure, mesoporous symmetry, pore size, proton conductivity and performance of the mesoporous Nafion membranes were studied in detail. Based on the *in situ* time-resolved SAXS data, the process of the silica-mediator assisted formation of mesoporous Nafion membrane was proposed in Fig. 3.5 for the typical 2D-H structured Nafion [40].

In the normal pH range, colloidal silica features a very low relative surface charge density and net zeta potential, which means that the proton adsorption degree on the SiOH surface groups is low. The addition of HCl promotes the protonation process, leading to the rapid increase in the zeta potential and formation of positively charged $SiOH^{2+}$ groups on the SiO_2 surface [41]. Driven by the electrostatic forces between the positively charged protonated colloidal silica and the SO^{3-} end groups of Nafion ionomers, the self-assembly of Nafion ionomers and silica occurs. Simultaneously, the as-formed Nafion-silica pair will anchor onto the ether block of the structure-directing block copolymers through hydrogen bonding forces. Due to the stabilizing functions of Nafion ionomers, further agglomeration of colloidal silica will be inhibited. The interfacial orientation between the Nafion–silica pair and the surfactant takes place through cooperative hydrogen bonding self-assembly, leading to the formation of highly ordered Nafion-silica-surfactant micelles in solution. Based on the time-resolved synchrotron SAXS results, the transformation from random micelles to the ordered mesoporous phase occurs at ~600 seconds. Mesopores with different structure symmetries such as 2D-H, 3D-FC, 3D-CB and 3D-BC can be achieved by adjusting the ratio of reagents and employing surfactants with different ether blocks, similar to that outlined for mesoporous silica [42]. The colloidal silica nanoparticles which attach both at Nafion ionomers and surfactants expand the Nafion ordered arrays, and create anelectrostatic-induction orientation of sulfonic groups towards the ordered silica-surfactant micelles. After removing both silica nanoparticles and surfactants, a highly ordered mesoporous Nafion with controlled structure symmetries is formed. The resulting mesoporous orientation would lead to the increasing exposure of sulfonic groups to the hydrophilic phase and the formation of highly ordered proton transporting channels, promoting fast proton conduction process.

2.2 Characterization and performance of mesoporous Nafion

The ordered meso-channels as observed by the AFM (Figs. 3.6a and 3.6b) are most likely induced during the recasting and drying of the Nafion membranes. On the other hand, the formation of ordered bicontinuous ionic hydrophilic channels embedded in a hydrophobic matrix is possible due to the well established phenomenon of microphase separation in block copolymers [43, 44]. Meso-Nafion shows a much higher water retention ability, higher proton conductivity and higher power output particularly under reduced RH and elevated temperatures, as compared to conventional Nafion 115 membranes (Figs. 3.6c and 3.6d). With the inherently high proton conductivity in hydrated conditions and excellent chemical and thermal stability, the meso-Nafion would be a much better PEM for the development of PEMFCs with high proton conductivity, excellent water retention properties, significantly reduced water management and thus a more compact design for the operation at reduced RH and elevated temperatures.

The morphology of Nafion synthesized by the colloidal silica mediated self-assembly process is characterized by parallel and well-organized pores and channels, as shown in Fig. 3.7 [40]. Nafion membranes retain the structural symmetries of the Nafion-silica precursor after removing the colloidal silica mediator. N_2 sorption measurement shows that the average pore diameter is 5.3, 3.8, 3.8 and 4.7 nm for Nafion membranes with 2D-H, 3D-FC, 3D-CB and 3D-BC symmetries, respectively, which is in agreement with the TEM results. In contrast, mesoporous silica prepared from the nonionic copolymer templates generally possess pore sizes ranging from 7 to 12 nm [45]. This suggests that the mesopores in Nafion shrink in size after removing the silica and surfactants. The archetypal parallel mesoporous channels of Nafion with 2D-H (Fig. 3.7a) and the grid-like framework of Nafion with 3D-FC (Fig. 3.7b) are in good agreement with the corresponding symmetrical structures, respectively, along the [110] and [100] directions of mesoporous silica, while the bicontinuous (Fig. 3.7c) and body-centered (Fig. 3.7d) structured Nafion appear to suffer from increased tortuousness of the mesopores as compared with their silica counterparts.

The polarization and power output performance of cells with mesoporous Nafion synthesized by the colloidal silica mediated self-assembly process and Nafion 112 membranes were measured at 60°C under various RH conditions (Fig. 3.8) [40]. The open circuit voltage of all five Nafion membranes was in the range of 0.95–0.98 V. The polarization performance of cells with mesoporous Nafion membranes is considerably

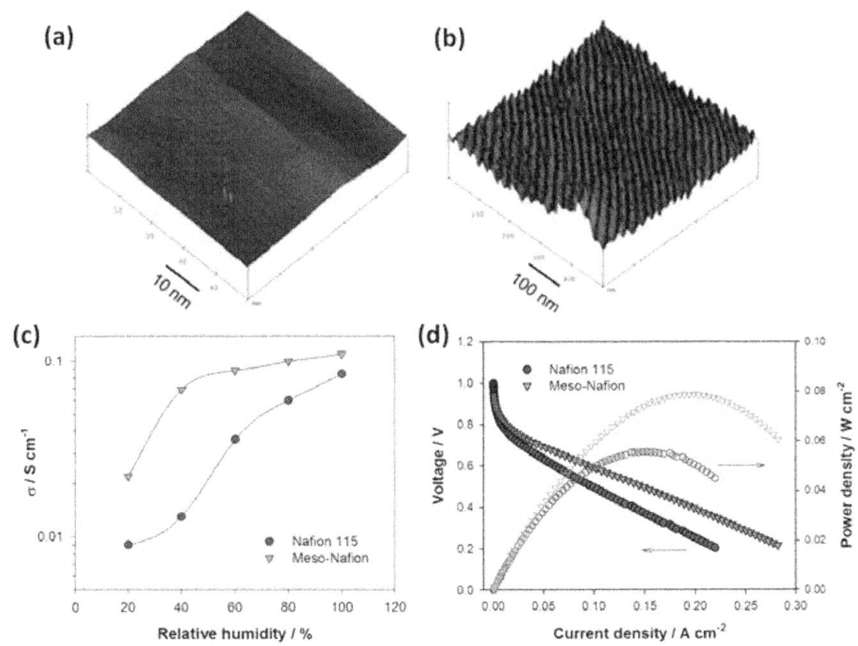

Figure 3.6. (a) AFM micrograph of Nafion 115, (b) AFM micrograph of meso-Nafion, (c) Conductivity plots of Nafion 115 and meso-Nafion membranes, (d) Performance of cell fabricated with Nafion 115 and meso-Nafion membranes, measured under dry H_2 and O_2 conditions at 25°C. (Reproduced from [37] with permission. Copyright (2011) The Royal Society of Chemistry.)

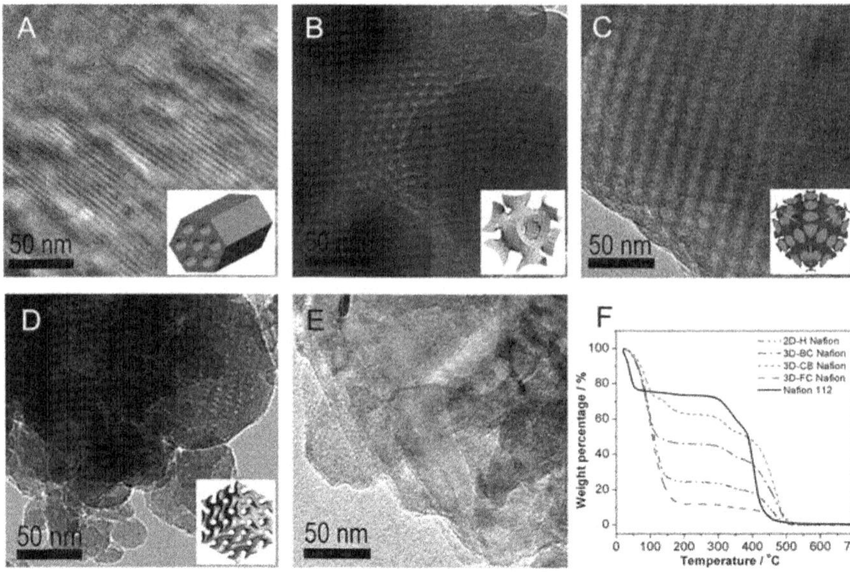

Figure 3.7. TEM morphologies of mesoporous Nafion with (A) 2D-H in the [110] direction, (B) 3D-BC in the [111] direction, (C) 3D-FC in the [100] direction, (D) 3D-CB mesoporous structures in the [311] direction and (E) pristine Nafion membrane. TGA profiles of mesoporous Nafion membranes with different space symmetries and pristine Nafion membrane as the control group are shown in (F). (Reproduced from [40] with permission. Copyright (2014) The Royal Society of Chemistry.)

higher than that of the cells with pristine Nafion membranes, particularly under reduced RH. For example, under 60% RH, the power density of the cells with 2D-H, 3D-CB, 3D-BC and 3D-FC Nafion membranes is 362 mW cm^{-2}, 224 mW cm^{-2}, 185 mW cm^{-2} and 203 mW cm^{-2}, higher than 115 mW cm^{-2} obtained on Nafion 112. The cell with a 2D-H Nafion membrane produced the highest power output. The peak power density of cells with 2D-H, 3D-BC, 3D-CB and 3D-FC mesoporous Nafion membranes shows a very different dependence on RH, as compared to the cell with pristine Nafion 112 (Fig. 3.8f). The peak power density of Nafion 112 reduced linearly with the decrease in RH. At 100% RH, the power density of the Nafion membrane cell is 228 mW cm^{-2} and decreases to 77 mW cm^{-2} when the RH is reduced to 40%, a 66.2% reduction in power output. The significant reduction in power output of Nafion 112 membrane cell is clearly due to the fact that the proton conductivity of Nafion membranes decreases significantly with decreasing RH. In the case of the cells with mesoporous Nafion membranes, the dependence of the power density on RH follows a S-type curve. The S-type dependence of the polarization performance on the RH has also been observed on functionalized mesoporous silica cells [46, 47]. The S curves indicate that the power output of the mesoporous Nafion membrane cells

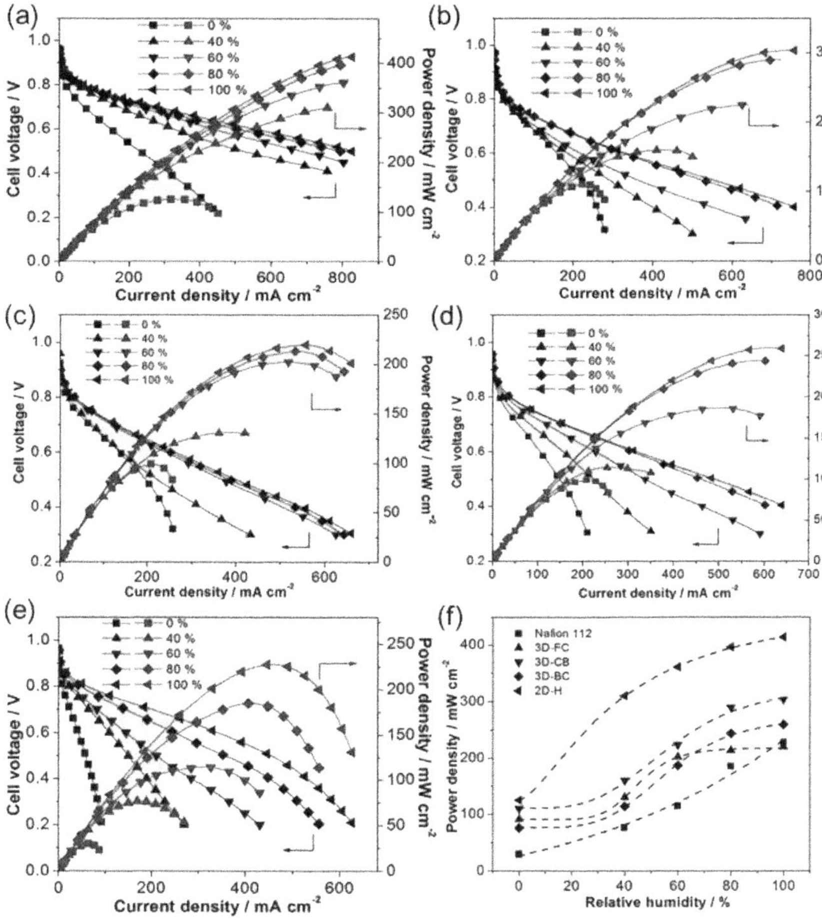

Figure 3.8. Polarization and power output of cells, measured at 60°C, H_2/O_2 with (a) 2D-H Nafion membrane, (b) 3D-CB Nafion membrane, (c) 3D-FC Nafion membrane, (d) 3D-BC Nafion membrane, (e) Nafion 112 membrane. Peak power density of mesoporous Nafion membranes against various relative humidity at 60°C is given in (f). (Reproduced from [40] with permission. Copyright (2014) The Royal Society of Chemistry.)

is less sensitive to the change in RH, which is likely due to the fact that ordered mesoporous channels have a much better water retention ability as compared to the random nanostructures associated with pristine Nafion membranes. The reduced sensitivity of the power performance of the cells to RH is an important factor for practical operations, as this will allow more stable power output under conditions of RH fluctuation. The cell with a 2D-H mesoporous Nafion membrane produces the best performance; 414 mW cm^{-2} under 100% RH and 310 mW cm^{-2} under 40% RH. Fukuda et al. [48] showed that variation of power density with the RH is consistent with the tendency of the amount of H_2O adsorbed on the Nafion membrane. Thus the low sensitivity or dependence on RH of mesoporous Nafion membrane cells is most likely due to the water capillary condensation effect of the mesoporous structure at reduced relative humidity. On the other hand, the hydrophilic clusters of the Nafion 112 membrane with $-SO_3H$ groups have to adsorb a large amount of water in order to form the continuous proton transportation channels, resulting in a high dependence on external humidification and RH [49]. Consequently, the highest performance of 2D-H mesoporous Nafion based fuel cells indicates the effectiveness of the highly ordered hexagonal cylinder structure for water retention.

3. Mesoporous Inorganic Proton Conducting Membranes

3.1 TiO_2, Al_2O_3 and other ceramic materials

Highly ordered mesoporous TiO_2 with uniform pore size (~7.6 nm) and high surface area (271 $m^2\ g^{-1}$) can be synthesized using titanium isopropoxide as a titania source and triblock copolymer P123 as a template [50]. In that work, the measurements clearly showed that the titania is made of highly crystalline anatase nanoparticles, which are uniformly embedded in the pore walls to form the "bricked-mortar" frameworks. The amorphous silica acts as a glue linking the TiO_2 nanocrystals and improves the thermal stability. Mesoporous TiO_2 was also prepared by the controlled drying of the precursor sol that was synthesized by the hydrolysis of titanium isopropoxide [34]. The conductivity of the mesoporous TiO_2 membrane is strongly affected by the chemistry of the pore walls. The conductivities of mesoporous TiO_2 increased with increasing RH. In this regard, the materials behave as typical protonic conductors. The conductivity versus RH curve for Nafion 117 has a different shape than those of mesoporous of TiO_2 as shown in Fig. 3.9, where the curves are S-shaped, while Nafion 117, in the same range of RH (30 to 80%) presents either a linear [51] or exponential [52] behavior. The large difference in conductivity among these four materials (Fig. 3.9a) can be attributed to their unequal water adsorption behavior, which should lead to differences in their conductivity values at a given RH. As demonstrated in Fig. 3.9b, the conductivities of samples Ti400, TiP4.0, TiP2.5 and Ti1.5 show a linear response to changes in their surface site density (*N*, number of water molecules per square nanometer). Variations in conductivity for a constant *N* should be associated with differences in the concentration of H^+ in the pore water, and also with potential differences in the mobility of the protons. The expected unequal conductivity at a given *N* between phosphated and non-phosphated samples will have additional sources, namely, differences in the surface pKa and changes in the structure of the first few layers of physisorbed water due to the presence of hydrated

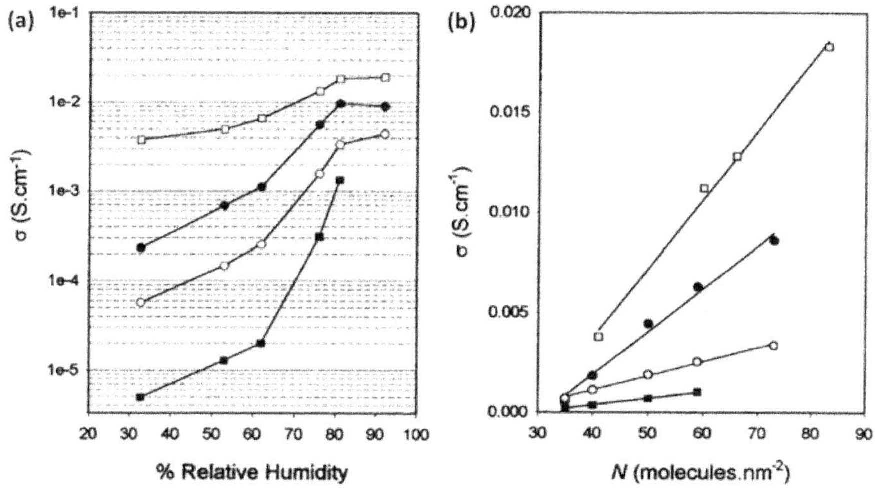

Figure 3.9. (a) Effect of RH on the proton conductivity and (b) effect of *N* on the proton conductivity of Ti400 (■), TiP4.0 (○), TiP2.5 (●) and Ti1.5 (□). (Reproduced from [34] with permission. Copyright (2000) American Chemistry Society.)

phosphate ions. The state of polarization of the water molecules hydrating the surface phosphate as well as their orientation in relation to the pore wall are expected to be different than that of the water hydrogen bonded to surface H_2O^+/OH groups.

Titanium (IV) oxides of various degrees of hydration, TiO_2 [53–55], $H_2Ti_3O_7$ [56] and $H_2Ti_4O_9$ [57], have been shown to conduct protons from room temperature to 150°C. It has been known that the surface of TiO_2 has both Levis acid (five-coordinated Ti^{4+} ions) and basic (two-coordinated O^{2-}) sites. Water can be adsorbed on the five-coordinated Ti^{4+} ions in a molecular and/or in a dissociated form [58]. Vichi et al. [53] investigated the protonic conductivity in a mesoporous TiO_2 membrane as a function of surface chemistry. Changing the *N* values from 5.5 to 5.7 leads to an increase in conductivity from 8.00×10^{-3} to 1.00×10^{-2} S cm^{-1} at 25°C and 81% RH. The results indicate that values in proton conductivity do not totally correlate with the water content in these materials, indicating that surface chemistry strongly affects the water uptake and proton conductivity of the nano-porous TiO_2. Further studies show that proton conductivity and water uptake of nanoporous TiO_2 is also affected by the pore structure (e.g., porosity, pore size and surface) [59].

Colomer [28] synthesized mesoporous TiO_2 thin film on glass substrates by a sol-gel technique. The films are mesoporous with an average pore diameter of 5.8 nm and the results show the conductivity improves with the increasing RH, exhibiting a sharp change at ~60% RH. The maximum conductivity is 8.71×10^{-3} S cm^{-1} at room temperature and 81%

RH and 3.78 × 10^{-2} S cm^{-1} at 80°C and 81% RH with activation energy of 0.23–0.37 eV [28]. Low conductivity in the range of 10^{-4} to 10^{-3} Scm^{-1} measured on pure TiO_2 membranes made by sol-gel methods was also reported by Tsuru et al. [60]. Titania nanotubes functionalized by 3-mercaptopropyltrimethoxysilane (MPTMS) as the sulfonic acid functional group precursor showed a proton conductivity of 0.08 S cm^{-1} at 80°C and 100% RH and has been shown to be a good additive to enhance the proton conductivity of Nafion membranes at elevated temperature and reduced RH [61].

Mesoporous alumina has also been synthesized and investigated as potential application as PEM for fuel cells. Shen et al. [62–64] studied the effect of pore size and salt doping on the proton conductivity of mesoporous alumina. Conductivity of mesoporous alumina increased with the pore size and best results were obtained with a pore diameter of 10.8 nm, 4.0 × 10^{-3} S cm^{-1} at 30°C and 90% RH. The surface chlorine has an important role in the change of the proton conductivity. The surface properties were modified by doping inorganic salts and the surface acidity was increased with doping chloride. Doping with chloride improves the proton conductivity by three to six times that of pure alumina. The results indicated that chlorine displaces some surface OH and increased the acidity of the mesoporous Al_2O_3, leading to the increase of the proton conductivity [64]. Mesoporous acid-free hematite ceramic membranes have also been studied as proton conductors by Colomer et al. [65]. This α-Fe_2O_3 ceramic membrane showed a sigmoidal dependence of the conductivity and the water uptake on the RH at a constant temperature. Further study showed that mesoporous α-Fe_2O_3 can also be synthesized from a hydrolytic ferric oxide polymer using a microwave-assisted sol-gel route [66]. Despite the unique acid-free property, the highest conductivity of this α-Fe_2O_3 ceramic membrane is 2.76 × 10^{-3} S cm^{-1} at 90°C and 81% RH [66]. Nd_5LnWO_{12} and Ln_6WO_{12} have been found to have satisfactory proton conductivity at high temperatures in hydrogen-containing atmosphere, however the coexisting electronic conductivity impedes the use in fuel cells [67].

Tsui et al. [68] studied the proton conductivity properties of ceramic membranes derived from ferroxane and alumoxane precursors. The ferroxane derived ceramics, FeOOH fired at 300°C showed highest proton conductivity from 1.29 to 2.65 × 10^{-2} S cm^{-1} at relative humidities of 33–100% and room temperature. The conductivity of ferroxane is comparable to that of the Nafion membrane with the advantages of lower methanol permeability and less sensitivity to humidity [69]. However, the ferroxane ceramic membranes are very brittle and can be easily broken to very small pieces. This is mainly due to the poor mechanical properties of the precursor, lepidocrocite as compared to other ceramic materials such as silica and alumina. For example, the hardness of lepidocrocite is 5 GPa, significantly lower than 30.6 GPa obtained on silica [70]. This makes it very hard to fabricate membranes from ferroxane-based ceramic powders.

The nature of the surface groups is expected to influence the protonic conductivity not only by affecting the water adsorption behavior of the material but also by influencing the mobility of protons in water clusters. Water structure, as determined by hydrogen bonding, mobility and polarization, is perturbed by the presence of the pore walls. The degree of perturbation is connected with the magnitude and distribution of the interfacial charge [71]. Proton mobility, on the other hand, is mainly governed by proton hopping and in some degree by the hydrodynamic proton mobility. Stronger hydrogen bonds result in a loss of hydrodynamic mobility (self-diffusion) and, up to a limit, in an increase of proton hopping frequency.

3.2 Silica and silica based glass

The ordered nanoporous or mesoporous silica materials with high structural order, large surface areas and pore volumes, and easy and variable surface functionalization offer great potential as porous frameworks not only for catalyst supports, protein separation, CO_2 capture [72, 73] but also for catalysts support and proton conductor as high temperature PEM applications [26–35, 74, 75]. Ye et al. [76] gave an excellent review on the recent applications of mesoporous materials as electrodes in solar cells, fuel cells and batteries. Due to their high hydrophilicity and capillary condensation effects in the nano-channels they can store and release water at elevated temperatures, facilitating water-assisted proton transport at temperatures above 100°C. The first reports on ordered mesoporous materials with well-defined pore sizes appeared in 1992 when researchers at Mobil reported for the first time on the synthesis of the M41S family [77, 78]. These mesoporous silica, synthesized by the self-assembly method using surfactants as a template, has a large surface area and an uniform pore diameter of several nm [42]. The water vapor condenses in pores through capillary force and this occurs at lower water vapor pressures for smaller pores. Thus, a porous structure with a high surface area and small pores is favorable for proton migration. The pore size and pore structure affect the proton conductivity [26]. The proton conductivity has been found to be associated with surface dehydroxylation of porous silica during the heat-treatment. However, high heat-treatment temperature would reduce the hydrophilic nature of the porous silica glass, leading to the reduced conductivity. Li and Kunitake [79] used the oxygen plasma technique to remove the surfactant template, obtaining porous silica thin films at reduced temperatures. The IR data showed that the plasma-treated film had a high retention of the silanol group on the pore wall, as compared to the heat-treated samples. The conductivity of plasma-treated porous silica is 4.8×10^{-4} S cm^{-1} at 90% RH and 50°C, an order of magnitude higher than that of the heat-treated porous silica under identical conditions [79]. On the other hand, high proton conductivity of 2.0 × 10^{-2} S cm^{-1} at 80°C and 81% RH was reported on mesoporous silica xerogels with average pore size of 3.7 nm [80]. Proton conductivity of mesoporous silica and silica xerogels showed a pronounced dependence on RH [79, 80].

Nogami et al. [81] studied in detail the proton conductivity in porous silica glass with an average pore size of ~1.7 nm and reported that 0.1 mm thick porous silica glasses sintered at 400°C exhibited proton conductivity of ~3 × 10^{-4} S cm^{-1} at 30°C under high RH. The activation energy for conduction decreases linearly with increasing logarithm of the product of proton and water concentration, and the conductivity increases with the water concentration, as shown in Fig. 3.10 [81]. Data for glasses containing small amount of water were from ref. [82]. The results indicate that proton conduction in porous silica glasses involves the protons dissociated from SiOH bonds and subsequently migrating or hopping from the initial site to a neighboring site, resulting in the conductivity, as shown below [81]:

$$SiOH\cdots H_2O \rightarrow SiO^- + H^+{:}H_2O \tag{4}$$

$$H_2O_{(I)}{:}H^+ \rightarrow H_2O_{(I)} + H^+ \rightarrow H_2O_{(II)}{:}H^+ \tag{5}$$

where the dotted line represents the hydrogen bonding between the proton and water molecule, $H_2O_{(I)}$ and $H_2O_{(II)}$ are the water molecules that the proton is hopping between. Thus the proton conductivity of porous silica glasses is highly dependent on the water concentration. However, the proton conductivity of pristine mesoporous silica is generally too low

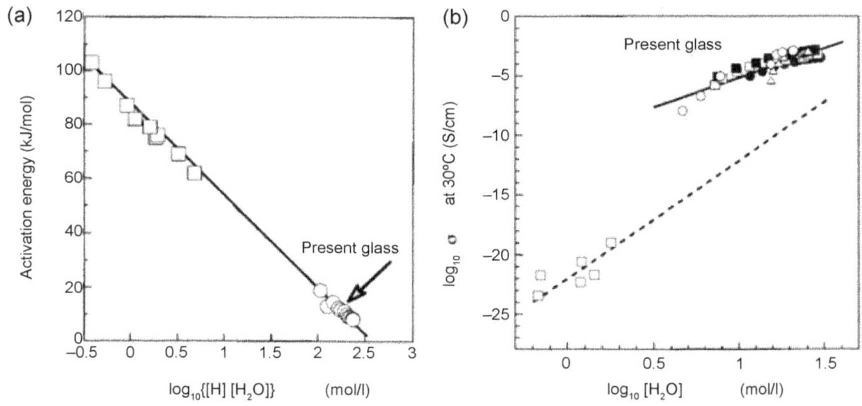

Figure 3.10. (a) Relationship between activation energy for conduction and concentration of protons and water, and (b) relationship between the conductivity and logarithm of concentration of water. (Reproduced from [81] with permission. Copyright (1998) American Chemistry Society.)

(e.g., 10^{-6} S cm^{-1} to 10^{-4} S cm^{-1} under 40 to 90% RH [26, 83]) to be applicable as PEM for fuel cells. Incorporation of aluminium in mesoporous silica framework can significantly enhance its proton conductivity [84], as the incorporation of aluminium increases the Brønsted acid sites, enhancing the proton conductivity.

As the protons are generated through the dissociation of adsorbed and condensed water, water dissociation and thus proton conduction can be enhanced by increasing the acidity of the material surface by introducing acid groups, including phosphoric acid (PA) [85, 86], sulfonic acids [75, 87] and HNO_3 [53, 88]. Strong acidity also contributes to retaining water molecules under low water vapor pressure [53]. The acid-functionalized mesoporous silica materials constitute an alternative and promising class of inorganic PEMs for high temperature operations due to their high charge carrier concentration, adjustable acid group density and oxidation resistance. Suzuki et al. [89] examined the proton conductivity of phosphorus incorporated mesoporous silica. Incorporation of phosphorus during mesoporous silica formation was found to be effective in improving proton conductivity. A sample with a low P/Si atomic ratio of 0.07 showed the highest proton conductivity above 10^{-2} S cm^{-1} at 100 to 120°C under saturated water vapor pressure. Nogami et al. [90] showed that incorporation of the P=O bond in the glass structure promotes proton dissociation from hydroxyl groups and thus enhances proton conductivity due to the stronger hydrogen bonds. A porous P_2O_5-SiO_2 glass exhibits a significantly high proton conductivity of 2×10^{-2} S cm^{-1} at room temperature and the proton conduction was suggested to be related to the proton hopping between hydroxyl groups and water molecules [90]. Although the phosphate unit improves proton conductivity, it becomes detached from silicate network upon immersion in water and the conductivity decreases significantly to values similar to those of pure silica [91]. In the case of silicate and phosphate glasses, the content of protons decreases with an increase in the sintering temperature. Thus it is desirable to prepare glasses with a large number of protons at low temperature, like the sol-gel process. The gelation time in the conventional sol-gel process can take as long as 1–6 months [92]. However, the sol-gel process can be accelerated by proper water/vapor management. Increasing the density of the surface group by immersing in protonic acids such as HNO_3 showed marginal improvement in proton conductivity of porous silica [53, 88].

There are many studies focusing on the sulfonic acid functionalization [75, 93–97]. Ioroi et al. [98] prepared nanoporous silica glass with 3-mercaptopropyltrimethoxysilance and subsequent oxidation of thiol group to sulfonic acid groups, achieving a proton conductivity of 1.0×10^{-2} S cm^{-1} at 40°C and 95% RH. However, operation at high temperatures results in the loss of grafted HSO_4^-. Marschall et al. [95] synthesized sulfonic acid

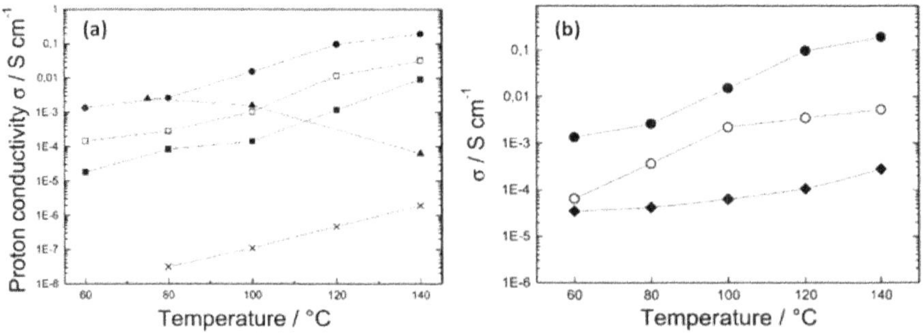

Figure 3.11. Proton conductivity measured (a) at 100% RH for 0% (×), 20% (■), 30% (□), and 40% (•) SO_3H-MCM-41 and for Nafion (▲) and (b) on 40% SO_3H-MCM-41 at 0% (♦), 50% (O) and 100% (•) RH. (Reproduced from [95] with permission. Copyright (2007) American Chemistry Society.)

functionalized MCM–41 by microwave assisted co-condensation method. Figure 3.11 shows the proton conductivity of SO_3H-MCM-41 synthesized by microwave assisted co-condensation method [95]. Microwave irradiation has the advantage of oxidation of thiol group and removal of the template within a short time. Pristine mesoporous silica, MCM-41 shows a negligible conductivity of ~10^{-6} S cm^{-1}, which is the result of a partial dissociation of water molecules in the presence of silanol group. However, SO_3H-MCM-41 composite samples very high proton conductivity at elevated temperatures, achieving 0.2 S cm^{-1} at 140°C and 100% RH (Fig. 3.11a). The proton conductivity of SO_3H-MCM-41 also increases significantly with increasing the RH. The proton transfer inside the sulfonic group functionalized mesopores is suggested to be the Grotthus-like mechanism [95]. Nevertheless, no cell performance was presented for the SO_3H-MCM-41 materials. It has been reported that mesoporous silica impregnated with 5.0M H_2SO_4 shows high conductivities in the order of 10^{-1} S cm^{-1} in a temperature range from 40 to 80°C at 60% RH [99]. However, the conductivity decreases drastically with time due to a rapid loss of the impregnated H_2SO_4 acid. Marschall et al. [100] synthesized imidazole functionalized mesoporous MCM-41 silica by immersing treatment and the highest conductivity was ~10^{-4} S cm^{-1} at 140°C and 100% RH. The proton conductivity of sulfonic acid and imidazole functionalized mesoporous silica composites was found to be very sensitive to RH [87, 100].

Qiao et al. [101, 102] studied in detail the proton conductivity properties of phosphoric or phosphonic acid functionalized mesoporous-structured silica as potential membranes for fuel cells. Owing to the amphoteric nature of the phosphonic acid, such solids can function as a proton donor (acidic) as well as a proton acceptor (basic) to form dynamic hydrogen bond networks in which protons are transported by breaking and forming of hydrogen bonds. The proton conductivity of phosphonic acid functionalized

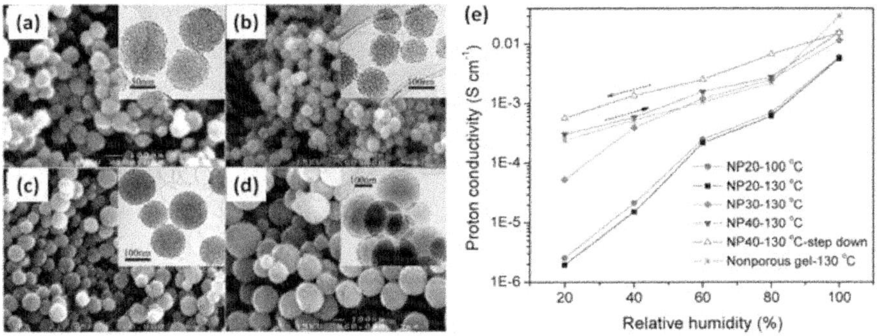

Figure 3.12. SEM and TEM (inset) images of (a) NP10, (b) NP20, (c) NP30, and (d) NP40, where NP10, NP20, NP30, and NP40 stands for the nominal P/Si molar ratio of 10, 20, 30 and 40% in the acidified product, respectively, (e) proton conductivity of phosphonic acid functionalized silica nanospheres as a function of RH at 100 and 130°C. (Reproduced from [101] with permission. Copyright (2009) American Chemistry Society.)

porous silica nanospheres synthesized by the co-condensation of diethylphosphatoethoxysilane (DPTS) and tetraethoxysilane using cetyltrimethylammonium bromide surfactant as a template, followed by acidification to PA in concentrated HCl, depends strongly on the nominal P/Si ratio, the morphology of the porous silica nanosphere and the RH (Fig. 3.12) [101]. The proton conductivity increases with the humidity and NP40 with most phosphonic acid shows the highest conductivity, 3.0×10^{-4} S cm^{-1} at 20% RH to 0.015 S cm^{-1} at 100% RH at 130°C. The results indicate that high surface area of the porous silica enhances the proton conductivity under low humidity conditions. However, no stability and cell performance were reported for the PA functionalized silica nanosphere composites.

Research has shown that replacing pure silica with composite silicate of strong acidity, e.g., the $Cs_3(H_2PO_4)(HSO_4)_2$-SiO_2 [103], CsH_2SO_4-SiO_2 [104], P_2O_5-SiO_2 [105], P_2O_5-TiO_2-SiO_2 [90] or P_2O_5-ZrO_2-SiO_2 [106] improves the proton conductivity. Due to the higher hydrophilicity of P_2O_5 as compared to SiO_2, the incorporation of P_2O_5 increases the amount of H_2O/OH hydrogen bonding. Tung et al. [105] indicated that increasing P_2O_5 content leads to the increased functional groups of P=O and P-OH on the pore surface and reduces the number of major pore size channels. The best conductivity is 1.14×10^{-2} S cm^{-1} at 80°C and 100% RH, obtained on $50P_2O_5$-$50SiO_2$ glass. Nogami et al. [32, 107–111] incorporated various heteropolyacids (HPAs) into silicaphosphate porous glass and yielded an electrolyte membrane with proton conductivity of 0.1 S cm^{-1} at 85°C and 85% RH. The cell based on phosphomolybdic acid (HPMo) incorporated silicaphosphate glass produced a maximum power density of 32 mW cm^{-2} at 29°C and 30% RH [108]. The characteristics of the polarization curves indicate that the cell performance is mainly dominated by the ohmic polarization losses most likely due to the high resistance of the HPMo/ZrO_2-P_2O_5-SiO_2 glass composite membrane. The power densities reported for silica-based glass PEMs is generally low, from 6 to 41.5 mW cm^{-2} on P_2O_5-SiO_2 and HPMo-P_2O_5-SiO_2 membranes [111, 112] to 45 mW cm^{-2} on surface modified silica glass membranes [98]. Using proton conducting H_3PO_4 as a binder, the power density of the P_2O_5-SiO_2 glass membrane cell was 70.3 mW cm^{-2} at 80°C and 75% RH, which is significantly higher than 28.3 mW cm^{-2} on the same membrane but with polytetrafluoroethylene as the binder [109]. Nakamoto et al. obtained maximum power densities of ~25 mW cm^{-2} at 130°C and 38 mW cm^{-2} at 150°C using the phosphosilicate gel/polyimidecomposite membrane [113]. Most recently, Ishiyama et al. reported the development of tungsten phosphate glass as PEM for fuel cells [114]. The proton was injected into sodium conductive oxide glass by electrochemical oxidation of hydrogen at the Pd anode and the maximum proton conductivity is 8×10^{-4} S cm^{-1} at 300°C. The maximum power density of a cell based on the proton injected tungsten phosphate glass membrane is 1.3 mW cm^{-2} at 300°C. Decreasing the membrane thickness as well as the interface resistance is essential to enhance the cellpower output of silica glass based membranes. Di et al. [115] introduced a thin Nafion layer to a phosphosilicate glass membrane and the addition of a thin Nafion layer reduced the glass membrane thickness to 500 mm, achieving a peak power density of 207 mW cm^{-2} at 70°C in H_2/O_2. However, the operation temperature of such Membrane Electrode Assemblies (MEAs) would be limited by the thermal stability of the inserted Nafion layer.

4. Mesoporous Silica based Inorganic/Organic Composite PEMs

Mesoporous inorganic materials have been extensively studied as potential PEM candidates because of their large surface area, chemical and mechanical stability. The addition of mesoporous materials into polymer matrix not only enhances the water uptake of polyelectrolyte membrane conductivity at high temperature and low RH as well as the cell performance [116]. Moreover, it improves the proton conductivity of the matrix by forming a proton conducting path via the functional proton conducting groups in the interior highly ordered meso-channels. Mesoporous silica is a typical mesoporous materials that are employed as fillers in PEMs [117]. The proton conductivity of pure mesoporous silica is too low to be used in PEM (10^{-5} S cm^{-1} at 90% RH) [26]. However, when it is functionalized by proton carrier or conductors, for instance, sulfonation, the proton conductivity of the material is significantly increased [118, 119]. Thus, functionalized mesoporous silica shows promising potential application as PEMs for fuel cells.

4.1 Sulfonated mesoporous silica as fillers for PEMs

Proton conducting groups in the highly ordered meso-channels form an effective proton conducting path to facilitate proton transportation. In order to improve water uptake and the proton conductivity of PEMs membranes, functionalized mesoporous silica with proton conducting groups is introduced into polymer matrix [120]. For instance, when 1 wt% of sulfonated mesoporous organosilicate is added into PA/PBI membrane, the proton conductivity of membrane was increased over one order of magnitude under 80°C, 40% RH [121]. Moreover, when 0.5 wt% sulfonated SBA-15 particles were added into sulfonated poly(phenylsulfone) (SPPSU), the composite membrane showed high proton conductivity of 5.9×10^{-3} S cm^{-1} after 130 minutes elapsed time under dehydrated condition (50% RH) at 80°C, higher than 3.6×10^{-3} S cm^{-1} of the pristine SPPSU membrane [122]. Besides, sulfonated cube mesoporous benzene-silica (SMBS) ($Im\ \bar{3}\ m$) was also added into sulfonated polyetheretherketon (SPEEK) with 10 wt% loading, the proton conductivity of the composite membrane increased up to 70°C at 40% RH. On the contrary, the proton conductivity of Nafion 117 membrane sharply dropped when temperature was higher than 60°C under the same RH [123].

In order to achieve high proton conductivity of the meso-silica/polymer composite membrane, high degree of sulfonation for the PEM is required. However, the increase of sulfonation degree substantially decreased the mechanical strength of PEMs. One practical approach to increase the mechanical strength of PEMs at high level of sulfonation degree is to employ interpenetrating polymer network (IPN) [124, 125]. When 10 wt% SMBS is added into the IPN structure with organosiloxane network (OSPN, 20 wt%) and SPEEK (65% in sulfonation degree), the ternary composite membrane showed a higher elongation before breaking than that of pristine SPEEK [126]. Furthermore, the ternary composite membrane demonstrated higher proton conductivity than that of pristine SPEEK membrane and the binary composite membrane both with OSPN and SMBS, especially at low RH [126]. This is due to the high water retention at low RH, resulting from the capillary condensation effect of the 2D periodic cylindrical channels of SMBS.

Besides sulfonated mesoporous silica, functionalized Hollow Mesoporous Silica (HMS) is developed as superior proton conductive filler under high temperature and low relative humidity [127]. It combines the traits of water reservoir of hollow materials and periodic proton transport path of mesopores. Hollow mesoporous silica is able to be synthesized via two protocols: self-templating etching method and template method. Self-templating etching contains surface-protected etching [128, 129], and structural difference-based selective etching [130]. Take structural difference-based selective etching as an example, the template, SiO_2 for instance, is dissolved in alkaline solution and redeposited into HMS via the assistance of surfactant micelles. The template method can be divided into hard template including SiO_2 [131], carbon [132], polystyrene (PS) [133] and soft template like emulsion droplets and vesicle [134]. The templates are then covered by mesoporous silica shell to form the core-shell composite materials. When the cores are etched by HF, NaOH and Na_2CO_3 solutions, etc. or calcined at high temperature, HMS is obtained. For instance, Jiang et al. synthesized HMS via the hard template method (SiO_2 as template, Fig. 3.13a) and it was functionalized by sulfonic acid, phosphoric acid and carboxylic acid groups via the connection of amine groups on the surface, as shown in Fig. 3.13d [135]. When 4.0 wt% the acid-base HMS particles were incorporated into Nafion matrix, the composite membranes indicated high proton conductivity of 1.02×10^{-2} S cm^{-1} at 80°C, 26.1% RH. That value was 11.1 times higher than that of recast Nafion membrane. That is due to the water reservoir of HMS microspheres with large volume of lumen, leading to the retention of water under low RH, as shown in Fig. 3.13e [135, 136]. The water uptake of Nafion/HMS (4.0 wt%) increased from 19.9 to 41.2% from 25°C to 80°C, while the swelling ratio slightly raised from 7.42 to 8.11%, respectively [135].

The high proton conductivity of sulfonated meso-silica/PEMs at high temperature and low hydration is due to capillary condensation effect of mesopores. There are two types of proton diffusion mechanism in PEMs including vehicle and Grotthuss mechanism. For vehicle mechanism, protons are diffused with the aid of water molecules. And the proton diffusion in the highly hydrated state of Nafion membrane is mainly considered as vehicle mechanism [137]. For Grotthuss mechanism, the protons transfer by hopping from one site to the neighboring site by the hydrogen bond construction and reorientation of the water molecules. Grotthuss mechanism will be a preferential pathway for proton transportation with low hydration water molecules [138]. In sulfonated *meso*-silica/Nafion composite membrane, the proton activation energy slightly increases from 7.1 to 7.6 kJ mol^{-1} when the RH decreases from 98 to 80%. Therefore, the proton diffusion in the sulfonated mesoporous silica based composite membrane seems to occur by the vehicle mechanism even at low RH. That is, the addition of sulfonated mesoporous silica into PEMs ensures sufficient water for proton diffusion via the vehicle pathway. That is due to the fact that mesoporous silica has a high propensity to retain water and inhibits the volatility of water in PEMs at high temperature and low RH.

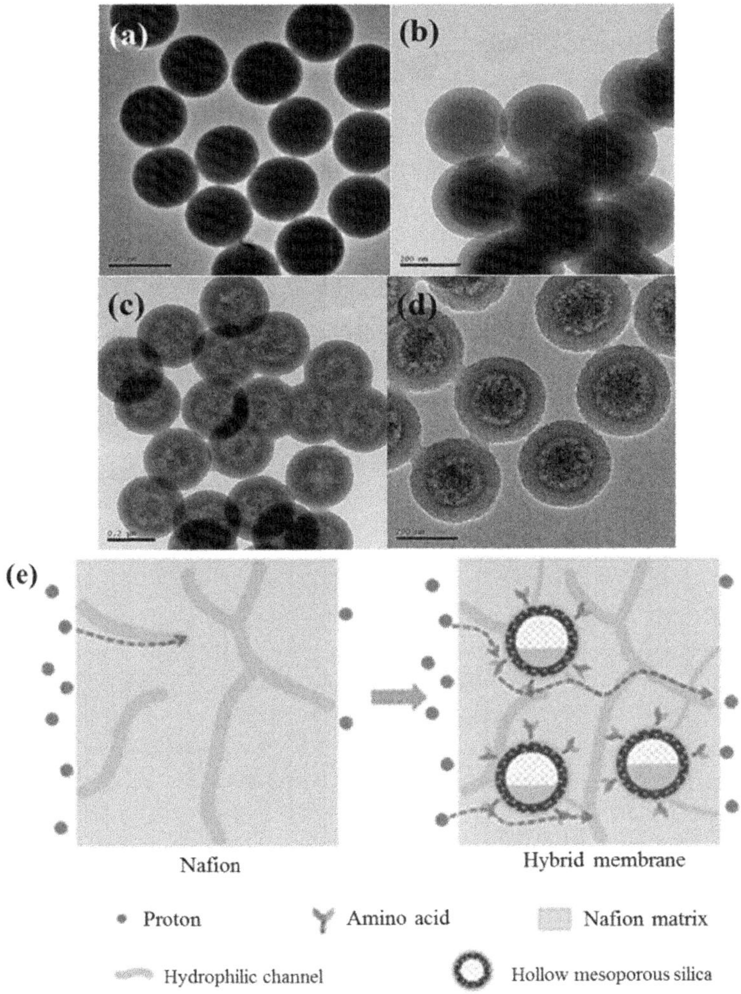

Figure 3.13. TEM images of (a) solid SiO_2; (b) core-shell SiO_2; (c) hollow mesoporous silica (HMS); (d) sulfonic acid functionalized HMS; (e) scheme for the water reservoir of functionalized HMS in Nafion composite membrane. (Reproduced from [135] with permission. Copyright (2015) The Royal Society of Chemistry.)

4.2 Sulfonation of mesoporous silica

There are two protocols for sulfonic groups functionalization of mesoporous silica including post-grafting method [95, 96] and co-condensation method [139]. Post-grafting functionalization method bases on the reaction of organic silanes with the framework of mesoporous silica [140]. Sulfonated organic silanes can be divided into two groups of thiol path [141] such as

3-(trihydroxysilyl)-1-propanesulfonic acid (TPS), mercaptopropyltrimethoxy silane (MPTMS) and sulfonic path [142, 143] such as 1,2,2-trifluoro-2-hydroxy-1-trifluoromethyl-ethanesulfonic acid sultone (FASA) under relatively severe conditions and long reaction time. This sulfonation strategy shows a slow deterioration effect on the periodical properties of silica support. However, the challenges for the post-grafting protocol are the precise control of the acid density and distributions of grafted sulfonated groups. The other protocol is the co-condensation one-step method that involves the self-assembly of hydrolyzed tetraalkoxysilanes ($(RO)_4Si$) with hydrolyzed trialkoxyorganosilanes ($(RO)_3SiR'$), where R and R′ are organic species, via structural-directing agent to functionalize mesoporous silica. In comparison with post grafting method, co-condensation method achieves higher functional sites with higher Ion Exchange Capacity (IEC) for mesoporous silica, larger uniform pores, higher surface area and better long-range order. Moreover, it can be conducted at mild and simple synthetic conditions, saving time and material in contrast to the post grafting method [94]. However, the loading and diversity of the functionalities are limited by the collective compatibility of various components under synthesis conditions. Furthermore, high concentration of the silane precursor species are detrimental for the mesoporous structural ordering of materials [144].

Compared with mesoporous silica, Periodic Mesoporous Organosilica (PMO) is easier to be sulfonated under relatively mild conditions [145]. PMO was first reported independently by Ozin et al. [146], Stein et al. [147], and Inagaki et al. [148] in 1999. Different from the mesoporous silica with pure inorganic framework, PMO has both organic and inorganic groups as integral parts of the porous framework. The organic groups are located within the pore walls as bridges between Si centers. For instance, TEOS and phenyltriethoxysilane (PTES) are co-condensed into phenyl-silica PMO and then it was sulfonated by concentrated sulfuric acid to form the sulfonated PMO [149]. When 3 wt% of sulfonated mesoporous phenyl-silica were added into Nafion matrix, the composite membrane showed two times higher water uptake than pristine Nafion 212 membrane. More importantly, it achieved power density of 414 mW cm^{-2} at 65°C under anhydrous condition, which is a milestone for the operation of PEMFC based on Nafion-based composite membrane without external humidification [149].

4.3 Properties of sulfonated mesoporous silica

Confinement of $-SO_3H$ group in the mesopores of silica materials reduces the dependence of proton conductivity on hydration for the composite membranes. That is due to the capillary condensation effect of mesopores, which increase the water uptake of composite membrane at low hydration.

And the water retention capability of the sulfonated mesoporous silica could be tuned by the pore structure, acidity density of the acid groups and content of the fillers.

Pore structure in terms of dimensionality and pore size affects proton transportation under conditions of similar acid loading. Mesoporous MCM-48 shows higher proton conductivity than that of zeolite beta with small pore and MCM-41 with one dimensional pore structure [150]. That is due to its large pore size and Three Dimensional (3D) interconnected pore structures. Another example is sulfonated KIT-6 with pore size of 8 nm and three dimensional interconnected pore structures. When it was incorporated into Nafion membrane, it demonstrated superior properties than pristine Nafion membrane in terms of proton conductivity and cell performance under 120°C, 72% RH [151].

Acidity of the organic groups in mesoporous silica materials is another critical factor that affects the activity of Nafion-based hybrid membranes. Aryl sulfonic acid mesoporous silica exhibits higher proton conductivity than that of propyl sulfonic acid containing samples, the phosphonic acid functionalized mesoporous silica, and carboxylic acid functionalized mesoporous silica, when the four types of organic functional groups have the similar acid loading in MCM-41 [150]. Tominaga et al. [152] synthesized three types of silica materials of SiO_2 sphere (p-SiO_2), mesoporous silica (Ne-MPSi) and sulfonated mesoporous silica (Su-MPSi) with acidity of 0.06, 0.38 and 0.46 mmol g^{-1}, respectively. When 5 wt% of the fillers embedded in Nafion matrix, Nafion/Su-MPSi composite showed the highest proton conductivity at 40°C, 95% RH. The high proton conductivity is due to strong acid sites anchored mainly on the internal surface of mesoporous silica material.

The effects of pore size and the acidity for the proton conductivity and the water sorption process in the mesopores are dominated by Kelvin equation [153]. Mesoporous electrolyte with small pore size and high acid density achieve high proton conductivity at low RH. Kelvin equation is:

$$\text{In}\left(\frac{p_W^{vap}}{p_{W,o}^{vap}}\right) = -\frac{2\alpha V_w \gamma^{\cos\theta}}{WRT} \tag{6}$$

where p_W^{vap}, $p_{W,o}^{vap}$, γ and V_W are the vapor press in the mesopores, normal vapor pressure, the surface tension and molar volume of water, respectively. R and T are the ideal-gas constant and the absolute temperature, respectively. θ is the contact angle between water and the pore surface, while w is the pore size (diameter of cylindrical pores and width of lamellar pores) [154].

Because of the capillary condensation effect of mesopores, the water in mesopores prefers to condense at low relative humidity because of the low relative vapor pressure in mesopores. When the relative humidity is lower than a certain value, water adsorption will decrease steeply, as shown

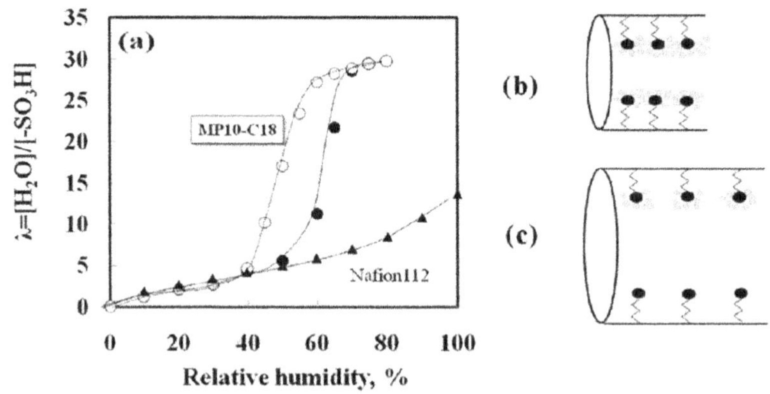

Figure 3.14. Water uptake (a) of sulfonic acid functionalized mesoporous silica films at 25°C against RH values. Solid and open circles show water adsorption and desorption processes, respectively. Water uptake of Nafion 112 and Nafion thin film show the adsorption process. Water sorption at high (b) and low (c) acid density at low RH and proton channel formation. (Reproduced from [155] with permission. Copyright (2013) American Chemistry Society.)

in the water vapor adsorption-desorption isotherms in Fig. 3.14a [155], leading to a steep drop of the conductivity of the mesoporous silica thin film. Moreover, according to Kelvin equation, the relative vapor pressure at the capillary condensation drops with the decrease of diameter of pores and contact angle between the pore surface and water molecule. The increase of acidity in the mesopores decreases the contact angle due to the hydrophilic property of the sulfonic acid group. Thus, the certain RH value at the steep change in proton conductivity shifts lower with the decrease in pore size and increase in acid density. For instance, Shannon and co-workers found that the proton conductivity of sulfonated mesoporous silica film was almost constantly down to as low as 20% RH, while the membrane with larger pores showed the decline beginning at a higher humidity level (50–60%) [153]. Consequently, small pore size and high density of sulfonic acid groups can effectively conduct protons in at low RH (Fig. 3.14b), while large pore size and low density of sulfonic acid groups need high level of hydration (Fig. 3.14c).

The filler content also significantly affects the performance of the composite membranes. The mesoporous structure of sulfonated mesoporous silica is favorable for water retention and proton diffusion derived from periodic proton transportation channels inside the mesopores. On the other hand, when excess filler is embedded into the matrix, filler prefers to aggregate and block the water uptake and decrease IEC of the composite membrane [157]. For instance, when 3 wt% of sulfonated mesoporous silica was added into Nafion membrane, the conductivity of the composite membrane was increased 20% at 60°C, 100% RH [156]. However, when large

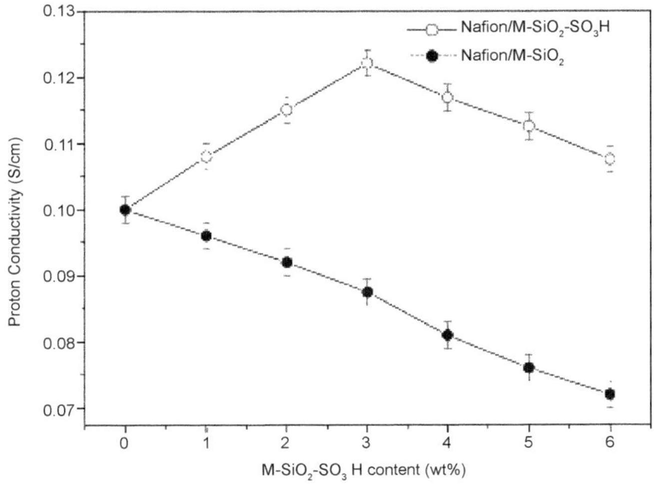

Figure 3.15. Proton conductivity of native and modified *meso*-silica in Nafion membranes under 100% RH, 60°C. (Reproduced from [156] with permission. Copyright (2007) Elsevier.)

amounts of sulfonated mesoporous silica (> 10 wt%) were added into Nafion matrix, the excessive fillers form the insulative phase by agglomerates, which prevents the proton transport in the conductive Nafion matrix (Fig. 3.15) [156]. Similarly to fillers in Nafion membrane, sulfonated mesoporous silica in SPI indicates the same bell shape for the proton conductivity against the filler content. Liu et al. indicated that 7.0 wt% filler was the optimum content for sulfonated mesoporous silica with particle size ~300 nm and pore size of 2.6 nm in SPI matrix for proton conductivity in the temperature range from 25°C to 80°C, 100% RH [158]. Nevertheless, by optimizing the synthesis method, the filler content in PEMs membrane can be further improved with little detrimental effect on the conductivity of PEMs membrane. Pereira et al. [159] synthesized perfluorosulfonic acid/mesoporous silica hybrid membranes via Evaporation-Induced Self-Assembly (EISA). Mesoporous silica nanoparticles with average size of 50 nm were homogenously dispersed in the Nafion matrix with a dense and smooth cross section image even when the filler content increased to 13 wt%. That is presumably due to the *in situ* growth of the inorganic phase entrapped in the polymeric network and well-dispersed at submicronic level creating strong interaction with Nafion. Besides, Dai et al. developed *in situ* synthesis of mesoporous silica in SPI [160]. In comparison with the blending method, the amount of mesoporous silica could reach up to 30 wt% in the SPI matrix by the *in situ* sol-gel and self-assembly approach, while the properties of the composite membrane in terms of water uptake and proton conductivity improved with the increase of filler content in SPI [160].

Table 3.1. The properties of mesoporous materials for proton exchange membranes.

Matrix	Fillers and content	Function groups	agents	Conductivity, mS cm^{-1}	Ref
Nafion	MCM-41, 3.0 wt%	__	__	~100 @100°C, 80%RH	[161]
	Mesoporous silica, 3.0 wt%	__	__	87@60°C, 100% RH	156]
	Mesoporous silica, 3.0 wt%	$-SO_3H$	FSAS	108@60°C, 100% RH	156]
	Mesoporous silica, 5.0 wt%	$-SO_3H$	TPS	94@40°C, 95%	[152]
	Mesopours silica, 13 wt%	$-SO_3H$	CSPTMS	22@RT, 100%	[159]
	KIT-6(Ia3d), 2.5 wt%	$-SO_3H$	FSAS	__	[151]
	MCM-41, MCM-48	$-SO_3H$	MPTMS	__	[150]
	Mesoporous silica, 5.0 wt%	$-SO_3H$	MPTMS		[120]
	Al-MCM-41, 0.50 wt%	$-SO_3H$	H_2SO_4	291@80°C, 100% RH	[162]
	phenyl-*meso*-silica, 3.0 wt%	$-SO_3H$	H_2SO_4	12.9@RT, 100% RH	
	HMS, 4.0 wt%	$-SO_3H$	Cysteine	10.2@80°C, 21.6%	[135]
SPI	*Meso*-silica, 7.0 wt%	$-SO_3H$	MPTMS	42.5@80°C, 100% RH	[158]
SPEEK	*Meso*-silica, 10.0 wt%	$-SO_3H$	H_2SO_4	17@70°C, 40% RH	[123]
SPPSU	SBA-15, 0.5 wt%	$-SO_3H$	MPTMS	5.9@80°C, 50% RH	[122]

Table 3.1 summarizes the proton conductivities of inorganic-organic hybrid membrane.

5. Functionalized Mesoporous Silica based PEM

5.1 Synthesis of PA and HPW functionalized meso-silica

In recent years, many efforts on the development of HPAs-functionalized meso-silica as new inorganic high temperature PEM have been done for fuel cell applications [163–165]. HPAs are known to be a Brønsted acid with unique nano-sized structures and a very strong acidity. Crystallized HPAs contain two types of protons in their structures: the water-combining proton which is dissociated and hydrated, the other is the unhydrated proton located on the bridge-oxygen in the HPAs [166]. Among HPAs,

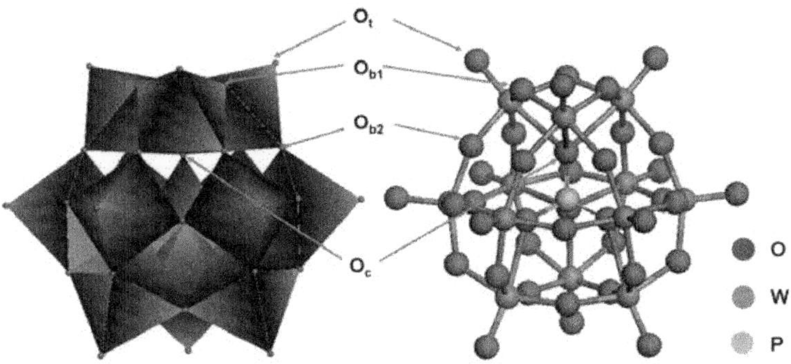

Figure 3.16. The Keggin structure of anion phosphotungsitc acid ($PW_{12}O_{40}^{3-}$): O_t, O_{b1}, O_{b2}, and O_c labeled the four types of oxygen in the structure (oxygen atoms in red; tungsten in cyan; phosphor in magenta).

phosphotungstic acid ($H_3PW_{12}O_{40}$, abbreviated as HPW) has the highest stability and strongest acidity with a high proton conductivity in fully hydrated state, 0.18 S cm^{-1} [167]. The most common structure is Keggin Unit (KU) as shown in Fig. 3.16. The central Phosphorus (P) atom in a tetrahedral environment is surrounded by 12 octahedral of composition WO_6. The oxygen atoms are shared by tungsten atoms, except for 12 terminal oxygen atoms (O_t) attached to only one addenda atom (W). Normally, protons coordinate to oxygen atoms on the exterior of the KU to construct the primary structure of the HPA. Then adsorbed water molecules interact with oxygen atoms of adjacent KUs via hydrogen bonds, linking the units together to form a secondary structure. The effect of water on proton conductivity of HPW is well known, which is pertinent to the use of HPW in the PEM fuel cells [168–171]. Another reason for the selection of HPW as proton carrier in the meso-silica matrix is the high thermal stability, upto 500°C [46].

There are two main approaches to anchor the acid onto the inner pore surface of the mesoporous silica: post synthesis by impregnation and one-pot synthesis. The disadvantage of the impregnation method is that neither the loading nor the surface distribution of the function groups as well as their accessibility can be easily controlled. The post-synthesis method allows for the incorporation of various proton conductors into mesopores of silica materials, such as PA and HPW. Zeng et al. [172] fabricated novel PA/*meso*-silica composite membrane based on the sintered mesoporous silica disk. They found that sintering treatment for the *meso*-silica disk significantly increased the mechanical strength of mesoporous silica disk and improved the operation temperature by getting rid of the polymeric binder. HPW molecules are homogenously dispersed in the nanochannels of *meso*-silica by the Vacuum Impregnation Method (VIM). By extending the vacuum treatment time in VIM method, the content of HPW in mesoporous silica improves from 2.5 wt to 80 wt% while the distance between HPW molecules decrease from 6 to 0 nm [46]. Besides, several types of mesoporous silica have been impregnated by HPW to form the HPW/*meso*-silica membrane including mesoporous silica in 2D hexagonal (*P6mm*), 3D face-centered cubic ($Fm\bar{3}m$), 3D body-centered ($Im\bar{3}m$), to 3D cubic bicontinuous ($Ia\bar{3}d$) symmetries [47]. On the other hand, a one-pot synthesis method has the advantages of precise control of the density and distribution of the function groups along the pores [95, 173, 174].

For the one-step synthesis method, when HPW molecules, silica precursor, TEOS are mixed in water, HPW forms negatively charged $PW_{12}O_{40}^{3-}$ ions and sufficient protons are adsorbed on SiOH groups to form positively charged $SiOH_2+$ in highly acidic aqueous solution [41]. Then self-assembly occurs between the positively charged silica and the negatively charged HPW by the electrostatic force. With the addition of structure-directing agent, P123, the tube-cumulated mesoporous HPW-silica with the template of P123 surfactant is formed through the cooperative hydrogen bonding and self-assembly between the HPW-silica structure and P123 surfactant [42]. With the phase separation of P123, the colloidal complex forms ordered mesoporous silica with the HPW self-assembled in the mesoporous structure. The surfactant template is then removed by the heat treatment at 400°C. Figure 3.17 shows the principles of the two synthesis approaches [46, 175].

The HPW loading in the mesoporous silica matrix depends on the synthesis method. In the case of impregnation route, the HPW loading can be as high as 80 wt%, while the maximum HPW content which can be incorporated into meso-silica structure without significant detrimental effect on the mesoporous structure is ~25 wt% in the case of one-pot synthesis route.

5.2 Conductivity and performance of PA and HPW functionalized meso-silica

After PA impregnation to the mesopores of sintered mesoporous silica disk (PA/*meso*-silica) with 27.7 wt% PA loading, superior conductivity was achieved in comparison with PA/PBI membrane as well as PA-porous-silica membrane at elevated temperature range [172]. Outstanding proton conductivity of 6.0×10^{-2} S cm^{-1} was achieved at 200°C under anhydrous condition while excellent power output of 689 and 200 mW cm^{-2} was obtained in H_2/O_2 and methanol/O_2 system without humidification at 190°C, respectively [172].

The microstructure of HPW functionalized mesoporous silica, HPW-meso-silica is related to the synthesis methods. In the case of HPW-meso-silica synthesized by impregnation method, the distribution of HPW inside the mesopores can be directly identified by TEM (Fig. 3.18) [46]. The mesoporous silica host was characterized by highly ordered nano-channels and the channel diameter of mesoporous silica is ~5.2 nm as shown in Fig. 3.18a. In the case of HPW-meso-silica nanocomposites, the distribution of the impregnated HPW nanoparticles in the mesoporous silica matrix is indicated by the ordered but not continuous black dots sandwiched between continuous channels (gray in color, see Figs. 3.18b and 3.18c). The thick line structure in gray color is siliceous pore walls and the phase contrast between HPW and silica becomes pronounced as the HPW content in the nanocomposite increased to 80 wt% (see the inset in Fig. 3.18c). Such phase contrast was also observed in the cesium substituted HPA-meso-silica system [176]. As the HPW content in meso-silica increased to 80 wt%, the density of HPW particles increases, indicating the distribution of HPW

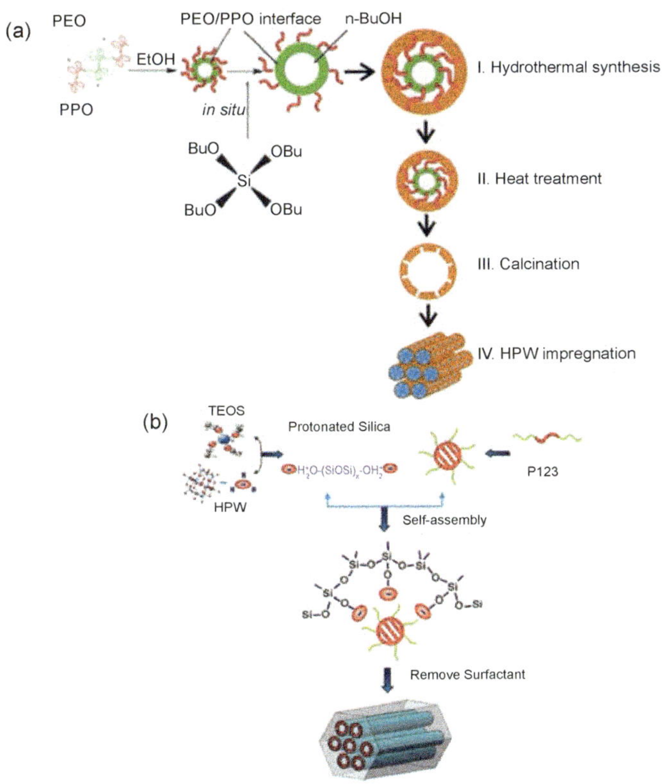

Figure 3.17. Schematic of synthesis of HPW-*meso*-silica via (a) two-step HPW impregnation process (Reproduced from [46] with permission. Copyright (2011) American Chemistry Society.) and (b) one-step self-assembly process. (Reproduced from [175] with permission. Copyright (2011) The Royal Society of Chemistry.)

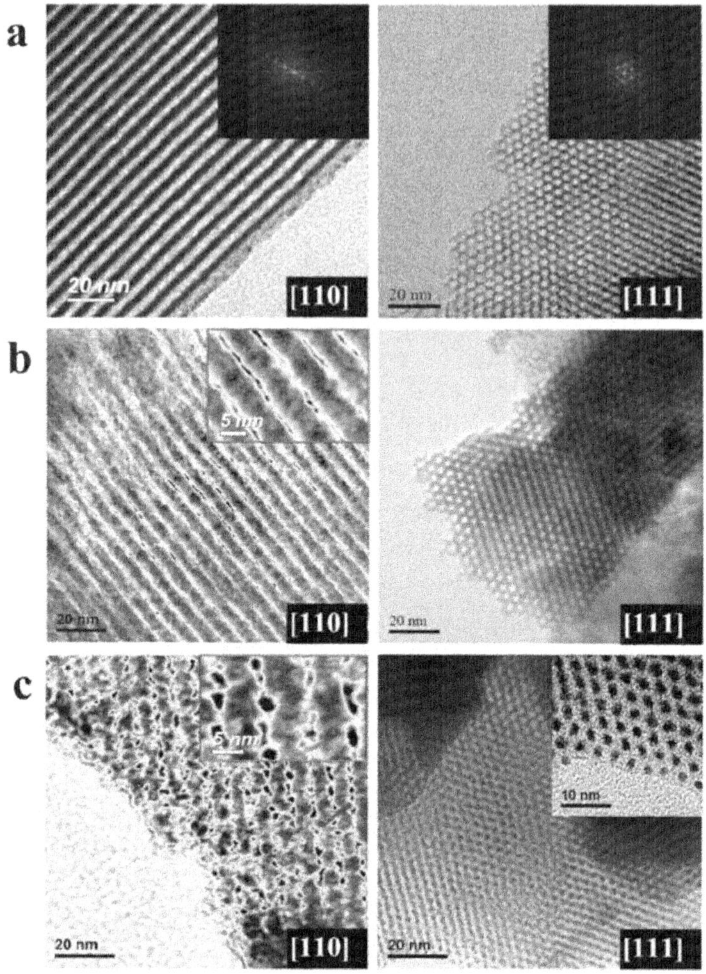

Figure 3.18. TEM images of (a) pure meso-silica and (b) 20% and (c) 80% HPW-meso-silica. The left images were viewed from the [110] direction, whereas the right images were viewed from the [111] direction. (Reproduced from [46] with permission. Copyright (2011) American Chemistry Society.)

particles in 80% HPW-meso-silica is much more uniform than that of 20% HPW-meso-silica. The higher intensity of HPW particles inside the meso-silica means shorter pathway of proton transportation, which would lead to the high proton conductivity and low activation energy barriers for the proton conductance in the meso-silica matrix. The average distances between the HPW particles in the HPW-meso-silica nanocomposites vary between 0 to 5 nm, and the shape and diameter of the interconnected channels of HPW-meso-silica nanocomposites are not as regular as that

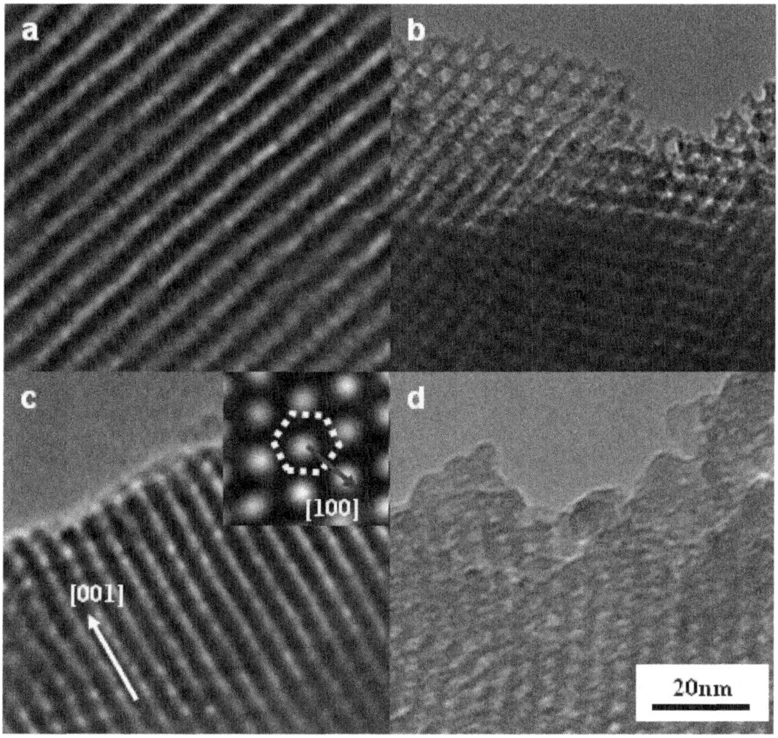

Figure 3.19. TEM micrographs of (a) pure SBA-15 silica; (b) 5 wt% HPW-meso-silica; (c) 25 wt% HPW-meso-silica and (d) 35 wt% HPW-meso-silica. The inset in (c) is the TEM image viewing along the pore axis. The scale bar applies to all TEM micrographs. (Reproduced from [175] with permission. Copyright (2011) The Royal Society of Chemistry.)

of the pristine meso-silica host (the inset in Fig. 3.18c), probably due to impregnation and insertion of HPW particles.

The structure and mesopore size of mesoporous silica by the one-spot method are highly dependent on the formation and size of micelle and also dependent on the size of the functional group. This is indicated by the limit of the HPW loading which can be added to the synthesis solution in order to have high density of HPW and at the same time to maintain the high degree of the mesoporous order of the functionalized mesoporous silica [164, 175]. Figure 3.19 is the TEM micrographs of the ordered mesoporous structure of meso-silica with different HPW contents, synthesized via one-pot self-assembly route [175]. The as-synthesized HPW-meso-silica structure is characterized by highly ordered 2D nanochannels. The center to center distance of the nanochannels is ~7.3 nm (Fig. 3.19a). From the TEM images viewing along the pore axis, the diameter of the hexagonal mesostructure is 4.3 nm for 5 wt% HPW-meso-silica (Fig. 3.19b) and 3.8 nm for 25 wt% HPW-meso-silica (Fig. 3.19c). The high-resolution TEM micrograph unambiguously demonstrates that the hexagonal packing of the nanochannels is well-aligned along the zone axis [001] direction. The uniformly ordered mesoporous arrays are still maintained when the HPW content is less than 25 wt%. The top-surface feature is reflective of long-range ordering of the mesochannels and their macroscopic alignment [177]. The results indicate the successful assembly of HPW molecules into the mesoporous SBA-15 structure. However, for the 35 wt% self-assembled HPW-meso-silica, the mesostructure becomes discontinuous and distorted as shown in Fig. 3.19d. The collapse of the ordered mesoporous structure is consistent with the observed significant shrinkage of the diffraction peak (100) of SAXS [175].

Similar to the sulfonic acid functionalized mesoporous silica materials, the proton conductivity of HPW-meso-silica nanocomposites depends strongly on the HPW content. In the case of HPW-meso-silica nanocomposites via impregnation route, the threshold of HPW content is ~10 wt%. When the HPW content reaches 65–67 wt%, the conductivity is almost independent of the HPW content and remains a constant value of 0.07 S cm^{-1} at 25°C and 100% RH, see Fig. 3.20 [46]. A similar trend was also observed on one-pot self-assembled HPW-meso-silica nanocomposites with the maximum conductivity on 25 wt% HPW-meso-silica [175]. The proton conductivity is also affected by the structure of the mesoporous silica matrix [178]. The best result was obtained with body-centered cubic ($Im\bar{3}m$)

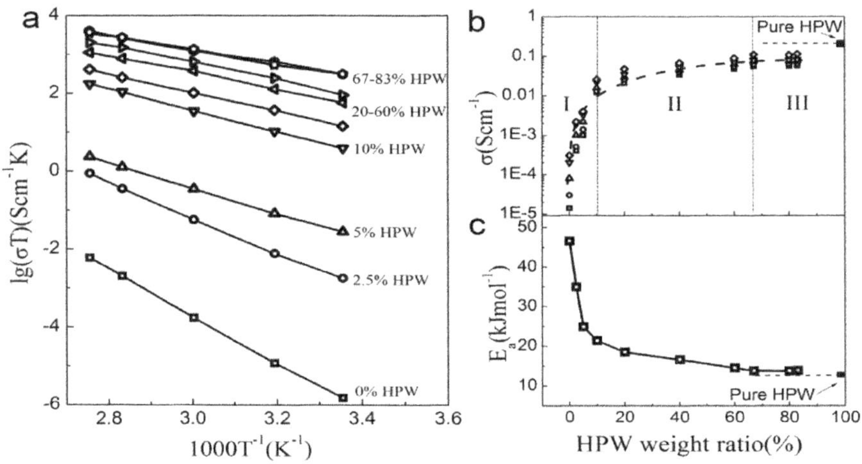

Figure 3.20. (a) Proton conductivity versus temperature of the HPW-meso-silica composite membranes as a function of HPW contents, (b) Proton conductivity versus HPW content measured at 25°C, and (c) Activation energy versus HPW content of HPW-meso-silica membrane. (Reproduced from [46] with permission. Copyright (2011) American Chemistry Society.)

HPW-meso-silica, showing proton conductivities of 0.061 S cm^{-1} at 25°C and 0.14 S cm^{-1} at 150°C under 100% RH, respectively and an activation energy of 10.0 kJ mol^{-1}. Most importantly, the proton conductivity of HPW-meso-silica has a much lower sensitivity toward the change in relative humidity.

Figure 3.21 shows the dependence of the conductivity of a HPW-meso-silica nanocomposite on RH measured at 80°C [179]. In the figure, the proton conductivity of Nafion 115 membrane was measured at 30°C as a function of RH. The initial rapid rise in conductivity originates from a change in the hydration state of HPW as the conductivity of HPW is strongly dependent on the number of water molecules contained in the HPW KU (noted as $H_3PW_{12}O_{40}{\cdot}nH_2O$, n = 4~29). However, the RH dependence of the conductivity is far less sensitive as compared to that of Nafion membranes [180, 181]. The much less sensitivity of the proton conductivity on the RH indicates the high water retention of the HPW KU-type nanoparticles anchored in the ordered meso-silica structure.

Leaching of acid function groups from functionalized mesoporous silica is a very important issue for the practical application of the materials as PEMs for fuel cells. The stability of the anchored HPW inside the mesopores was assessed by measuring the stability of proton conductivity of HPW-meso-silica under an accelerated durability test with a constant water flow. In the case of a 80% HPW-meso-silica, the conductivity value of the nanocomposites drops rather rapidly during the first few hours and reaches a constant value after tested for 6 hours under an accelerated durability test at 80°C with a constant water flow rate [163]. For example, the conductivity decreased from 0.108 and stabilized at 0.075 S cm^{-1} after a 6 hour test and the

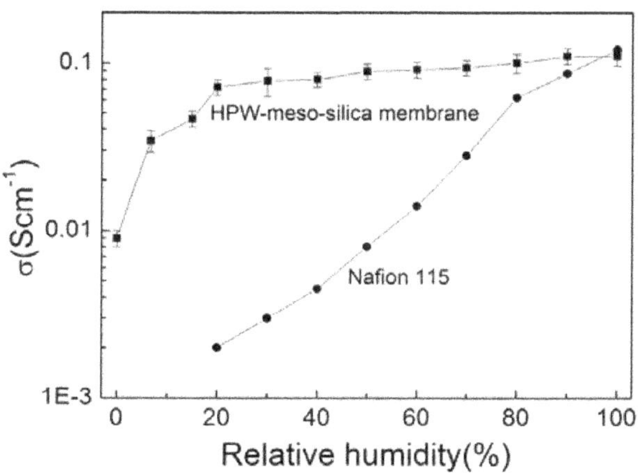

Figure 3.21. The conductivity of a HPW-meso-silica nanocomposite measured at 80°C and reference Nafion 115 membrane measured at 30°C as a function of relative humidity. (Reproduced from [179] with permission. Copyright (2011) The Royal Society of Chemistry.)

loss in proton conductivity is 28% [163]. The initial decay in the conductivity is most likely due to the leaching of HPW during the flushing of water. The stability of the impregnated HPW also depends on the pore size. The mesoporous silica with pore size of ~5 nm shows a much higher retention ability for impregnated HPW as compared to that with larger pore size [179]. The stabilized conductivity indicates the successful immobilization of water soluble HPW in the mesoporous silica framework. On the other hand, the HPW-meso-silica synthesized by one-pot synthesis method shows very stable proton conductivity under elevated temperature and reduced RH [175, 182]. The stability of HPW inside the mesoporous silica is most likely due to the formation of ($\equiv SiOH_2^+$) ($H_2PW_{12}O_{40}^-$) species [182, 183].

The applicability of HPW-meso-silica nanocomposite PEMs has been demonstrated in both H_2 and alcohol fuels such as methanol and ethanol on small button cells [163, 175, 178, 182]. Figure 3.22 is the typical cell performance for direct methanol and ethanol fuel measured at different temperatures on a cell with a 165 μm-thick 25 wt% HPW-meso-silica nanocomposite membrane synthesized via one-pot synthesis method and conventional Pt-based electrocatalysts [182]. The open circuit potential is low, 0.6–0.65V probably due to the electrolyte membrane which is not 100% dense. The maximum power density is 16.8 and 43.4 mW cm^{-2} for direct ethanol and methanol at 80°C, respectively, indicating low electrocatalytic activity of Pt electrocatalysts for the electrochemical oxidation reactions of ethanol and methanol at low temperatures. As the temperature is raised to 200°C, the maximum power output is 112 mW cm^{-2} for direct ethanol fuel cell and 128.5 mW cm^{-2} for direct methanol fuel cell. The DMFC performance based on HPW-meso-silica nanocomposite membrane and low Pt catalyst loading (1 mg cm^{-2}) is close to the advanced DMFCs (100–200 mW cm^{-2}) [184]. The preliminary cell stability was tested under

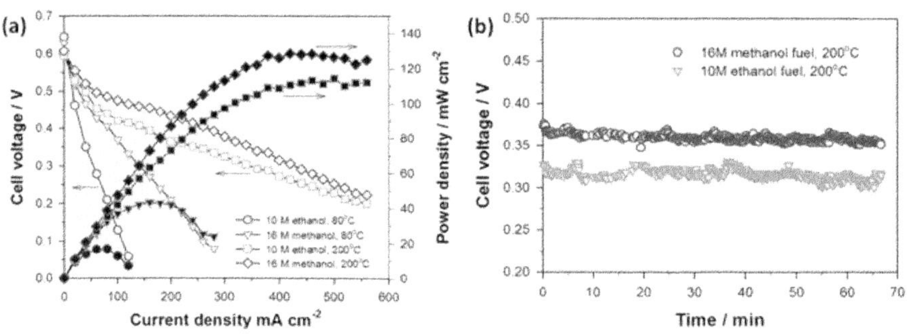

Figure 3.22. Performance and stability of single cells assembled by a 165 μm-thick 25 wt% HPW-meso-silica nanocomposite membrane for direct alcohol fuel cells in the absence of external humidification. The stability of the cell was measured at a constant current of 300 mA cm^{-2} at 200°C. Pt black (1.0 mg cm^{-2}) was used as the anode and cathode electrocatalysts. (Reproduced from [182] with permission. Copyright (2011) The Royal Society of Chemistry.)

a constant current density of 300 mA cm^{-2} at 200°C for ~1 hour. The cell performance indicates the elimination of poisoning effect of the alcohol oxidation reaction on the Pt black catalysts at elevated high temperatures (Fig. 3.22b), very different from that commonly observed significant drop in performance for the DAFCs at low temperatures [185]. Further study showed that maximum power density of the cell can be as high as ~240 mW cm^{-2} at 150°C in 10% RH or under no external humidification in methanol fuel [175, 178]. The studies show that HPW-meso-silica nanocomposites have a much lower methanol crossover as compared to the Nafion membranes. For example, the crossover currents of HPW-meso-silica nanocomposite PEMs are in the range of 12–21 mA cm^{-2} in the temperature range of 100–160°C in a 10 M methanol solution [178]. In the case of a Nafion 117 membrane (178 μm in thickness), the methanol crossover current is 70 mA cm^{-2}, as measured in 1 M methanol at an ambient temperature [186] and 277 mA cm^{-2} in 2 M methanol at 80°C [187]. The much lower methanol crossover current on the HPW-meso-silica membrane could also be related to the gas phase diffusion of methanol and dimensional stability of the inorganic nanocomposite.

Demonstration of the performance in stack cells rather than small button cells is an essential step towards the practical application of inorganic PEMs like HPW-meso-silica. However, very different from polymeric membrane such as Nafion and PBI/PA membranes [188–192], the fabrication of homogeneous, crack-free and mechanically strong membranes with a good working surface area is a significant technological challenge for acid functionalized inorganic materials. The large size HPW-meso-silica membranes were assembled in fuel cell stacks for testing in H_2/O_2 without external humidification [193]. HPW-meso-silica nanocomposite membranes were fabricated by a modified hot-pressing process so as to increase the mechanical strength and increase the size of membrane. Figure 3.23 shows the power output of the 10-cell stack assembled with a 0.76 mm-thick HPW-meso-silica membrane and Pt/C as both catalyst anode and cathode in H_2/O_2 [193]. The OCV of the stack is 7.9 V. The stack produced a maximum power output of 74.4 W, corresponding to a power density of 372.1 mW cm^{-2} at 150°C in H_2/O_2 based on the total effective cell area of 200 cm^2. This power output is quite high compared to values reported on alternate inorganic membrane fuel cells such as In^{3+} doped SnP_2O_7 (264 mW cm^{-2} at 250°C) [194], CsH_2PO_4-based electrolyte (48.9 mW cm^{-2} at 235°C) [195] and HPMo/HPW doped glass electrolytes (32 mW cm^{-2} at 29°C) [108]. The 10-cell stack demonstrates high stability during a 50 hour test and generated a near-constant power output of ~32 W in H_2/air (viz., power density of 160 mW cm^{-2}) at 150°C without external humidification (see Fig. 3.23). The results indicate that the HPW-meso-silica nanocomposites can be applied in PEM fuel cell stacks operated at elevated temperature in the absence of external humidification.

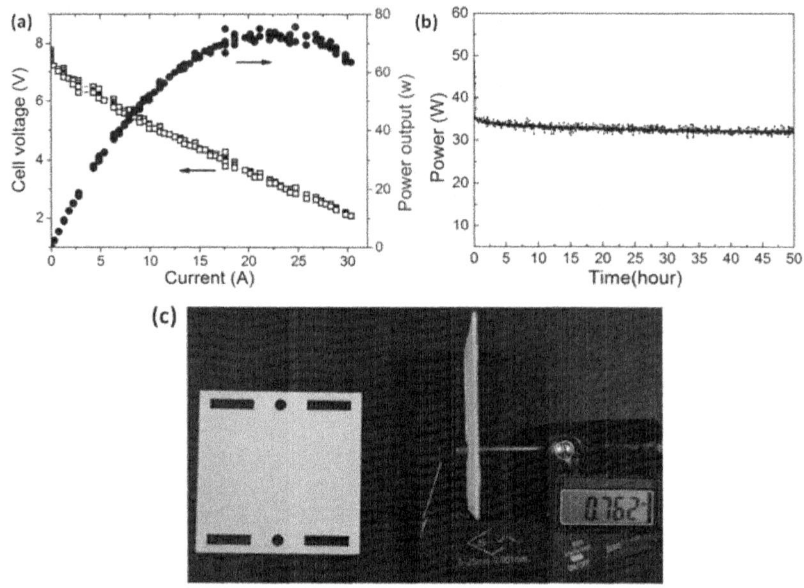

Figure 3.23. (a) Polarization and power density of a 10-cell stack measured in H_2/O_2 (1/1 atm) at 150°C; (b) Stability of a 10-cell stack measured at a stack voltage of 6.0 V and 150°C in H_2/air (1/1 atm). The image in (c) shows a HPW-meso-silica membrane with a thickness of 762 μm and dimension of 62 mm × 62 mm. Note, Pt/C was used as catalyst with a loading of 0.4 mg cm^{-2} for both anode and cathode, and the effective area of the stack was 200 cm^2. (Reproduced from [193] with permission. Copyright (2013) Elsevier.)

5.3 Proton hopping within HPW-meso-silica nanocomposite

Within HPW-meso-silica membrane, the negative charged heteropolyacid $PW_{12}O_{40}^{3-}$ are anchored inside the mesoporous silica channels most likely due to the formation of ($\equiv SiOH_2^+$) ($H_2PW_{12}O_{40}^-$) species [182]. As shown in Fig. 3.18, TEM has provided a direct observation of the morphology and distribution of HPW nanoparticles in the highly ordered mesoporous silica channels, in which the distances between HPW vary according to the weight percentage of HPW. When the content of HPW increases, the average distance between HPWs is accordingly decreased, which leads to the decrease of the energy barrier of proton transfer, so does the increase of the proton conductivity (see Fig. 3.20) [46, 179]. For pristine mesoporous silica, the activation energy for proton transfer is 47–55 kJ mol^{-1} and it is 10–14 kJ mol^{-1} for the proton conductivity on HPW-meso-silica [46, 164, 178]. Clearly, the average distances between HPWs, determined by the HPW loading, affect both the proton conductivity and energy barrier of proton transfer. Figure 3.24 depicts the effective proton transport pathways in the HPW functionalized mesoporous silica. There are mainly two effective pathways determining the rate of proton transfer. One is the intramolecular

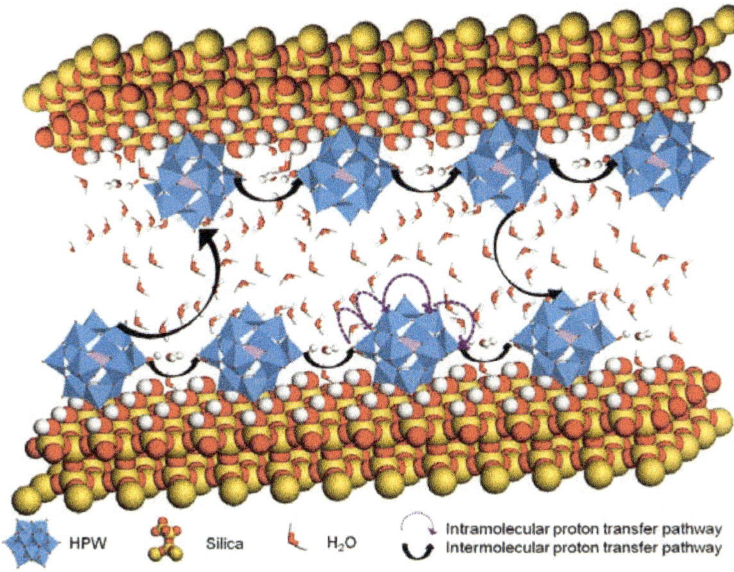

Figure 3.24. The proposed proton transportation in HPW-meso-silica nanocomposites via intramolecular and intermolecular proton hopping. (Reproduced from [36] with permission. Copyright (2014) The Royal Society of Chemistry.)

proton transfer pathway, in which the proton-hopping happens on an isolated HPW. The other is the intermolecular proton transfer pathway, which differs from the intra-pathway since the proton-exchange process is composed of a series of "hops" from one HPW to the neighboring HPW along the water-assisted hydrogen bond. Obviously, the pathway with higher energy barrier will be the rate-limited step, and determines the overall proton conductivity of HPW-meso-silica nanocomposite.

Table 3.2 summarizes the properties and performance of typical mesoporous structured materials as PEMs applying in PEMFCs. As shown in Table 3.2, the highly ordered mesoporous structures normally have excellent water retention ability and can significantly enhance proton conductivity of PEMs at high temperatures and low humidity conditions. Furthermore, the doping of proton conducting agents such as phosphate and sulfuric acids is effective to enhance the proton conductivity of the mesoporous materials. For the polymeric PEMs, the highest power output was observed on the cell with meso-Nafion (modified with 10% F108) at 80°C in H_2/O_2 [38]. For the inorganic PEMs, the highest power output was observed on the cell with PA functionalized sintered meso-silica membranes, 570 mW cm^{-2} at 190°C in H_2/O_2 [172, 196].

Table 3.2. Properties, conductivities and cell performance of mesoporous structured materials as PEM in PEMFCs.

Mesoporous materials	Dopant or functionalization	Properties			Notes	Ref.
		BET, pore size/nm	Conductivity	Cell performance		
Meso-Nafion /10% F108			0.11 S cm^{-1}, 100°C, 100% RH	1046 mW cm^{-2} at 80°C, 100% RH, H_2/O_2 0.633 W cm^{-2} at 80°C, 60% RH, H_2/O_2	σ is less sensitive to RH as compared to pristine Nafion	[37, 38]
2D-H Nafion		609 m^2 g^{-1}, 5.3 nm	0.08 S cm^{-1}, 60°C, 100% RH 0.062 S cm^{-1}, 60°C, 40% RH	414 mW cm^{-2} at 60°C, 100% RH, H_2/O_2 310 mW cm^{-2} at 60°C, 40% RH, H_2/O_2	σ is less sensitive to RH as compared to pristine Nafion	[39, 40]
TiO_2	None	134 m^2 g^{-1}, ~5 nm	0.01 S cm^{-1}, 25°C, 81% RH	n.a.	σ is less sensitive to RH	[34]
TiO_2	Phosphate	134 m^2 g^{-1}, ~5 nm	9.7×10^{-3} S cm^{-1}, 25°C, pH 2.5	n.a.		[34]
Al_2O_3	$LaCl_3$		3.0×10^{-2} S cm^{-1}, 25°C, 90% RH	n.a.		[64]
Al_2O_3	None	43.1 m^2 g^{-1}, 10.2 nm	4.0×10^{-3} S cm^{-1}, 25°C, 90% RH	n.a.		[62, 63]
5% P_2O_5-95% SiO_2	5% P_2O_5	631 m^2 g^{-1}, 2 nm	1.3×10^{-2} S cm^{-1}, 80°C, 80% RH	70.3 mW cm^{-2} at 80°C, 75% RH, H_2/O_2	10 wt% H_3PO_4 was used as the binder	[197]
5%P_2O_5-95% SiO_2	5% P_2O_5	631 m^2 g^{-1}, 2 nm	1.0×10^{-3} S cm^{-1}, 80°C, 80% RH	28.3 mW cm^{-2} at 80°C, 75% RH, H_2/O_2	10 wt% PTFE was used as the binder	[197]
ZrO_2-P_2O_5 -SiO_2	HPM	~350 m^2 g^{-1}, 2.4 nm	10^{-3}–10^{-1} S cm^{-1}, 30–90°C, 80% RH	32 mW cm^{-2} at 29°C, 30% RH, H_2/O_2	Active cell area: 0.25 cm^2	[198]
P_2O_5-SiO_2 /Nafion	None			207 mW cm^{-2} at 70°C, humidified, H_2/O_2	Nafion layer: 800 nm	[115]

Si-MCM-41	None	1030 $m^2\ g^{-1}$, 2.7 nm	1×10^{-6} S cm^{-1}, 140°C, 100% RH	n.a.		[95]
Si-MCM-41	$-SO_3H$	1030 $m^2\ g^{-1}$, 2.7 nm	0.2 S cm^{-1}, 140°C, 100% RH	n.a.	σ is sensitive to RH	[95]
Si-MCM-41	H_3PO_4	~950 $m^2\ g^{-1}$, ~2.0 nm	0.015 S cm^{-1}, 130°C, 100% RH	n.a.	σ is sensitive to RH	[101]
Si-MCM-41	HPW	~3.0 nm	0.045 S cm^{-1}, 150°C, 100% RH	95 mW cm^{-2} at 100°C, 100% RH, H_2/O_2 90 mW cm^{-2} at 150°C, 1% RH, CH_3OH/O_2	Membrane thickness: 0.8 mm	[163]
SiO_2	HPW	~950 $m^2\ g^{-1}$, 5–8 nm	0.076 S cm^{-1}, 100°C, 100% RH; 0.05 S cm^{-1}, 200°C, 0% RH	128.5 mW cm^{-2} at 200°C, CH_3OH/O_2 112 mW cm^{-2} at 200°C, CH_3CH_2OH/O_2	Membrane thickness: 0.16 mm; active area: 4 cm^2	[179, 182]
SiO_2	HPW		0.13 S cm^{-1}, 100°C, 100% RH	74.4 W at 150°C in H_2/O_2	10 cell stack; membrane thickness: 0.75 mm; active area: 200 cm^2	[193]
SiO_2, sintered at 650°C SiO_2, sintered at 650°C	H_3PO_4	948 $m^2\ g^{-1}$, 8.2 nm	0.04 S cm^{-1}, 200°C, 0% RH	570 mW cm^{-2} at 190°C, 0% RH, H_2/O_2 700 mW cm^{-2} at 190°C, 3% RH, H_2/O_2	Membrane thickness: 0.5 mm	[172, 196]

6. Prospect and Summary

For a PEMFC with capability to operate at elevated high temperatures above 100°C it essentially requires a PEM with adequate proton conductivity and stability. To achieve high proton conductivity, the mesoporous structured materials based PEMs must have the capability to retain water under conditions of high temperatures and low RH. As the temperature increases, the total pressure required to maintain the same RH increases dramatically. For example, to maintain 100% RH at 180°C, it would require a total pressure of > 10 atm even without considering the partial pressure of fuel and oxidant gas [199]. This essentially places a requirement of the adequate proton conductivity of the mesoporous structured materials under very low RH at high temperatures in order to achieve adequate power output under normal operation (i.e., low pressure) conditions. The degree of structural order and size of the mesopores of the porous matrix as well as the structure of the anchored proton carrier have a significant effect on the water retention properties of the system. On the other hand, the durability of PEMs depends largely on the stability of proton carriers in the mesopores, which in turn is related to the nature of the interaction between the dopant and host mesoporous materials. If no chemical bonds are formed between the proton carrier and the exterior surface of the mesoporous matrix, there might be a considerable leaching out of the proton carrier due to water produced in a running fuel cell. Mesoporous materials without doping of the proton carrier generally show significant dependence on the RH. In the case of mesoporous zirconium phosphates, σ is 4.1×10^{-6} at 22°C and 84% RH and when RH decreases to 30%, σ is ~3.0×10^{-8} at the same temperature [200], resulted in a conductivity decrease by more than two orders of magnitude. Thus, the nature of the proton carrier or conductor is critical. For example, phosphates are lacking chemical stability as P^{5+} ions are easily dissolved in water and become detached from the silicate network [91]. Rapid decrease in conductivity was also observed on the sulfuric acid impregnated mesoporous silica glass membrane due to a partial or total loss of the impregnated acid [95]. Membranes based on phosphate-silicate glasses show the performance at a temperature as high as 300°C, but the conductivity and cell performance ($\sigma = 8 \times 10^{-4}$ S cm^{-1} and P = 1.3 mW cm^{-2} at 300°C) [114] are probably too low for practical applications. The proton conductivity of phosphate functionalized mesoporous silica also shows high sensitivity to humidity [101]. One of the most fundamental reasons for the leaching of proton carriers such as phosphate and sulfuric acid is most likely due to the lack of strong chemical bonding between the proton carriers and the ZrO_2,TiO_2 and SiO_2 based mesoporous matrixes.

It should be highly desirable if PEMFCs can be operated without using any external humidification system. Mesoporous structures that possess ordered and interconnected networks, high surface areas and high water retention ability have shown advantages as the support matrix for the proton carrier when used as PEMs of fuel cells. However, the control of crystallinity, pore size, porosity and ordering in the mesoporous materials is critical to obtain optimum performance as supports for the proton carrier. For example, meso-Nafion membranes with 2D-H and 3D-CB show much higher conductivities than their counterparts and the cell with 2D-H meso-Nafion membrane shows the best performance under the testing conditions indicating the effectiveness of the highly ordered hexagonal cylinder structure for water retention [39, 40]. HPW-meso-silica nanocomposites with 3D mesostructures display a significantly higher proton conductivity and higher stability as a function of relative humidity in comparison to 2D mesostructures [47]. For unsupported mesoporous materials, e.g., a-Fe_2O_3 and P_2O_5-SiO_2 glass, high specific surface areas, high pore volume and narrow pore size distribution in the range of 2–3 nm would be required to maintain high water uptake and thus adequate conductivity [66, 81, 197]. In this case, water molecules within the mesopores play a major role in proton transport by forming protonic charge carriers through hydrogen bonding. On the other hand, size of nano or mesopores is related to the proton carrier or the dopant. In the case of HPW as the proton carrier, pore size in the range of ~6 nm would be sufficient to maintain high proton conductivity and stability [179]. Large pores will lead to a significant loss of the impregnated HPW from the HPW-meso-silica nanocomposites. Thermal stability and sintering ability of the mesoporous structure is also an important parameter if the fabrication of the mesoporous materials based membrane involves the high temperature sintering process. For example, the mesoporous structure of TiO_2 is thermally stable at 450°C [50], but sintering ability of TiO_2 at this temperature is not known. As shown here, functionalized mesoporous silica materials show the most promising properties as alternative PEMs for high temperature operation. The proton conductivity of the acid functionalized mesoporous silica is comparable with that of Nafion membranes and under conditions of high temperature and reduced humidity, it is significantly better than Nafion membranes [46, 95, 164].

Among acid functionalized mesoporous silica materials, HPAs such as HPW functionalized mesoporous silica, HPW-meso-silica, appears to be most promising, achieving a proton conductivity of 0.14 S cm^{-1} at 150°C and under anhydrous conditions [47]. The most significant aspect of the HPW-meso-silica nanocomposites is the formation of ($\equiv SiOH^{2+}$) ($H_2PW_{12}O_{40}^{-}$) species between the positively charged siliceous surface of mesoporous silica and negatively charged HPW in the presence of water [182]. This ensures the immobilization and high stability of impregnated or assembled HPW within a highly ordered mesoporous silica structure. The good performance and preliminary stability data of the small button cells and stacks with HPW-meso-silica nanocomposite membranes demonstrate that the inorganic PEMs based on highly ordered HPW-meso-silica nanocomposites can be applied to the fuel cell operated at temperatures up to 200°C. However, silica based materials are inherently inflexible and brittle. Thus it is difficult to fabricate MEAs with good interface contact between the electrode and electrolyte under pressurized conditions similar to that of the polymeric material based membranes. Mechanical strength of the silica based membrane is also a problem and it is very difficult to fabricate MEAs with high density and adequate mechanical strength for handling and operation. Tezuka et al. [201] used glass paper as a support for hybrid gel membranes made from 3-glycidoxypropyltrimethoxysilane, tetramethoxysilane and orthophosphoricacid, achieving a power density of 85 mW cm^{-2} at 130°C and 7% RH. The utilization of glass paper reduces the thickness of the hybrid gel membranes. It is also possible to use the thin film technique to reduce the thickness of the inorganic material based membranes. However, in practice the quality of the thin film is critically dependent on the morphology and porous structure of the substrates, which also needs to be catalytically active and electronic conducting as electrodes for fuel cells. As the ceramic materials are generally in a powdered form and thus the membranes fabricated by conventional die pressing techniques need to be in the thickness range of 0.5 to 1.0 mm in order to have sufficient mechanical strength. Nakamoto et al. [113] used carbon paper to sandwich phosphosilicate gel and reduced the thickness of the membrane to ~100 mm after hot-press at 130°C under 11 MPa. However, the use of carbon paper and the polyimide binder limits the thermal stability of the membrane at high temperatures. Thick membranes result in high ohmic losses in the PEMFC. Furthermore, the mechanical strength of the die-pressed disks is still low and brittle in nature, making it very difficult to scale up the system to fuel cell stacks. Very different from the polymeric materials, such difficulty appears to be one of the major obstacles in the development of inorganic material based PEM fuel cells.

Another issue is the testing a PEMFC at high temperatures as the temperature increases, the total pressure required to maintain the same RH increases dramatically. This essentially places a requirement of the adequate proton conductivity of the inorganic materials under very low RH at high temperatures in order to utilize the conventional fuel cell test stations. The technical hurdles appear to be indirectly indicated by the observations that majority of the open literature report the excellent proton conductivity of acid functionalized nano or mesoporous ceramic materials such as mesoporous TiO_2, FeOOH, silica or P_2O_5-SiO_2 glass materials but with little or no cell performance [34, 66, 81, 95, 202, 203]. Additionally, the use of polymeric binders in the die-pressed disks also limits operation temperatures (200°C or below) of the PEM due to degradation and decomposition of the polymeric binder. One of the solutions to substantially increase the operation temperature as well as the mechanical strength is to get rid of the polymeric binder.

It was recently found that mesoporous silica powder can form dense membranes after sintering at 650°C with the ordered mesoporous structure still intact [196]. The mechanical strength of sintered meso-silica is 51 MPa, significantly higher than ~5 MPa on hot-pressed meso-silica membranes [182]. The feasibility of utilization of sintered mesosilica based PEMs has been demonstrated by functionalization of the sintered meso-silica with PA and the results indicate that the PA/sintered meso-silica PEM exhibits excellent conductivity under non-humidified conditions, producing a maximum power density of 689 mW cm^{-2} in H_2 at 190°C in the absence of external humidification [172]. The mesoporous structure of the sintered meso-silica is thermally stable at temperatures up to 650°C [196]. If HPW instead of PA is used to functionalize sintered mesoporous silica, the HPW/sintered meso-silica nanocomposite could be used for PEMFCs operating at temperatures as high as 450°C, noting that HPW is thermally stable up to 500°C [46]. The operation of PEMFCs in the temperature range of 300–450°C would fill an important temperature gap in current fuel cell technologies and demonstrate the promising potential in the development of new types of fuel cells based on liquid fuels such as methanol and ethanol. Conventional high surface area carbon-supported Pt based electrocatalysts developed for low temperature fuel cells would not be applicable due to the significant agglomeration of nano-sized Pt catalysts at such high temperatures and oxidation of carbon based supports. And it is necessary to develop and optimize new classes of electrocatalysts for the catalysts/meso-silica electrolyte interface. However, high operation temperatures in the range of 300–450°C open a window for the development of non-precious metal and metal oxide based electrocatalysts for fuel cells. Nano-scaled and nano-structured electrode concept developed for high temperature solid oxide fuel cells [204] may be applicable for the operation of a PEM fuel cell in the temperatures of 300 to 450°C. In the meantime, considerably more efforts should be made to fundamentally understand the mesoporous structure-property relationships and mechanism and kinetics of the interfacial reactions and phenomena of fuel cells at temperatures up to 450°C.

In summary, mesoporous structured materials as new PEMs for fuel cell applications have been attracting more and more attention, and displaying a lot of very important effects on the development of PEMFCs. Definitely, mesoporous structured materials are very promising and would play a more important role for the development of PEMFCs.

Acknowledgement

This project was supported by the Australian Research Council under Discovery Project Scheme (project number: DP150102025 and DP150102044).

References

[1] Grant, P.M. 2003. Hydrogen lifts off - With a heavy load. *Nature,* 424: 129–30.

[2] Steele, B.C.H. and Heinzel, A. 2001. Materials for fuel-cell technologies. *Nature,* 414: 345–52.

[3] Carrette, L., Friedrich, K.A. and Stimming, U. 2001. Fuel cells—fundamentals and applications. *Fuel Cells,* 1: 5–39.

[4] Biyikoglu, A. 2005. Review of proton exchange membrane fuel cell models. *Int. J. Hydrog. Energy,* 30: 1181–212.

[5] Mauritz, K.A. and Moore, R.B. 2004. State of understanding of Nafion. *Chem. Rev.,* 104: 4535–85.

[6] Galperin, D., Khalatur, P.G. and Khokhlov, A.R. 2009. Morphology of Nafion Membranes: Microscopic and Mesoscopic Modeling, S.J. Paddison and K.S. Promislow, Editors. p. 453–82.

[7] Hamrock, S.J. and Yandrasits, M.A. 2006. Proton exchange membranes for fuel cell applications. *Polym. Rev.,* 46: 219–44.

[8] Ciureanu, M. 2004. Effects of Nafion (R) dehydration in PEM fuel cells. *J. Appl. Electrochem.,* 34: 705–14.

[9] Zawodzinski, T.A., Springer, T.E., Davey, J. et al. 1993. Comparative-study of water-uptake by and transport through ionomeric fuel-cell membranes. *J. Electrochem. Soc.,* 140: 1981–5.

[10] Buchi, F.N. and Scherer, G.G. 2001. Investigation of the transversal water profile in nafion membranes in polymer electrolyte fuel cells. *J. Electrochem. Soc.,* 148: A183–A8.

[11] Haubold, H.G., Vad, T., Jungbluth, H. et al. 2001. Nano structure of NAFION: A SAXS study. *Electrochim. Acta,* 46: 1559.

[12] Kreuer, K.D. 2001. On the development of proton conducting polymer membranes for hydrogen and methanol fuel cells. *J. Membr. Sci.,* 185: 29–39.

[13] Schuster, M.F.H. and Meyer, W.H. 2003. Anhydrous proton-conducting polymers. 233–61.

[14] Li, Q., He, R., Jensen, J.O. et al. 2003. Approaches and recent development of polymer electrolyte membranes for fuel cells operating above 100°C. *Chem. Mater.,* 15: 4896–915.

[15] Li, Q., He, R., Jensen, J.O. et al. 2004. PBI-based polymer membranes for high temperature fuel cells - Preparation, characterization and fuel cell demonstration. *Fuel Cells,* 4: 147–59.

[16] Asensio, J.A., Borrós, S. and Gómez-Romero, P. 2004. Polymer electrolyte fuel cells based on phosphoric acid-impregnated poly(2,5-benzimidazole) membranes. *J. Electrochem. Soc.,* 151.

[17] Roziére, J. and Jones, D.J. 2003. Non-fluorinated polymer materials for proton exchange membrane fuel cells. *Mater. Res.,* 33: 503–55.

[18] Sopian, K. and Wan Daud, W.R. 2006. Challenges and future developments in proton exchange membrane fuel cells. *Renew. Energy,* 31: 719–727.

[19] Rubatat, L. and Diat, O. 2007. Stretching effect on Nafion fibrillar nanostructure. *Macromolecules,* 40: 9455–62.

[20] Jang, S.S., Molinero, V., Cagin, T. et al. 2004. Nanophase-segregation and transport in Nafion 117 from molecular dynamics simulations: Effect of monomeric sequence. *J. Phys. Chem. B,* 108: 3149–57.

[21] Schmidt-Rohr, K. and Chen, Q. 2008. Parallel cylindrical water nanochannels in Nafion fuel-cell membranes. *Nat. Mater.,* 7: 75–83.
[22] Zawodzinski, Jr. T.A., Derouin, C., Radzinski, S. et al. 1993. Water uptake by and transport through Nafion(R) 117 membranes. *J. Electrochem. Soc.,* 140: 1041–7.
[23] Armatas, G.S., Salmas, C.E., Louloudi, M. et al. 2003. Relationships among pore size, connectivity, dimensionality of capillary condensation, and pore structure tortuosity of functionalized mesoporous silica. *Langmuir,* 19: 3128–36.
[24] Dourdain, S. and Gibaud, A. 2005. On the capillary condensation of water in mesoporous silica films measured by x-ray reflectivity. *Appl. Phys. Lett.,* 87: 1–3.
[25] Bocquet, L., Charlaix, E., Ciliberto, S. et al. 1998. Moisture-induced ageing in granular media and the kinetics of capillary condensation. *Nature,* 396: 735–737.
[26] Li, H.B. and Nogami, M. 2002. Pore-controlled proton conducting silica films. *Adv. Mater.,* 14: 912–4.
[27] Athens, G.L., Ein-Eli, Y. and Chmelka, B.F. 2007. Acid-functionalized mesostructured aluminosilica for hydrophilic proton conduction membranes. *Adv. Mater.,* 19: 2580–2587.
[28] Colomer, M.T. 2006. Nanoporuous anatase thin films as fast proton-conducting materials. *Adv. Mater.,* 18: 371–374.
[29] Yamada, M., Li, D.L., Honma, I. et al. 2005. A self-ordered, crystalline glass, mesoporous nanocomposite with high proton conductivity of 2 x 10(-2) S cm(-1) at intermediate temperature. *J. Am. Chem. Soc.,* 127: 13092–3.
[30] Daiko, Y., Kasuga, T. and Nogami, M. 2002. Proton conduction and pore structure in sol-gel glasses. *Chem. Mater.,* 14: 4624–7.
[31] Li, H.B. and Nogami, M. 2003. Ordered mesoporous phosphosilicate glass electrolyte film with low area specific resistivity. *Chem. Commun.,* 236–7.
[32] Uma, T. and Nogami, M. 2007. Structural and transport properties of mixed phosphotungstic acid/phosphomolybdic acid/SiO2 glass membranes for H-2/O-2 fuel cells. *Chem. Lett.,* 19: 3604–10.
[33] Xiong, L.M. and Nogami, M. 2006. Proton-conducting ordered mesostructured silica monoliths. *Chemistry Letters,* 35: 972–3.
[34] Vichi, F.M., Tejedor-Tejedor, M.I. and Anderson, M.A. 2000. Effect of pore-wall chemistry on proton conductivity in mesoporous titanium dioxide. *Chem. Mater.,* 12: 1762–70.
[35] Halla, J.D., Mamak, M., Williams, D.E. et al. 2003. Meso-SiO2-C12EO10OH-CF3SO3H - A novel proton-conducting solid electrolyte. *Adv. Funct. Mater.,* 13: 133–8.
[36] Jiang, S.P. 2014. Functionalized mesoporous structured inorganic materials as high temperature proton exchange membranes for fuel cells. *J. Mater. Chem. A,* 2: 7637–55.
[37] Lu, J.L., Lu, S.F. and Jiang, S.P. 2011. Highly ordered mesoporous Nafion membranes for fuel cells. *Chem. Commun.,* 47: 3216–8.
[38] Lu, J., Tang, H., Xu, C. et al. 2012. Nafion membranes with ordered mesoporous structure and high water retention properties for fuel cell applications. *J. Mater. Chem.,* 22: 5810–9.
[39] Li, J.R., Tang, H.L., Chen, L.T. et al. 2013. Highly ordered and periodic mesoporous Nafion membranes via colloidal silica mediated self-assembly for fuel cells. *Chem. Commun.,* 49: 6537–9.
[40] Zhang, J., Li, J.R., Tang, H.L. et al. 2014. Comprehensive strategy to design highly ordered mesoporous Nafion membranes for fuel cells under low humidity conditions. *J. Mater. Chem. A,* 2: 20578–87.
[41] Pettersson, A. and Rosenholm, J.B. 2002. Streaming potential studies on the adsorption of amphoteric alkyldimethylamine and alkyldimethylphosphine oxides on mesoporous silica from aqueous solution. *Langmuir,* 18: 8447–54.
[42] Wan, Y. and Zhao, D.Y. 2007. On the controllable soft-templating approach to mesoporous silicates. *Chem. Rev.,* 107: 2821–60.

[43] Leibler, L. 1980. Theory of microphase separation in block copolymers. *Macromolecules,* 13: 1602–17.
[44] Khandpur, A.K., Forster, S., Bates, F.S. et al. 1995. Polyisoprene-polystyrene diblock copolymer phase diagram near the order-disorder transition. *Macromolecules,* 28: 8796–806.
[45] Zhao, D., Huo, Q., Feng, J. et al. 1998. Nonionic triblock and star diblock copolymer and oligomeric sufactant syntheses of highly ordered, hydrothermally stable, mesoporous silica structures. *J. Am. Chem. Soc.,* 120: 6024–36.
[46] Zeng, J. and Jiang, S.P. 2011. Characterization of high-temperature proton-exchange membranes based on phosphotungstic acid functionalized mesoporous silica nanocomposites for fuel cells. *J. Phys. Chem. C,* 115: 11854–63.
[47] Zeng, J., Shen, P.K., Lu, S. et al. 2012. Correlation between proton conductivity, thermal stability and structural symmetries in novel HPW-meso-silica nanocomposite membranes and their performance in direct methanol fuel cells. *J. Membr. Sci.,* 397: 92–101.
[48] Fukada, S., Ohba, K. and Nomura, A. 2013. Relation between water adsorption in polymer-electrolyte fuel cell and its electric power. *Energy Conv. Manag.,* 71: 126–30.
[49] Hwang, G.S., Kaviany, M., Gostick, J.T. et al. 2011. Role of water states on water uptake and proton transport in Nafion using molecular simulations and bimodal network. *Polymer,* 52: 2584–93.
[50] Dong, W., Sun, Y., Lee, C.W. et al. 2007. Controllable and repeatable synthesis of thermally stable anatase nanocrystal–silica composites with highly ordered hexagonal mesostructures. *J. Am. Chem. Soc.,* 129: 13894–904.
[51] Sumner, J.J., Creager, S.E., Ma, J.J. et al. 1998. Proton conductivity in Nafion (R) 117 and in a novel bis (perfluoroalkyl)sulfonyl imide ionomer membrane. *J. Electrochem. Soc.,* 145: 107–10.
[52] Sone, Y., Ekdunge, P. and Simonsson, D. 1996. Proton conductivity of Nafion 117 as measured by a four-electrode AC impedance method. *J. Electrochem. Soc.,* 143: 1254–9.
[53] Vichi, F.M., Colomer, M.T. and Anderson, M.A. 1999. Nanopore ceramic membranes as novel electrolytes for proton exchange membranes. *Electrochem. Solid State Lett.,* 2: 313–6.
[54] Kasuga, T. 2006. Formation of titanium oxide nanotubes using chemical treatments and their characteristic properties. *Thin Solid Films,* 496: 141–5.
[55] Ekstrom, H., Wickman, B., Gustavsson, M. et al. 2007. Nanometer-thick films of titanium oxide acting as electrolyte in the polymer electrolyte fuel cell. *Electrochim. Acta,* 52: 4239–45.
[56] Thorne, A., Kruth, A., Tunstall, D. et al. 2005. Formation, structure, and stability of titanate nanotubes and their proton conductivity. *J. Phys. Chem. B,* 109: 5439–44.
[57] Krogh Andersen, E., Krogh Andersen, I.G. and Skou, E. 1988. Proton conduction in H2Ti4O9, 1.2 H2O. *Solid State Ionics,* 27: 181–7.
[58] Bredow, T. and Jug, K. 1995. Theoretical investigation of water adsorption at rutile and anatase surfaces. *Surf. Sci.,* 327: 398–408.
[59] Tejedor-Tejedor, M.I., Vichi, F.M. and Anderson, M.A. 2005. Effect of pore structure on proton conductivity and water uptake in nanoporous TiO2. *J. Porous Mat.,* 12: 201–14.
[60] Tsuru, T., Yagi, Y., Kinoshita, Y. et al. 2003. Titanium phosphorus oxide membranes for proton conduction at intermediate temperatures. *Solid State Ionics,* 158: 343–50.
[61] Jun, Y., Zarrin, H., Fowler, M. et al. 2011. Functionalized titania nanotube composite membranes for high temperature proton exchange membrane fuel cells. *Int. J. Hydrog. Energy,* 36: 6073–81.
[62] Shen, H., Maekawa, H., Kawamura, J. et al. 2008. Effect of pore size and salt doping on the protonic conductivity of mesoporous alumina. *Solid State Ionics,* 179: 1133–7.
[63] Shen, H.Y., Maekawa, H., Kawamura, J. et al. 2006. Development of high protonic conductors based on amorphous mesoporous alumina. *Solid State Ionics,* 177: 2403–6.

[64] Shen, H., Maekawa, H., Wang, L. et al. 2009. Effect of chloride on the acid properties of proton-conducting mesoporous Al2O3. *Electrochem. Solid State Lett.*, 12: B18–B21.
[65] Colomer, M.T. 2011. Proton transport, water uptake and hydrogen permeability of nanoporous hematite ceramic membranes. *J. Power Sources*, 196: 8280–5.
[66] Colomer, M.T. and Zenzinger, K. 2012. Mesoporous alpha-Fe2O3 membranes as proton conductors: Synthesis by microwave-assisted sol-gel route and effect of their textural characteristics on water uptake and proton conductivity. *Microporous Mesoporous Mat.*, 161: 123–33.
[67] Escolastico, S., Solis, C. and Serra, J.M. 2011. Hydrogen separation and stability study of ceramic membranes based on the system Nd(5)LnWO(12). *Int. J.*, 36: 11946–54.
[68] Tsui, E.M., Cortalezzi, M.M. and Wiesner, M.R. 2007. Proton conductivity and methanol rejection by ceramic membranes derived from ferroxane and alumoxane precursors. *J. Membr. Sci.*, 306: 8–15.
[69] Zhang, L.W., Chae, S.R., Hendren, Z. et al. 2012. Recent advances in proton exchange membranes for fuel cell applications. *Chem. Eng. J.*, 204: 87–97.
[70] Gao, F.M., He, J.L., Wu, E.D. et al. 2003. Hardness of covalent crystals. *Phys. Rev. Lett.*, 91.
[71] Marrink, S.J., Berkowitz, M. and Berendsen, H.J.C. 1993. Molecular dynamics simulation of a membrane/water interface: the ordering of water and its relation to the hydration force. *Langmuir*, 9: 3122–31.
[72] Kaithwas, A., Prasad, M., Kulshreshtha, A. et al. 2012. Industrial wastes derived solid adsorbents for CO2 capture: A mini review. *Chem. Eng. Res. Des.*, 90: 1632–41.
[73] Zhu, J.J., Wang, T., Xu, X.L. et al. 2013. Pt nanoparticles supported on SBA-15: Synthesis, characterization and applications in heterogeneous catalysis. *Appl. Catal. B-Environ.*, 130: 197–217.
[74] Azizi, S.N., Ghasemi, S. and Chiani, E. 2013. Nickel/mesoporous silica (SBA-15) modified electrode: An effective porous material for electrooxidation of methanol. *Electrochim. Acta*, 88: 463–72.
[75] Jiang, B.Y., Tang, H.L. and Pan, M. 2012. Well-ordered sulfonated silica electrolyte with high proton conductivity and enhanced selectivity at elevated temperature for DMFC. *Int. J. Hydrog. Energy*, 37: 4612–8.
[76] Ye, Y., Jo, C., Jeong, I. et al. 2013. Functional mesoporous materials for energy applications: solar cells, fuel cells, and batteries. *Nanoscale*, 5: 4584–605.
[77] Beck, J.S., Vartuli, J.C., Roth, W.J. et al. 1992. A new family of mesoporous molecular sieves prepared with liquid crystal templates. *J. Am. Chem. Soc.*, 114: 10834–43.
[78] Kresge, C.T., Leonowicz, M.E., Roth, W.J. et al. 1992. Ordered mesoporous molecular sieves synthesized by a liquid-crystal template mechanism. *Nature*, 359: 710–2.
[79] Kulikovsky, A.A., Schrnitz H., Wippermann, K. et al. 2006. DMFC: galvanic or electrolytic cell? *Electrochem. Commun.*, 8: 754–60.
[80] Colomer, M.T., Rubio, F. and Jurado, J.R. 2007. Transport properties of fast proton conducting mesoporous silica xerogels. *J. Power Sources*, 167: 53–7.
[81] Nogami, M., Nagao, R. and Wong, C. 1998. Proton conduction in porous silica glasses with high water content. *J. Phys. Chem. B*, 102: 5772–5.
[82] Nogami, M. and Abe, Y. 1997. Evidence of water-cooperative proton conduction in silica glasses. *Phys. Rev. B*, 55: 12108–12.
[83] Daiko, Y., Kasuga, T. and Nogami, M. 2004. Pore size effect on proton transfer in sol-gel porous silica glasses. *Microporous Mesoporous Mat.*, 69: 149–55.
[84] Sharifi, M., Marschall, R., Wilkening, M. et al. 2010. Proton conductivity of ordered mesoporous materials containing aluminium. *J. Power Sources*, 195: 7781–6.
[85] Nogami, M., Li, H.B., Daiko, Y. et al. 2004. Proton-conducting phosphosilicate films prepared using template for pore structure. *J. Sol-Gel. Sci. Technol.*, 32: 185–8.
[86] Li, H.B., Jin, D.L., Kong, X.Y. et al. 2011. High proton-conducting monolithic phosphosilicate glass membranes. *Microporous Mesoporous Mat.*, 138: 63–7.

[87] Marschall, R., Bannat, I., Feldhoff, A. et al. 2009. Nanoparticles of mesoporous SO3H-functionalized Si-MCM-41 with superior proton conductivity. *Small*, 5: 854–9.

[88] Colomer, M.T. and Anderson, M.A. 2001. High porosity silica xerogels prepared by a particulate sol-gel route: pore structure and proton conductivity. *J. Non-Cryst. Solids*, 290: 93–104.

[89] Suzuki, S., Nozaki, Y., Okumura, T. et al. 2006. Proton conductivity of mesoporous silica incorporated with phosphorus under high water vapor pressures up to 150 degrees C. *J. Ceram. Soc. Jpn.*, 114: 303–7.

[90] Nogami, M., Nagao, R., Wong, G. et al. 1999. High proton conductivity in porous P2O5-SiO2 glasses. *J. Phys. Chem. B*, 103: 9468–72.

[91] Nogami, M., Goto, Y., Tsurita, Y. et al. 2001. Effect of phosphorus ions on the proton conductivity in the sol-gel-derived porous glasses. *J. Am. Ceram. Soc.*, 84: 2553–6.

[92] Wang, C. and Nogami, M. 2000. Effect of formamide additive on protonic conduction of P2O5-SiO2 glasses prepared by sol-gel method. *Mater. Lett.*, 42: 225–8.

[93] Mikhailenko, S., Desplantier-Giscard, D., Danumah, C. et al. 2002. Solid electrolyte properties of sulfonic acid functionalized mesostructured porous silica. *Microporous Mesoporous Mat.*, 52: 29–37.

[94] Margolese, D., Melero, J.A., Christiansen, S.C. et al. 2000. Direct syntheses of ordered SBA-15 mesoporous silica containing sulfonic acid groups. *Chem. Mat.*, 12: 2448–59.

[95] Marschall, R., Rathousky, J. and Wark, M. 2007. Ordered functionalized silica materials with high proton conductivity. *Chem. Mater.*, 19: 6401–7.

[96] Marschall, R., Bannat, I., Caro, J. et al. 2007. Proton conductivity of sulfonic acid functionalised mesoporous materials. *Microporous Mesoporous Mater.*, 99: 190–6.

[97] Supplit, R., Sugawara, A., Peterlik, H. et al. 2010. Supported and free-standing sulfonic acid functionalized mesostructured silica films with high proton conductivity. *Eur. J. Inorg. Chem.*, 3993–9.

[98] Ioroi, T., Kuraoka, K., Yasuda, K. et al. 2004. Surface-modified nanopore glass membrane as electrolyte for DMFCs. *Electrochem. Solid State Lett.*, 7: A394–A6.

[99] Matsuda, A., Kanzaki, T., Tadanaga, K. et al. 2002. Sol-gel derived porous silica gels impregnated with sulfuric acid - Pore structure and proton conductivities at medium temperatures. *J. Electrochem. Soc.*, 149: E292–E7.

[100] Marschall, R., Sharifi, M. and Wark, M. 2009. Proton conductivity of imidazole functionalized ordered mesoporous silica: Influence of type of anchorage, chain length and humidity. *Microporous Mesoporous Mat.*, 123: 21–9.

[101] Jin, Y.G., Qiao, S.Z., Xu, Z.P. et al. 2009. Porous silica nanospheres functionalized with phosphonic acid as intermediate-temperature proton conductors. *J. Phys. Chem. C*, 113: 3157–63.

[102] Jin, Y.G., Qiao, S.Z., Xu, Z.P. et al. 2009. Phosphonic acid functionalized silicas for intermediate temperature proton conduction. *J. Mater. Chem.*, 19: 2363–72.

[103] Ponomareva, V.G. and Shutova, E.S. 2005. Composite electrolytes Cs-3(H2PO4)(HSO4)(2)/SiO2 with high proton conductivity. *Solid State Ionics*, 176: 2905–8.

[104] Ponomareva, V.G. and Shutova, E.S. 2007. High-temperature behavior of CsH2PO4 and CsH2PO4-SiO2 composites. *Solid State Ionics*, 178: 729–34.

[105] Tung, S.P. and Hwang, B.J. 2005. Synthesis and characterization of hydrated phosphor-silicate glass membrane prepared by an accelerated sol-gel process with water/vapor management. *J. Mater. Chem.*, 15: 3532–8.

[106] Nogami, M., Daiko, Y., Akai, T. et al. 2001. Dynamics of proton transfer in the sol-gel-derived P2O5-SiO2 glasses. *J. Phys. Chem. B*, 105: 4653–6.

[107] Inoue, T., Uma, T. and Nogami, M. 2008. Performance of H-2/O-2 fuel cell using membrane electrolyte of phosphotungstic acid-modified 3-glycidoxypropyl-trimethoxysilanes. *J. Membr. Sci.*, 323: 148–52.

[108] Uma, T. and Nogami, M. 2009. PMA/ZrO2-P2O5-SiO2 glass composite membranes: H-2/O-2 fuel cells. *J. Membr. Sci.*, 334: 123–8.

[109] Uma, T. and Nogami, M. 2007. A novel glass membrane for low temperature H-2/O-2 fuel cell electrolytes. *Fuel Cells*, 7: 279–84.
[110] Uma, T. and Nogami, M. 2007. Synthesis and characterization of glasses as an electrolyte for low-temperature H-2/O-2 fuel cells. *J. Electrochem. Soc.*, 154: B32–B8.
[111] Uma, T. and Nogami, M. 2008. Proton-conducting glass electrolyte. *Anal. Chem.*, 80: 506–8.
[112] Nogami, M., Matsushita, H., Goto, Y. et al. 2000. Sol-gel-derived glass as a fuel cell electrolyte. *Adv. Mater.*, 12: 1370–2.
[113] Nakamoto, N., Matsuda, A., Tadanaga, K. et al. 2004. Medium temperature operation of fuel cells using thermally stable proton-conducting composite sheets composed of phosphosilicate gel and polyimide. *J. Power Sources*, 138: 51–5.
[114] Ishiyama, T., Suzuki, S., Nishii, J. et al. 2013. Electrochemical substitution of sodium ions in tungsten phosphate glass with protons. *J. Electrochem. Soc.*, 160: E143–E7.
[115] Di, Z., Li, H., Li, M. et al. 2012. Improved performance of fuel cell with proton-conducting glass membrane. *J. Power Sources*, 207: 86–90.
[116] Tang, H., Wan, Z., Pan, M. et al. 2007. Self-assembled Nafion-silica nanoparticles for elevated-high temperature polymer electrolyte membrane fuel cells. *Electrochem. Commun.*, 9: 2003–8.
[117] Zhao, D., Wan, Y. and Zhou, W. 2013. Synthesis Approach of Mesoporous Molecular Sieves, in Ordered Mesoporous Materials. Wiley-VCH Verlag GmbH & Co. KGaA. p. 5–54.
[118] Jin, Y.G., Qiao, S.Z., Xu, Z.P. et al. 2009. Porous silica nanospheres functionalized with phosphonic acid as intermediate-temperature proton conductors. *J. Phys. Chem. C*, 113: 3157–63.
[119] Matsuda, A., Nono, Y., Kanzaki, T. et al. 2001. Proton conductivity of acid-impregnated mesoporous silica gels prepared using surfactants as a template. *Solid State Ionics*, 145: 135–40.
[120] Choi, Y., Kim, Y., Kim, H.K. et al. 2010. Direct synthesis of sulfonated mesoporous silica as inorganic fillers of proton-conducting organic-inorganic composite membranes. *J. Membr. Sci.*, 357: 199–205.
[121] Tominaga, Y. and Maki, T. 2014. Proton-conducting composite membranes based on polybenzimidazole and sulfonated mesoporous organosilicate. *International Journal of Hydrogen Energy*, 39: 2724–30.
[122] Won, J.-H., Lee, H.-J., Yoon, K.-S. et al. 2012. Sulfonated SBA-15 mesoporous silica-incorporated sulfonated poly(phenylsulfone) composite membranes for low-humidity proton exchange membrane fuel cells: Anomalous behavior of humidity-dependent proton conductivity. *Int. J. Hydrogen Energy*, 37: 9202–11.
[123] Xie, L., Cho, E.-B. and Kim, D. 2011. Sulfonated PEEK/cubic (Im3m) mesoporous benzene-silica composite membranes operable at low humidity. *Solid State Ionics*, 203: 1–8.
[124] Delhorbe, V., Thiry, X., Cailleteau, C. et al. 2012. Fluorohexane network and sulfonated PEEK based semi-IPNs for fuel cell membranes. *J. Membr. Sci.*, 389: 57–66.
[125] Luu, D.X. and Kim, D. 2011. Semi-interpenetrating polymer network electrolyte membranes composed of sulfonated poly(ether ether ketone) and organosiloxane-based hybrid network. *J. Power Sources*, 196: 10584–90.
[126] Han, S.Y., Park, J. and Kim, D. 2013. Proton-conducting electrolyte membranes based on organosiloxane network/sulfonated poly(ether ether ketone) interpenetrating polymer networks embedding sulfonated mesoporous benzene-silica. *J. Power Sources*, 243: 850–8.
[127] Zhao, Y., Yang, H., Wu, H. et al. 2014. Enhanced proton conductivity of hybrid membranes by incorporating phosphorylated hollow mesoporous silica submicrospheres. *J. Membr. Sci.*, 469: 418–27.
[128] Zhang, T., Ge, J., Hu, Y. et al. 2008. Formation of hollow silica colloids through a spontaneous dissolution–regrowth process. *Angew. Chem. Int. Ed.*, 47: 5806–11.

[129] Zhang, Q., Zhang, T., Ge, J. et al. 2008. Permeable silica shell through surface-protected etching. *Nano Lett.*, 8: 2867–71.

[130] Chen, Y., Chen, H., Guo, L. et al. 2010. Hollow/rattle-type mesoporous nanostructures by a structural difference-based selective etching strategy. *ACS Nano.*, 4: 529–39.

[131] Fang, X., Chen, C., Liu, Z. et al. 2011. A cationic surfactant assisted selective etching strategy to hollow mesoporous silica spheres. *Nanoscale*, 3: 1632–9.

[132] Deshmukh, R. and Schubert, U. 2013. Synthesis of CuO and Cu3N nanoparticles in and on hollow silica spheres. *Eur. J. Inorg. Chem.*, 2013: 2498–504.

[133] Qi, G., Wang, Y., Estevez, L. et al. 2010. Facile and scalable synthesis of monodispersed spherical capsules with a mesoporous shell. *Chem. Mater.*, 22: 2693–5.

[134] Li, Y. and Shi, J. 2014. Hollow-structured mesoporous materials: Chemical synthesis, functionalization and applications. *Adv. Mater.*, 26: 3176–205.

[135] Yin, Y., Deng, W., Wang, H. et al. 2015. Fabrication of hybrid membranes by incorporating acid-base pair functionalized hollow mesoporous silica for enhanced proton conductivity. *J. Mater. Chem. A*, 3: 16079–88.

[136] Feng, K., Tang, B. and Wu, P. 2015. A "H2O donating/methanol accepting" platform for preparation of highly selective Nafion-based proton exchange membranes. *J. Mater. Chem. A.*

[137] Kreuer, K.-D., Rabenau, A. and Weppner, W. 1982. Vehicle mechanism, a new model for the interpretation of the conductivity of fast proton conductors. *Angew. Chem. Int. Ed.*, 21: 208–9.

[138] Bernard, L., Fitch, A., Wright, A.F. et al. 1981. Proceedings of the international conference on fast ionic transport in solids mechanisms of hydrogen diffusion and conduction in DUO2AsO4·4D2O as inferred from neutron diffraction evidence. *Solid State Ionics*, 5: 459–62.

[139] Yang, Q., Liu, J., Zhang, L. et al. 2009. Functionalized periodic mesoporous organosilicas for catalysis. *J. Mater. Chem.*, 19: 1945–55.

[140] Cho, E.-B., Luu, D.X. and Kim, D. 2010. Enhanced transport performance of sulfonated mesoporous benzene-silica incorporated poly(ether ether ketone) composite membranes for fuel cell application. *J. Membr. Sci.*, 351: 58–64.

[141] Wilhelm, M., Jeske, M., Marschall, R. et al. 2008. New proton conducting hybrid membranes for HT-PEMFC systems based on polysiloxanes and SO3H-functionalized mesoporous Si-MCM-41 particles. *J. Membr. Sci.*, 316: 164–75.

[142] Athens, G.L., Ein-Eli, Y. and Chmelka, B.F. 2007. Acid-functionalized mesostructured aluminosilica for hydrophilic proton conduction membranes. *Adv. Mater.*, 19: 2580–7.

[143] Alvaro, M., Corma, A., Das, D. et al. 2004. Single-step preparation and catalytic activity of mesoporous MCM-41 and SBA-15 silicas functionalized with perfluoroalkylsulfonic acid groups analogous to Nafion[registered sign]. *Chem. Commun.*, 956–7.

[144] Hoffmann, F., Cornelius, M., Morell, J. et al. 2006. Silica-based mesoporous organic-inorganic hybrid materials. *Angew. Chem. Int. Ed.*, 45: 3216–51.

[145] Cho, E.-B., Kim, H. and Kim, D. 2009. Effect of morphology and pore size of sulfonated mesoporous benzene-silicas in the preparation of poly(vinyl alcohol)-based hybrid nanocomposite membranes for direct methanol fuel cell application. *J. Phys. Chem. B*, 113: 9770–8.

[146] Asefa, T., MacLachlan, M.J., Coombs, N. et al. 1999. Periodic mesoporous organosilicas with organic groups inside the channel walls. *Nature*, 402: 867–71.

[147] Melde, B.J., Holland, B.T., Blanford, C.F. et al. 1999. Mesoporous sieves with unified hybrid inorganic/organic frameworks. *Chem. Mater.*, 11: 3302–8.

[148] Inagaki, S., Guan, S., Fukushima, Y. et al. 1999. Novel mesoporous materials with a uniform distribution of organic groups and inorganic oxide in their frameworks. *J. Am. Chem. Soc.*, 121: 9611–4.

[149] Tsai, C.-H., Lin, H.-J., Tsai, H.-M. et al. 2011. Characterization and PEMFC application of a mesoporous sulfonated silica prepared from two precursors, tetraethoxysilane and phenyltriethoxysilane. *Int. J.*, 36: 9831–41.

[150] McKeen, J.C., Yan, Y.S. and Davis, M.E. 2008. Proton conductivity of acid-functionalized zeolite beta, MCM-41, and MCM-48: Effect of acid strength. *Chem. Mater.*, 20: 5122–4.

[151] Chen, Z. and Hsu, R.S. 2009. Nafion/acid functionalized mesoporous silica nanocomposite membrane for high temperature PEMFCs. *Proton Exchange Membrane Fuel Cells 9*, 25: 1151–7.

[152] Tominaga, Y., Hong, I.-C., Asai, S. et al. 2007. Proton conduction in Nafion composite membranes filled with mesoporous silica. *J. Power Sources*, 171: 530–4.

[153] Moghaddam, S., Pengwang, E., Jiang, Y.-B. et al. 2010. An inorganic-organic proton exchange membrane for fuel cells with a controlled nanoscale pore structure. *Nature Nanotech.*, 5: 230–6.

[154] Park, M.J., Downing, K.H., Jackson, A. et al. 2007. Increased water retention in polymer electrolyte membranes at elevated temperatures assisted by capillary condensation. *Nano. Lett.*, 7: 3547–52.

[155] Fujita, S., Koiwai, A., Kawasumi, M. et al. 2013. Enhancement of proton transport by high densification of sulfonic acid groups in highly ordered mesoporous silica. *Chem. Mater.*, 25: 1584–91.

[156] Lin, Y.-F., Yen, C.-Y., Ma, C.-CM. et al. 2007. High proton-conducting Nafion (R)/-SO3H functionalized mesoporous silica composite membranes. *J. Power Sources*, 171: 388–95.

[157] Geng, L., He, Y., Liu, D. et al. 2012. New organic-inorganic hybrid membranes based on sulfonated polyimide/aminopropyltriethoxysilane doping with sulfonated mesoporous silica for direct methanol fuel cells. *J. Appl. Polym. Sci.*, 123: 3164–72.

[158] Liu, D., Geng, L., Fu, Y. et al. 2011. Novel nanocomposite membranes based on sulfonated mesoporous silica nanoparticles modified sulfonated polyimides for direct methanol fuel cells. *J. Membr. Sci.*, 366: 251–7.

[159] Pereira, F., Vallé, K., Belleville, P. et al. 2008. Advanced mesostructured hybrid silica–Nafion membranes for high-performance PEM fuel cell. *Chem. Mater.*, 20: 1710–8.

[160] Geng, L., He, Y., Liu, D. et al. 2012. Facile *in situ* template synthesis of sulfonated polyimide/mesoporous silica hybrid proton exchange membrane for direct methanol fuel cells. *Microporous Mesoporous Mat.*, 148: 8–14.

[161] Jin, Y., Qiao, S., Zhang, L. et al. 2008. Novel Nafion composite membranes with mesoporous silica nanospheres as inorganic fillers. *J. Power Sources*, 185: 664–9.

[162] Meenakshi, S., Sahu, A.K., Bhat, S.D. et al. 2013. Mesostructured-aluminosilicate-Nafion hybrid membranes for direct methanol fuel cells. *Electrochim. Acta*, 89: 35–44.

[163] Lu, S.F., Wang, D.L., Jiang, S.P. et al. 2010. HPW/MCM-41 phosphotungstic acid/mesoporous silica composites as novel proton-exchange membranes for elevated-temperature fuel cells. *Adv. Mater.*, 22: 971–6.

[164] Tang, H.L., Pan, M., Lu, S.F. et al. 2010. One-step synthesized HPW/meso-silica inorganic proton exchange membranes for fuel cells. *Chem. Commun.*, 46: 4351–3.

[165] Jiang, S.P. 2014. Functionalized mesoporous materials as new class high temperature proton exchange membranes for fuel cells. *Solid State Ionics*, 262: 307–12.

[166] Amirinejad, M., Madaeni, S.S., Rafiee, E. et al. 2011. Cesium hydrogen salt of heteropolyacids/Nafion nanocomposite membranes for proton exchange membrane fuel cells. *J. Membr. Sci.*, 377: 89–98.

[167] Yang, J., Janik, M.J., Ma, D. et al. 2005. Location, acid strength, and mobility of the acidic protons in Keggin 12-$H_3PW_{12}O_{40}$: A combined solid-state NMR spectroscopy and DFT quantum chemical calculation study. *J. Am. Chem. Soc.*, 127: 18274–80.

[168] Wang, D.L., Lu, S.F. and Jiang, S.P. 2010. Pd/HPW-PDDA-MWCNTs as effective non-Pt electrocatalysts for oxygen reduction reaction of fuel cells. *Chem. Commun.*, 46: 2058–60.

[169] Xiang, Y., Yang, M., Zhang, J. et al. 2011. Phosphotungstic acid (HPW) molecules anchored in the bulk of Nafion as methanol-blocking membrane for direct methanol fuel cells. *J. Membr. Sci.*, 368: 241–5.

[170] Kukino, T., Kikuchi, R., Takeguchi, T. et al. 2005. Proton conductivity and stability of Cs2HPW12O40 electrolyte at intermediate temperatures. *Solid State Ionics,* 176: 1845–8.
[171] Cui, Z.M., Xing, W., Liu, C.P. et al. 2010. Synthesis and characterization of H5PMo10V2O40 deposited Pt/C nanocatalysts for methanol electrooxidation. *J. Power Sources,* 195: 1619–23.
[172] Zeng, J., He, B., Lamb, K. et al. 2013. Phosphoric acid functionalized pre-sintered meso-silica for high temperature proton exchange membrane fuel cells. *Chem. Commun.,* 49: 4655–7.
[173] Corriu, R.J.P., Hoarau, C., Mehdi, A. et al. 2000. Study of the accessibility of phosphorus centres incorporated within ordered mesoporous organic-inorganic hybrid materials. *Chem. Commun.,* 71–2.
[174] Macquarrie, D.J. 1996. Direct preparation of organically modified MCM-type materials. Preparation and characterisation of aminopropyl-MCM and 2-cyanoethyl-MCM. *Chem. Commun.,* 1961–2.
[175] Lu, J.L., Tang, H.L., Lu, S.F. et al. 2011. A novel inorganic proton exchange membrane based on self-assembled HPW-meso-silica for direct methanol fuel cells. *J. Mater. Chem.,* 21: 6668–76.
[176] Rao, P.M., Goldberg-Oppenheimer, P., Kababya, S. et al. 2007. Proton enriched high-surface-area cesium salt of phosphotungstic heteropolyacid with enhanced catalytic activity fabricated by nanocasting strategy. *J. Mol. Catal. A-Chem.,* 275: 214–27.
[177] Fan, R., Huh, S., Yan, R. et al. 2008. Gated proton transport in aligned mesoporous silica films. *Nat. Mater.,* 7: 303–7.
[178] Zeng, J., Shen, P.K., Lu, S.F. et al. 2012. Correlation between proton conductivity, thermal stability and structural symmetries in novel HPW-meso-silica nanocomposite membranes and their performance in direct methanol fuel cells. *J. Membr. Sci.,* 397: 92–101.
[179] Zeng, J., Zhou, Y.H., Li, L. et al. 2011. Phosphotungstic acid functionalized silica nanocomposites with tunable bicontinuous mesoporous structure and superior proton conductivity and stability for fuel cells. *Phys. Chem. Chem. Phys.,* 13: 10249–57.
[180] Ye, Y.S., Chen, W.Y., Huang, Y.J. et al. 2010. Preparation and characterization of high-durability zwitterionic crosslinked proton exchange membranes. *J. Membr. Sci.,* 362: 29–37.
[181] Yan, X.M., Mei, P., Mi, Y.Z. et al. 2009. Proton exchange membrane with hydrophilic capillaries for elevated temperature PEM fuel cells. *Electrochem. Commun.,* 11: 71–4.
[182] Tang, H.L., Pan, M. and Jiang, S.P. 2011. Self assembled 12-tungstophosphoric acid-silica mesoporous nanocomposites as proton exchange membranes for direct alcohol fuel cells. *Dalton Trans.,* 40: 5220–7.
[183] Lefebvre, F. 1992. 31P MAS NMR study of H3PW12O40 supported on silica: Formation of (≡SiOH2 +)(H 2PW12O40 -). *J. Chem. Soc., Chem. Commun.,* 756–7.
[184] Thomas, S.C., Ren, X.M., Gottesfeld, S. et al. 2002. Direct methanol fuel cells: progress in cell performance and cathode research. *Electrochim. Acta,* 47: 3741–8.
[185] Wasmus, S. and Kuver, A. 1999. Methanol oxidation and direct methanol fuel cells: a selective review. *J. Electroanal. Chem.,* 461: 14–31.
[186] Jiang, S.P., Liu, Z.C. and Tian, Z.Q. 2006. Layer-by-layer self-assembly of composite polyelectrolyte-nafion membranes for direct methanol fuel cells. *Adv. Mater.,* 18: 1068–1072.
[187] Ren, X., Springer, T.E. and Gottesfeld, S. 2000. Water and methanol uptakes in nafion membranes and membrane effects on direct methanol cell performance. *J. Electrochem. Soc.,* 147: 92–8.
[188] Cho, E.A., Jeon, U.S., Hong, S.A. et al. 2005. Performance of a 1 kW-class PEMFC stack using TiN-coated 316 stainless steel bipolar plates. *J. Power Sources,* 142: 177–83.

[189] Alcaide, F., Alvarez, G., Blazquez, J.A. et al. 2010. Development of a novel portable-size PEMFC short stack with electrodeposited Pt hydrogen diffusion anodes. *Int. J. Hydrog. Energy*, 35: 5521–7.
[190] Siracusano, S., Baglio, V., Di Blasi, A. et al. 2010. Electrochemical characterization of single cell and short stack PEM electrolyzers based on a nanosized IrO(2) anode electrocatalyst. *Int. J. Hydrogen Energy*, 35: 5558–68.
[191] Thirumalai, D. and White, R.E. 1997. Mathematical modeling of proton-exchange-membrane fuel-cell stacks. *J. Electrochem. Soc.*, 144: 1717–23.
[192] Weng, F.B., Jou, B.S., Su, A. et al. 2007. Design, fabrication and performance analysis of a 200 W PEM fuel cell short stack. *J. Power Sources*, 171: 179–85.
[193] Zeng, J., Jin, B.Q., Shen, P.K. et al. 2013. Stack performance of phosphotungstic acid functionalized mesoporous silica (HPW-meso-silica) nanocomposite high temperature proton exchange membrane fuel cells. *Int. J. Hydrog. Energy*, 38: 12830–7.
[194] Nagao, M., Takeuchi, A., Heo, P. et al. 2006. A proton-conducting In3+-doped SnP2O7 electrolyte for intermediate-temperature fuel cells. *Electrochem. Solid State Lett.*, 9: A105–A9.
[195] Boysen, D.A., Uda, T., Chisholm, C.R.I. et al. 2004. High-performance solid acid fuel cells through humidity stabilization. *Science*, 303: 68–70.
[196] Zeng, J., He, B.B., Lamb, K. et al. 2013. Anhydrous phosphoric acid functionalized sintered mesoporous silica nanocomposite proton exchange membranes for fuel cells. *Acs. Appl. Mater. Interfaces*, 5: 11240–8.
[197] Nogami, M., Tanaka, K. and Uma, T. 2009. Preparation and characterisation of pelletised glass electrolytes for fuel cells. *Fuel Cells*, 9: 528–33.
[198] Uma, T. and Nogami, M. 2009. PMA/ZrO2–P2O5–SiO2 glass composite membranes: H2/O2 fuel cells. *J. Membr. Sci.*, 334: 123–8.
[199] Zhang, J., Xie, Z., Zhang, J. et al. 2006. High temperature PEM fuel cells. *J. Power Sources*, 160: 872–91.
[200] Hogarth, W.H.J., da Costa, J.C.D., Drennan, J. et al. 2005. Proton conductivity of mesoporous sol-gel zirconium phosphates for fuel cell applications. *J. Mater. Chem.*, 15: 754–8.
[201] Tezuka, T., Tadanaga, K., Matsuda, A. et al. 2005. Utilization of glass papers as a support for proton conducting inorganic-organic hybrid membranes from 3-glycidoxypropyltrimethoxysilane, tetraalkoxysilane and orthophosphoric acid. *Solid State Ionics*, 176: 3001–4.
[202] Mishra, A.K., Kuila, T., Kim, D.Y. et al. 2012. Protic ionic liquid-functionalized mesoporous silica-based hybrid membranes for proton exchange membrane fuel cells. *J. Mater. Chem.*, 22: 24366–72.
[203] Ponomareva, V.G., Kovalenko, K.A., Chupakhin, A.P. et al. 2012. Imparting high proton conductivity to a metal-organic framework material by controlled acid impregnation. *J. Am. Chem. Soc.*, 134: 15640–3.
[204] Jiang, S.P. 2012. Nanoscale and nano-structured electrodes of solid oxide fuel cells by infiltration: Advances and challenges. *Int. J. Hydrog. Energy*, 37: 449–70.

CHAPTER 4

Mesoporous Structured Electrocatalysts for Fuel Cells

*Kamel Eid, Liang Wang** and *Hongjing Wang**

ABSTRACT

Mesoporous materials, especially mesoporous metallic nanocrystals (NCs), are promising electrode materials for fuel cell applications due to their catalytic properties with highly accessible surface areas, low density and stability. These structural and functional characteristics are not only required to improve the catalytic activity and durability in electrode reactions for fuel cells, but also favorably provide low cost for commercialization. Control of the mesoporous structures (e.g., pore structure and pore size) is very favorable for tailoring their catalytic properties to achieve high performance for fuel cells. There has been a lot of interest in synthesis of mesoporous materials with designed compositions and mesoporous structures. This chapter presents the synthetic methods for mesoporous structured electrocatalysts and their fuel cell applications with special emphasis on methanol oxidation reaction, oxygen reduction reaction, formic acid oxidation and ethanol oxidation reaction.

Keywords: Mesoporous structures, Electrode materials, Electrocatalysts, Electrode reaction, Fuel cells

College of Chemical Engineering, Zhejiang University of Technology, Hangzhou, Zhejiang 310014, P.R. China.

* Corresponding authors: wangliang@zjut.edu.cn; hjw@zjut.edu.cn.

1. Introduction

Porous nanostructures including microporous (< 2 nm), mesoporous (2 ~ 50 nm), and macroporous (> 50 nm) are of great importance due to their unique physicochemical properties such as quantum size, surface effect and low density [1–4]. Mesoporous materials (e.g., metallic NCs) stand out among the most intriguing porous materials due to their aligned pores, uniform porous structure and high available surface area enabling their utilization in various applications including medical and catalytic applications [1, 2]. Evidently, particle size, morphology, composition and pore structure are pivotal factors that govern the applicability of mesoporous NCs [1, 2, 5]. In fact, there are several studies in literature that are dedicated to the synthesis of various types of mesoporous NCs including organic, inorganic and hybrid NCs [1–5]. Mesoporous metals are a subset of porous materials which exhibit fascinating catalytic activities towards various reactions such as reduction of NO_x into N_2 and oxidation of CO into CO_2 [1, 5–7]. Notably, to date, mesoporous Pt-based NCs are the most active candidates as catalysts for both direct methanol (DMFCs) and proton exchange membrane (PEMFCs) [8, 9]. This is ascribed to the large accessible surface area, durability, high mass activity and low density of the mesoporous Pt-based NCs [8].

The synthesis of highly ordered mesoporous metals is greatly challenged due to the fact that metals at nanoscale (< 5 nm) are highly energetic, preferring growth into solid particles to minimize their interfacial free energy [1, 2, 4]. Accordingly, massive efforts were devoted to the fabrication of mesoporous metals NCs through template-based, dealloying, self-assembly and surfactant-based strategies [2, 8–14]. This yielded a surge of studies to tailor pore size, volume and distribution for better catalytic performance and even for further exploration for newer applications. Due to the expeditious progress in this field, it is of great importance to provide timely updates of the recent advances in the fabrication methods, properties, challenges and applications. The application of porous Pt-based NCs in fuel cells will be particularly emphasized. This will be concluded by revealing the future prospective for synthesis of mesoporous metals and their further applications.

2. Fabrication Strategies for Mesoporous Metals

2.1 Dealloying

Dealloying is the primary synthetic approach for the fabrication of porous metal NCs. It was used for the first time in 1920s to obtain porous Raney Ni heterogeneous catalysts by chemical dealloying of $NiAl_3$ [15]. Later in the

1970s, with the great advances in the microscopic equipment, dealloying was used as an effective route for the preparation of wide varieties of nanoporous metals [14, 16, 17]. The porosity originated from selective dissolution or corrosion of more active metal from alloy phase under corrosive conditions [10, 18–20]. This caused a decomposition process and subsequent formation of porous structure resulting from contiguous of the noble metals residuals into interconnected clusters [10, 21]. Hence, the alloy elements should be dissimilar in their activity in the presence of an appropriate corrosive agent, allowing only dissolution of the active metal. Meanwhile, the other metal remains intact to prevent formation of any stable bulk oxide under corrosive conditions. The dealloying technique is categorized into electrochemical dealloying and chemical dealloying.

2.1.1 Electrochemical dealloying

In this route, the active metal is corroded away electrochemically in an electrolyte solution under electrochemical conditions [9, 22–26]. The alloy elements should exhibit distinct electrochemical activity during the electrochemical dealloying. Several studies are concerned with investigation of the electrochemical dealloying process and its correlated factors. Regarding this, the parting limit (inert metal concentration should be suitable to be aggregated into interconnected clusters in the porous morphology) and critical potential (E_c) (electrode potential required for dissolution the active element only) are the main factors in the electrochemical dealloying [18, 27]. Specifically, electrochemical dealloying processes ubiquitously occur solely under the parting limit, whereas, it takes place above and/or lower the E_c [27, 28]. It is noteworthy that the dealloying rate at potentials over the E_c is faster than at lower E_c that is time dependent. This is clearly evident in the dealloying of Cu_3Au into porous Cu-Au under E_c, the Cu dissolve slowly forming pits onto the alloys surface that are converted to deeper pores with time upon increasing the potential ($< E_c$) (Fig. 4.1) [28]. Other factors also should be considered including electrolyte concentration, less noble element dissolution rate, noble element diffusivities and capillary forces [27, 28].

The formation mechanism of the pores was ambiguous till Erlebacher and coworkers elucidated it deeply throughout simulation on dealloying Ag-Au that is the typical alloy structure [10, 26]. This due to the fact that, both metals have the same face-centered cubic *fcc* lattice structure, similar atomic radii, while Ag is more vulnerable to electrochemical dissolution relative to Au [10, 19, 27]. Specifically, the mechanism is attributed to the inherent dynamical pattern; Ag atoms solely dissolved away from the first layer, facilitating dissolution of its adjacent Ag atoms leaving cavities among Au atoms. That diffuses simultaneously in the form of 2-D islands

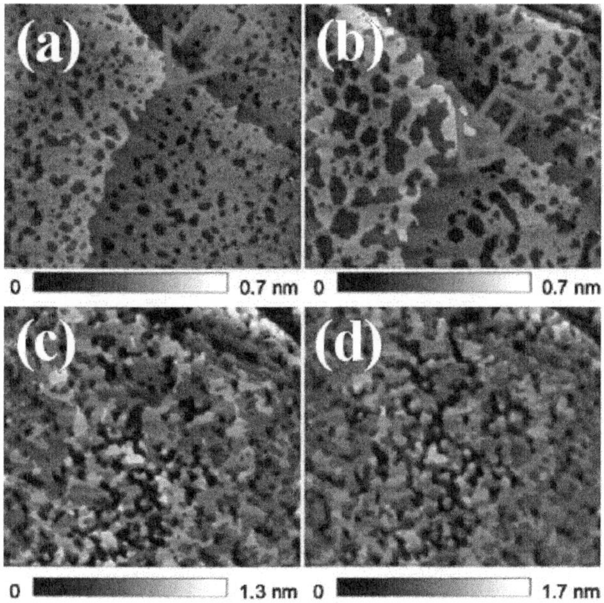

Figure 4.1. STM images of selective electrochemical dissolution of a Cu_3Au (111) surface in 0.1 M H_2SO_4 at different times under various potentials in reference to Ag/AgCl electrode (a) 270 mV, t = 174 minutes (b) 330 mV, t = 183 minutes (c) 470 mV, t = 211 minutes and (d) 510 mV, t = 263 minutes. Reprinted with permission from ref. [28]. Copyright 2011 American Chemical Society.

entailed patterns of Au clusters to lessen the surface energy. Ligaments structure with various pore sizes formed within further dissolution of Ag atoms and Au diffusion. The balance between dissolution rate (k_{diss}) of the Ag atoms and the surface diffusion (k_{SD}) of Au atoms tune the ligament size (λ) based on the equation (1)

$$\lambda \ \alpha \ (k_{SD}/k_{diss})^{1/6} \quad (1)$$

Notably, the diffusion rate of noble metals has an unambiguous effect on the dealloying process [10, 26]. In particular, the less noble metal could not be dealloyed completely resulting from the sluggish diffusion rate of more noble metals. This is indicated by detection of less noble metal fractions after electrochemical dealloying. Several studies are dedicated to optimize the E_c for tailoring the pore size in the electrochemical dealloying [10, 26, 29, 30]. Various porous metallic NCs ranging from mono-metallic to multi-metallic were fabricated by the electrochemical dealloying method such as Pt, Pd, Ag, Ni, Au, Cu-Pt, Ag-Au, Pt-Co, Pt-Cu [1, 9, 19, 25, 29, 31–39]. Intriguingly, the particle sizes, morphology and compositions are clearly correlated to the electrochemical dealloying as indicating through dealloying of PtCu, PtCO, and PtFe (Fig. 4.2a–b) [40, 41]. Recently, Sieradzki and coworkers fabricated mesoporous AgAu by electrochemical dealloying of $Ag_{0.73}Au_{0.27}$ nanoparticles with optimizing the potentials and dealloying time. The results demonstrate that dealloying occurs at a potential above E_c that shapes the particle porosity (Fig. 4.3a–i) [42].

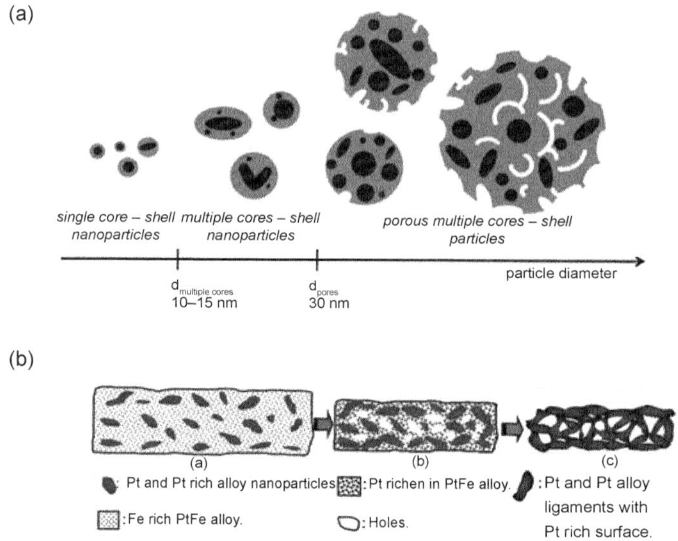

Figure 4.2. (a) The relation among size, morphology and composition of Pt-Cu and Pt-Co prepared by the electrochemical dealloying. Reprinted with permission from ref. [40]. Copyright 2012 American Chemical Society. (b) Scheme shows the formation of nanoporous $PtFe_5$ nanowires by electrospinning coupled with chemical dealloying. Reprinted with permission from ref. [41]. Copyright 2011 Wiley.

2.1.2 Chemical dealloying

Chemical dealloying is more facile, and free of corrosion route relative to electrochemical dealloying, which predominately executes on using acid or base solution to dissolve the less noble elements [14, 43, 44]. It was utilized by Andean in the past for decorative arts; specifically they used salts and/or some plant juices to dissolve Cu and Ag from the surfaces of Au-Cu-Ag alloy to obtain a shiny golden appearance [45]. Dealloying of Ag-Au in HNO_3 to form various nanoporous Ag-Au is a typical example, the pore sizes are dealloying time dependent [18, 20]. Subsequently, various nanoporous metals such as Cu, Pd, Ag, Au and Pt on dealloying of their alloys Al-Cu, Ni-Cu, Cu-Pd, Zn-Ag, Mn-Au, Cu-Au, Zr-Au, Cu-Pt and Zn-Pt [43, 46–54]. In spite of the great advances in this field, the pore

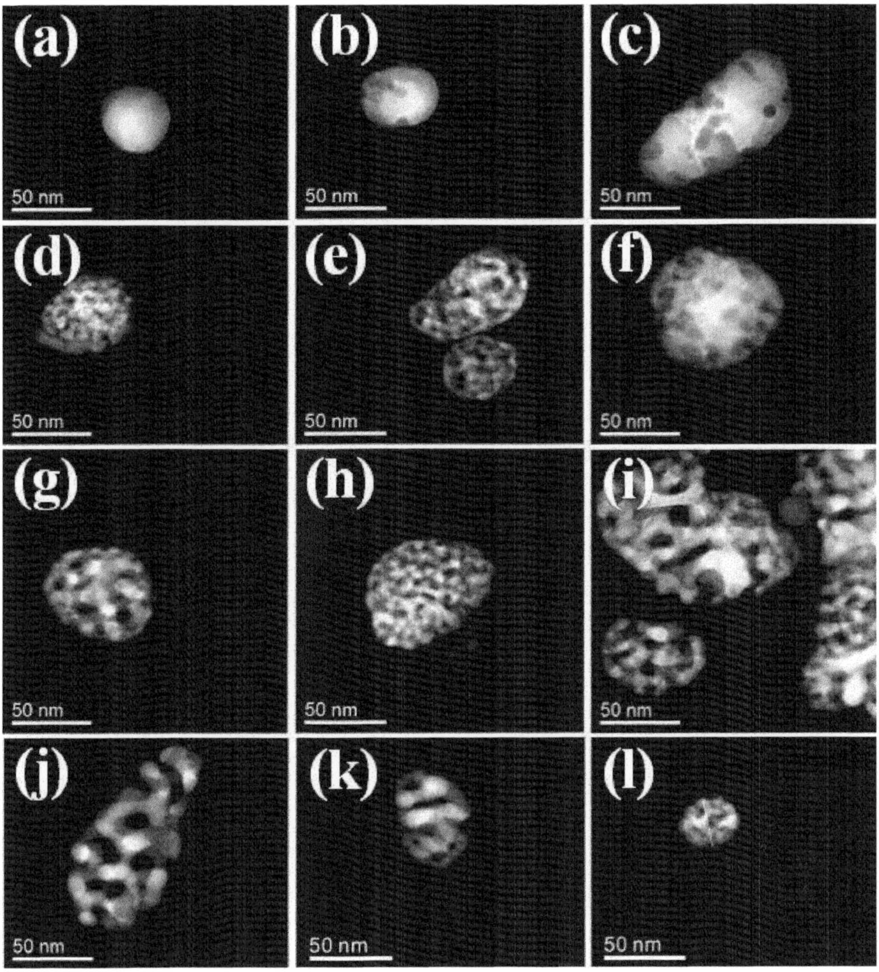

Figure 4.3. (a) High-angle annular dark-field (HAADF) images of undealloyed AgAu nanoparticles and (b–i) nanoporous $Ag_{0.73}Au_{0.27}$ prepared by electrochemical dealloying in 0.1 M H_2SO_4 with optimizing the potentials and dealloying time. Conditions (b) 0.54 V, 6 hours; (c, d) 0.64 V, 6 hours; (e, f) 0.74 V, 6 hours; (g) 0.84 V, 30 minutes; (h) 0.94 V, 2 minutes; (i) 0.94 V, 30 minutes; (j) 0.94 V, 6 hours; (k); 1.1 V, 30 minutes; (I) 1.3 V, 15 minutes. Reprinted with permission from ref. [42]. Copyright 2014 American Chemical Society.

size and pore distribution is barely controlled during chemical dealloying ascribed to the drastic power of the etching agent enabling indiscriminate etching sites of less noble metal. These barriers can be avoided by adjusting the concentration of etching agent, dealloying time and hiring thermal annealing. On this basis, various mesoporous metals such as Cu, Pd, Ag, Pt, Pt-Cu, Pt-Co and Ag-Au are obtained with controlled pore size [54].

For example, mesoporous Au-Pt NCs are obtained on chemical dealloying of $Pt_{10}Au_{10}Cu_{80}$ in HNO_3, because both Au and Pt are more noble than Cu metal (Fig. 4.4a–b) [36]. Interestingly, the pore size of Au-Pt increased from 6 nm to 50 nm by using thermal annealing during chemical dealloying (Fig. 4.4c–d). Recently, Yamauchi and coworker synthesized Pt-Pd nanocages with dendrite shells on dealloying of Pt-Pd nanodendrites in HNO_3 solution [55]. The pore size is tuned by using a proper concentration of HNO_3 (Fig. 4.5a–c). Most recently, mesoporous $Pt_{60}Cu_{40}$ microwires were obtained via chemical dealloying of Pt_3Cu_{97} alloy ribbons in diluted HNO_3 solution at room temperature (Fig. 4.5d–e) [56].

Combining the chemical dealloying with other approaches such as template-based, surfactant-based and seed mediated growth is a decisive strategy for dominating pore size, composition and morphology. For instance, mesoporous Pt-Ni is obtained on sequential dealloying of PtNiAl in NaOH and HNO_3 [57]. Similarly, PtFe nanowires are obtained by combined electrospinning method with chemical dealloying as shown in

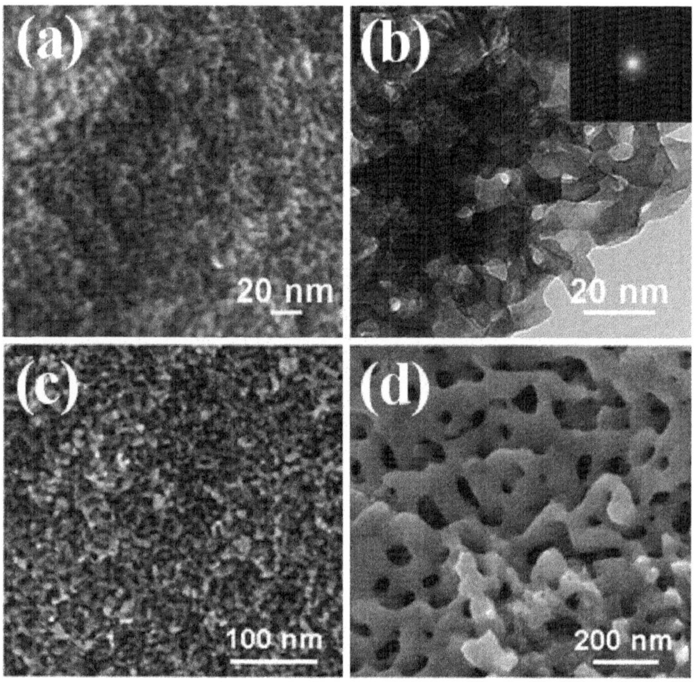

Figure 4.4. (a) SEM and (b) TEM images of nanoporous alloys by electrochemical dealloying of $Au_{10}Pt_{10}Cu_{80}$ without thermal annealing. (c) SEM image of a nanoporous $Au_{10}Pt_{10}Cu_{80}$ dealloyed chemically in HNO_3 for 48 hours, and (d) the $Au_{20}Pt_{80}$ NPA after annealing at 500°C for 15 minutes. Reprint with permission from ref. [36]. Copyright 2010 The Royal Society of Chemistry.

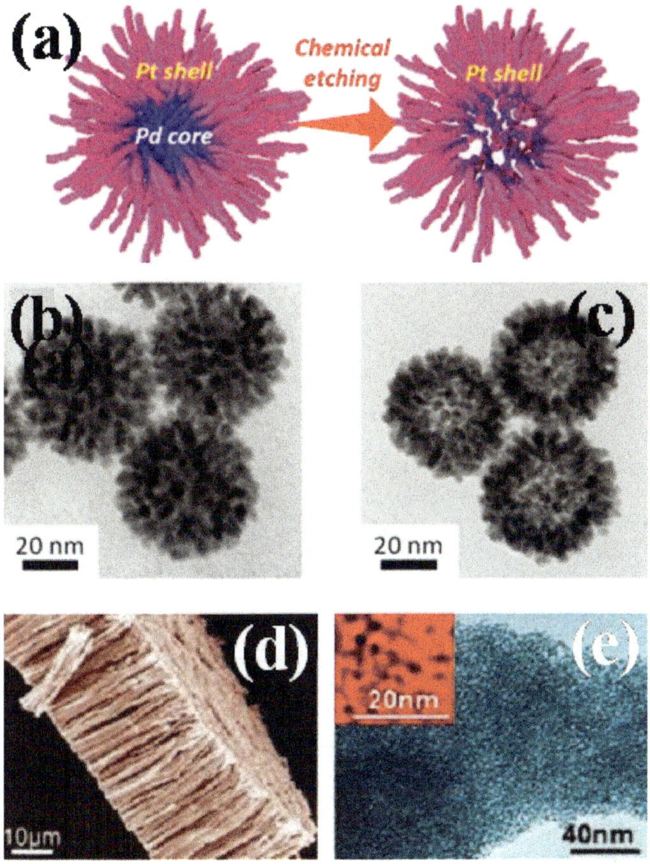

Figure 4.5. (a) Scheme presents the preparation process of PtPd nanocages with porous dendritic shell by selective chemical etching of Pd core in Pt-Pd nanodendrites by concentrated HNO_3 solution. (b) TEM image of Pt-Pd nanodendrites before chemical etching and (c) TEM of Pt-Pd nanocages after chemical etching. Reprinted with permission from ref. [55]. Copyright 2013 American Chemical Society (d) SEM image of nanoporous $Pt_{60}Cu_{40}$ microwires and (d) TEM of porous microwires $Pt_{60}Cu_{40}$ with its HAADF shown as inset in (b). Samples are prepared by the chemically dealloyed of Pt_3Cu_{97} in a diluted HNO_3 solution at room temperature. Reprinted with permission from ref. [56]. Copyright 2015 American Chemical Society.

(Fig. 4.2b) [41]. Most recently, a combination of advanced powder metallurgy, template and dealloying approaches was used to fabricate highly ordered mesoporous Au-Ag and Ni-Mn nanostructure [58]. Particularly, metal precursors and metal alloy are infiltrated into channels of cellulosic fiber in the presence of water-soluble binder followed by thermal sintering at high temperature to decompose template and then chemical dealloying in a concentrated HNO_3 solution to form mesoporous structure (Fig. 4.6a) [58]. The pure metal and/or metal alloy can be changed in order to control the

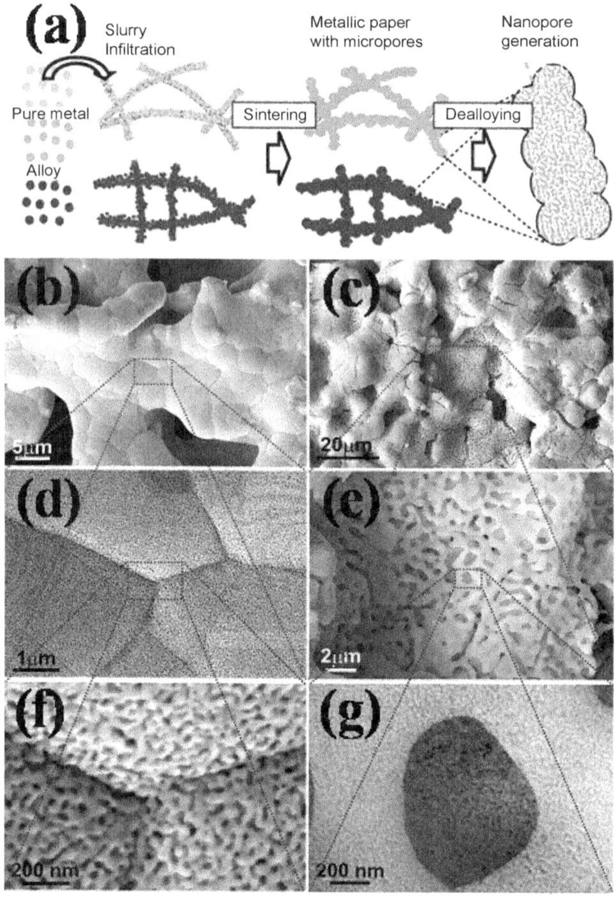

Figure 4.6. (a) scheme illustrates the fabrication method of a hierarchical nanoporous Au structure via sintering slurry of metal powder and water-soluble binder on a paper sheet. (b–d) SEM images of bimodal nanoporous $Au_{35}Ag_{65}$ sheet prepared by one-step dealloying at room temperature in concentrated HNO_3 solution and (e–g) trimodal nanoporous Au_5Ag_{95} created through dealloying/annealing/redealloying [58].

composition of the obtained mesoporous structure. Interestingly, bimodal mesoporous $Au_{35}Ag_{65}$ nanostructure is obtained on dealloying at room temperature (Fig. 4.6b–d), while trimodal mesoporous Au_5Ag_{95} is formed via redealloying process (Fig. 4.6f–g) [58].

Selective use of an appropriate etching solution to remove less unstable metal is another effective strategy for engineering NCs porosity. Based on this various mono, bimetallic and even tri-metallic NCs are obtained on using Ag NPs as a template followed by selective etching by bis(p-sulfonatophenyl) phenylphosphane (BSPP) (Fig. 4.7a–m) [59]. Ag

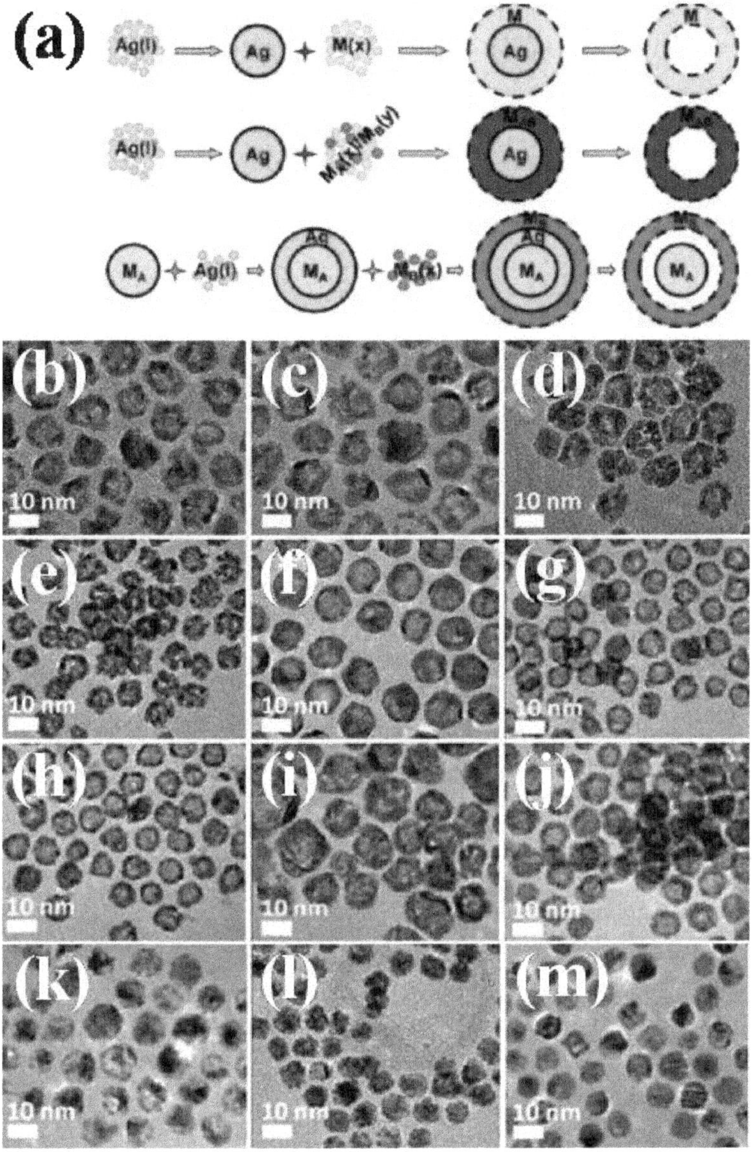

Figure 4.7. (a) Scheme shows the preparation method of various nanoporous single, binary and ternary metallic NCs by using Ag NPs as a template followed by chemical etching. TEM images of hollow structured (b) Ru, (c) Rh, (d) Os, (e) Ir and (f) Pt nanoparticles. TEM images of hollow (g) PtRu alloy, (h) PtRh alloy, (i) PtOs alloy, and (j) PtRuOs alloy. Cage-bell structured of (k) Pt-Ru, (l) Pt-Os, and (m) Pt-Pt nanoparticles. Reprinted with permission from ref. [59]. Copyright 2012 American Chemical Society.

atoms bind strongly with BSPP and etch away forming various porous morphologies due to instability of Ag atoms in the acidic solutions [59]. This mechanism allows control dissolution of Ag atoms to form metal alloys with different mesostructures (Fig. 4.7b–m). This method opens a new frontier for synthesis of various mesoporous structure. Although there is great progress in the chemical dealloying method, it stills remains a great challenge to control pore size and pore volume of the final product, due to the drastic power of the etching solution. Furthermore, the surfaces atoms of the active metal in the alloy system are eroded randomly, which yield a non-uniform mesoporous structure.

2.2 Template-based method

Template-based synthesis is among the most effective strategies for fabrication various porous structure ranging from macroporous, mesoporous to microporous. After understanding the dealloying process, template-based emerged as a decisive route for synthesis of porous materials started from hard templates in the early 1970s [60]. The synthetic process entails (1) selecting the suitable templates with well-defined pore structures, (2) filling void spaces of the templates with the metal precursors before solidification (converting them into solid phase), and (3) eroding the template (if essential) for creation porous structure [2, 13, 61–63]. Based on the template nature, there are three main categories of hard, soft and hybrid templates. Those templates are naturally such as minerals, biological molecular assemblies and organs, whereas other synthesized by chemical or physical methods [11, 12]. The physical properties of the template mainly vary by their nature whereas; their chemical properties are determined by their role during chemical transformation step in the synthesis process [11]. In particular, template involved in the chemical reaction namely chemical templates while, physical templates are uninvolved. There are various methodologies for fabrication of porous physical template-based such as channel replication and surface coating wherein, the chemical templates-based synthesis include isomerization, addition and substitution [11, 12].

Various factors should be considered during template-based synthesis including, the ability of templates to reserve their porous structure throughout filling their cavities with the metal precursors and also after template removal without demolition [11, 12]. Added to that, choosing the appropriate method is highly desired for facilitating deposition of metal precursors inside the template pores. This is generally executed by chemical or electrochemical reduction approaches [11]. The template can be removed physically by dissolution or chemically through calcinations or etching which are determined by the templates nature. Based on the template synthesis method, myriad porous materials are synthesized with different structure ranging from zero-dimensional (0D) to 3D and hierarchical nanostructures [11, 12, 62].

2.2.1 Hard template

In this route, the metal precursors are deposited inside the voids of porous solid structure followed by template removal. There are multifarious hard-templates including natural ones such as minerals and biological molecules and organs which have been used for fabrication of porous materials for the past several decades [11, 60]. The synthesis of nanoporous structures by hard-template is a great challenge due to the limited availability of natural templates with tunable pores. In the early 1990s, artificial templates including honeycomb-like anodic alumina and highly ordered colloidal crystals stood out as an attractive source for fabrication of porous metallic nanostructures [64, 65].

There are multifarious synthetic hard templates available such as Anodized Aluminium Oxide membranes (AAO) and track-etched polymeric membranes such as polycarbonate (PC-TE) and silica membranes, whereas Meldrun and Seshadri used sea-urchins as hard templates [11, 66–70]. Both AAO and PC-TE templates are the most prominent templates for fabrication of various porous materials with well-defined porous structures [11, 66, 67]. This is because of their commercial availability and ease of preparation with various pore densities. Particularly, the AAO is predominately synthesized electrochemically from Al metal which possesses a highly uniform cylindrical pores structure 10–200 nm and with pore densities of 10^9–10^{11} pores per cm^2 [66–69]. Meanwhile, polymeric templates are generally prepared by the track-etch method where it depends on bombarding the rigid material via nuclear fission fragments for wrecking in the nonporous materials. Usually polymeric membranes have irregular porous structure with larger pore diameter 10 nm–10 μm and with less pore density up to 10^8 pores per cm^2 relative to AAO.

Interestingly, the pore size and the thickness of the porous metal films can be carefully controlled by adjusting the mother structures of the AAO templates [66–69]. The most common synthetic approaches used for the fabrication of porous structures based on hard templates are electrochemical deposition and solution-based synthesis methods sol gel and layer by layer assembly (LbL) [11]. Martin et al. synthesized quantum dots metal oxide with different morphologies including nanotubes and fibrils by using sol gel method in the presence of AAO template [71]. In particular, for the preparation of TiO_2 AAO a template is immersed in a solution of titanium isopropoxide precursor, ethanol and water in the presence of HCl as a catalyst. This is followed by drying, heating and polishing the

AAO to remove the excess amount of TiO_2 before eroding the templates by alkaline etching for obtaining TiO_2 nanotubes or fibrils. The same method is vulnerable for the synthesis of other 1D semiconductor metal oxides such as MnO_2, V_2O_5, Co_3O_4, ZnO, WO_3 and SiO_2.

Besides the efficiency of these approaches for fabrication of porous metal oxide, the obtained particle suffers from inevitable inconsistency. This is due to low loading density of the metal precursor resulting from the insufficient driven force of the template capillaries, which led to production of non-uniform porous metal oxide NCs. Unfortunately, by increasing the driven force by using electrophoresis with the sol-gel method highly uniform TiO_2, SiO_2 and $BaTiO_3$, solid nanowires are obtained but with the absence of a porous structure [72]. This issue can be overcome by repeating replications of AAO. For example, honeycomb structures of Au and Pt with highly ordered pores are obtained by repeating replication of AAO [64].

Electrochemical deposition is one of the most effective techniques for the fabrication of porous metallic and/or metallic oxide NCs in presence of AAO templates [73, 74]. Specifically, in this approach, the metal working electrode is initially attached to the templates surface by using the evaporation technique. This is followed by reduction of the metal precursors with a negative potential in the presence of an electrolyte solution for depositing the metal ions inside the templates channels under electrochemical conditions [73, 74]. This approach is utilized for the synthesis of various metallic NCs including noble metals, transition metals and even semiconductors. For instance, Co, Ni, Zn and Cu nanotubes are synthesized using electrochemical deposition in the presence of AAO templates [75–77]. Particularly, AAO is functionalized with silane layer by using 3-aminopropyltriethoxysilane in toluene before drying. Following that, the salinized AAO is polished by refined sand paper before sensitizing and activating with $SnCl_2$ and $PdCl_2$ in the presence of HCl respectively. Afterwards, the activated AAO is immersed in a solution of metal precursors containing Co^{+2} or Cu^{+2} or Ni^{+2} and the reducing agent for electrochemically deposition of metal ions inside the AAO channels to obtain nanotubes.

Similarly, various semiconductor nanowires such as CdSe are obtained throughout sequential deposition of cationic and anionic components of the semiconductor by reverse potential sweeping [76]. Specifically, Se seeds are initially deposited before deposition Cd during the reverse potential sweep. It should be noted that, the electroless plating is possibly used for coating metal seeds nanoparticles onto the surface of membrane template. Various mono, bi-metallic and even tri-metallic Pt-based NCs with control porous morphology, particle size and composition are fabricated based on using AAO templates. In this regard, Liu et al. successfully used electrochemical deposition for the preparation of mesoporous Pt-CO nanowire with crystalline ligaments in the presence of AAO templates [78]. The fabrication procedure starting from electrodepositing of Pt-Co precursors into the porous of AAO templates followed by chemical dealloying in H_3PO_4 (Fig. 4.8a–d) [78]. The pore size, morphology and composition of Pt-Co are tailored by adjusting dealloying time in H_3PO_4 (Fig. 4.8e–j).

The LbL assembly method is generally used for the fabrication of porous organic polymer capsules or organic-inorganic hybrid nanotubes, biomaterials such as DNA and proteins [79, 80]. It should be noted that, prior modification of the template walls by the linker such as poly (ethylenimine) followed by sequentially depositing oppositely charged polyelectrolytes poly(acrylic acid) (PAA) and poly(allylamine hydrochloride) and 3-amino propylphosphonic acid facilitate the deposition of desired materials [79–81].

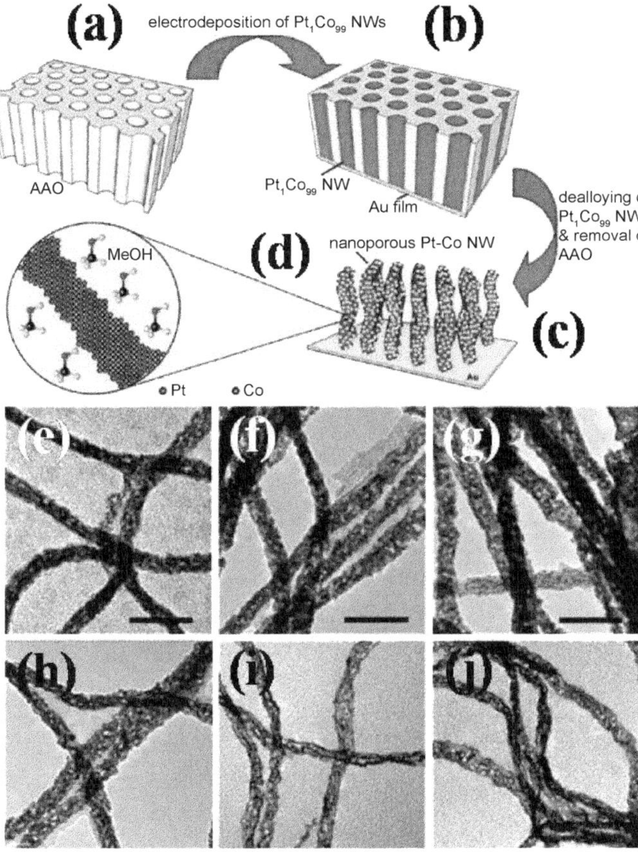

Figure 4.8. (a–d) Scheme represents the fabrication of porous Pt_1Co_{99} nanowires and (e–j) TEM images of as prepared samples by chemical dealloying in 10 wt% H_3PO_4 at 45°C for different times. (e) 10 minutes; (f) 30 minutes; (g) 2 hours; (h) 5 hours; (i) 15 hours; and (j) 5 days. The scale bars are 50 nm. Reprinted with permission from ref. [78]. Copyright 2009 American Chemical Society.

Physical gas phase routes such Chemical Vapor Deposition (CVD), Atomic Layer Deposition (ALD), thermal evaporation and epitaxial growth are also utilized for fabrication of various mesoporous metals. For example, Knez et al. used ALD with AAO template for the synthesis of crescent-shaped Au nanotubes [82]. This was achieved by initial coating of AAO with ZnO and Al_2O_3 layers by ALD followed by ion milling and selective chemical acidic etching of ZnO which led to formation of circular nanochannels. Later annealing is used for the transformation of nanochannels into crescent channels to facilitate displacement of the Al_2O_3 nanowires attached to the pore walls. Then, electroplating is used for depositing Au inside the formed crescent nanochannels followed by alkaline eroding both Al_2O_3 nanowires and AAO templates to form Au crescent nanotubes. Similarly, Whitesides et al. prepared Au and indium-tin oxide nanotubes based on AAO shadow evaporation approach. Gösele et al. synthesized high-density epitaxial Si (100) nanowire arrays on a Si (100) substrate using an AAO template [84]. They initially functionalized the inner channels of AAO templates with Al or Al_2O_3 before selective chemical etching of Al_2O_3. This is followed by the deposition of Au NCs inside the vertical pores to facilitate binding and growing the Si nanowires inside the AAO channels by using of pulse electrodeposition followed by a vapor-liquid-solid growth process. This is followed by the selective acidic chemical etching of AAO templates by hydrofluoric acid (HF) solution. The use of AAO templates during epitaxy growth led to the formation of Si (100) nanowire instead of (111), (112) and (110). In spite of the great progress in the synthesis of porous materials with hard-templates, most of the reported materials possess, only 1D structure with absence of 2D and 3D architecture which are highly desired for various catalytic applications especially for the fuel cells. In pursuit of these goals, artificial close-packed colloidal crystals templates such as silica or polymer (e.g., polystyrene or polymethylmethacrylate) were explored as active routes for the synthesis of 2D and/or 3D porous materials in the early 1999s [13, 65]. This is attributed to their ability to self-assemble onto solid substrates to form highly ordered colloidal crystals in 3D. This subsequently, led to the creation of wide ranges of porous materials within the replicating of the porous structures of colloidal crystal templates and their inverse opals into durable solid matrices [63].

There are three main pathways to obtain a highly uniform porous structure by using colloidal crystals templates, including infiltration the colloidal crystal templates with metal ions, co-deposition of colloidal spheres templates with metal NPs and are assembled by core-shell structures into periodic arrays (Fig. 4.9a–c) respectively. In all approaches, the template should be removed to create mesoporous structure [63].

In particular, the metal nanoparticles (NPs) can infiltrate into the interstices of the colloidal crystal templates by filtration enabling production

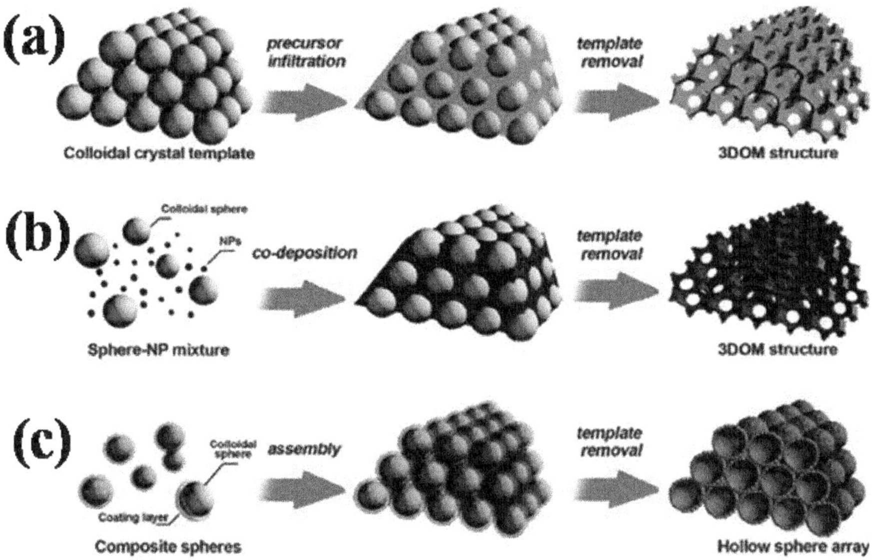

Figure 4.9. The three main methods for preparation of 3D porous nanostructure based on using colloidal crystal templating. (a) Metal precursors infiltration with colloidal crystal followed by template removal. (b) Co-deposition of nanoparticles with templates. (c) Assembly of templates with nanoparticles into periodic arrays. Reprinted with permission from ref. [63]. Copyright 2007 American Chemical Society.

of 3D porous metals with mesoporous structure after template removal (Fig. 4.9a). The etchings of colloidal crystal templates can be archived by several methods depending on the type of templates. For instance, silica is usually removed by immersion in dilute HF and NaOH, whereas polystyrene is eroded with various solvents such as trichloromethane. Meanwhile, the pyrolysis at high temperature is also used to remove polymer templates which somehow lead to shrinkage of the pores. The adjusting of the colloidal crystal features (e.g., particle size, pore size and porous volume) and metal NPs prior to templating determine the porous morphology, composition and properties (e.g., surface area) of resulted metal NCs product [11].

During co-deposition of metal NPs with the colloidal crystal templates, the voids of the colloidal template should be completely filled with the metal precursors (Fig. 4.9b). This can be controlled through electrodeposition by adjusting the deposition parameters such as voltage, pH value and concentration of the solution to facilitate reduction of metal ions from an electrolyte solution and templating into the interstices of the colloidal crystal templates. However, nonuniform deposition is obtained due to the fact that, the electrodeposition is merely vulnerable to the conductive surfaces and the tortuous pores of templates are curved, narrow and are not accessible to metal ions. Several attempts have been dedicated to solve these barriers culminating in using a reducing agent to decrease the metal ions to enable the electrodeposition process with the absence of conductive substrates for creation mesoporous metal structures. Wide varieties of porous metals such as Au, Pt, Ni, Cu, Co and Ag are obtained upon coupling between the electroless deposition and colloidal crystal templates [14, 85–87].

Furthermore, 3D porous metal structures are also obtained by concurrent assembly among both colloidal spheres and metal NPs (Fig. 4.9c). The filtration, sedimentation and/or solvent evaporation are the main approaches for the uniform arrangement of the metal NPs as thin layer of core-shell metal spheres. Then on removing the colloidal spheres by calcination and dissolution lead to creation of porous morphology due to interconnection between the metallic NPs. Which consequently produce hollow sphere arrays of metallic shells (Fig. 4.9c). Notably, the regulation assembly process for producing periodic arrays and preserving the porous structure after templates removal are the key factors determining the uniformity of the resulted porous structure.

Among the most intriguing hard templates for the synthesis of diverse mesoporous metal structures are silica templates due to the ease of preparation into a 3D porous structure and its fascinating chemical and mechanical stabilities. In particular, wide varieties of mesoporous silicas such as MCM-41 (p6mm), SBA-15 (p6mm), KIT-6 (Ia3d) and MCM-48 (Ia3d) have been used as templates for preparing mesoporous metal powders or thin films via chemical or electrochemical reduction [88]. For instance, mesoporous rod-like $MoSe_2$ was synthesized using mesoporous silica SBA-15 through a nanocasting strategy [89]. MCM-48 template was used for fabrication of Pt nanowires with a 3D interconnected porous structure [90]. Various mesoporous morphologies such as polyhedral- and olive-shaped of Pt-based NCs nanoparticles can be obtained by using KIT-6 and SBA-15 templates. The results warrant that, the reducing agent power determines the deposition rate of Pt in the confined hollows of silica templates to obtain highly uniform 3D or 2D mesoporous structure.

Lu and coworkers developed a facile approach for preparing of porous metal nanowire by using electrochemical deposition in the presence of silica templates [91]. Specifically, silicate and surfactant molecules simultaneously co-assembled prior to the surfactant removal in form of mesoporous templates on the conductive substrates to be ready for filling with the metals or semiconductors NPs via electrodeposition. Then a mesoporous nanowire network was produced by removing the silica template.

Metals NPs can be synthesized in desired morphologies and subsequently used as a template for preparing porous structures based on the galvanic replacement reaction due to the dissimilarity in the equilibrium potentials. This approach possesses advantages over the other template approaches owing to the absence of additional template removal steps. Various porous Au NCs such as nanoboxes, nanocages and nanotubes were obtained by galvanic replacement reaction between Ag NCs and aqueous solution of chloroauric acid [92]. Puntes and coworkers synthesized various porous multi-metallic NCs such as Ag-Au, Au-Ag-Pd, Pd-Ag Au-Pd-Ag with controlled composition and morphologies (e.g., spherical, cubic and cylindrical) by using galvanic replacement between Ag NPs and other metal precursors at room temperature [93]. Similarly, hollow nanocages Fe_2O_3 were obtained by galvanic replacement among Mn_3O_4 NCs and Iron (11) perchlorate [94]. This method is generalized to synthesis other porous metals including Co_3O_4/SnO_2 and Mn_3O_4/SnO_2. Recently, Stamenkovic and coworkers obtained highly ordered 3D porous Pt_3Ni nanoframe with Pt-skin by interior erosion of $PtNi_3$ NCs [95]. The thermal treatment has a substantial effect on decreasing the conversion time from two weeks to only 12 hours. The advantages of template synthesis are that, the morphology, composition and properties are controllable in the structured products. However, there are some drawbacks in the template synthesis, including shrinkage of template pores after calcinations or etching. Furthermore, multi-steps are required to increase the loading density of metal precursor inside template pores to overcome the insufficient driven force of the capillaries.

2.2.2 Soft template method

There are various types of soft templates such as self-assembled, biological and artificial structures including micelles, microemulsions, liposomes and vesicles, biological macromolecules, and viruses used for creating highly uniform porous NCs [11]. This approach is advantageous over other templates due to the ease of preparing various porous NCs with different morphologies and compositions under mild reaction conditions. Among the most intriguing soft templates is Lyotropic Liquid Crystalline (LLC) that is a phase of amphiphilic molecules like block copolymers and involve long-range spatially periodic nanostructures with various lattice parameters [1, 2, 96–98]. Notably, the concentration of LLCs plays a paramount role in directing the morphology process through changing their phases. Particularly the structures of LLCs alter easily from a micellar solution (L_I) through micellar cubic (II), hexagonal (H_I), bicontinuous cubic (V_I), lamellar (L_α) and even to inverse micellar (L_2) phases by increasing LLCs concentration. Therefore, reduction of metal precursors in the presence of LLCs templates enables production of myriad well defined ordered mesoporous metals NCs. The reduction of metal ions can be performed electrochemically and/or chemically in the confined LLCs.

2.2.2.1 Electrochemical reduction

In this route the metal precursors are reduced electrochemically in the presence of soft templates. Various mesoporous metallic NCs are prepared with different morphologies and compositions by using this method. For example, Attard and coworkers used electrochemical reduction approaches for preparing mesoporous Pt film with pore size in the range of 1.7–10 nm by reducing Pt precursor in an aqueous solution of LLCs templates [97]. Inspired by this, various efforts have been made to exploit the role of LLCs in the synthesis of mesoporous metallic NCs. The results warrant that the mesoporous structures including pore size diameter and pore volume are mainly dependent on the length of the alkyl chain of surfactants [98]. Meanwhile, the production of ordered continuous mesoporous metal film require the formation of a highly stable mesophases of LLCs in the presence of metal ions. Also, electrochemical reaction parameters such as deposition potential, precursor concentrations and temperature have substantial effects on the mesoporous film properties and morphology especially for both surface and uniformity [98]. The LLCs templates-based approach is scaled up for production various mesoporous mono and multi-metallic NCs such as Pd, Co, Ni, Sn, Cd, Sn, Pt-Ru, Pt-Pd and Pt-Au [11, 12, 99].

For instance, Yamauchi and coworkers synthesized mesoporous Pt-Au without phase segregation by using 2D hexagonally LLCs templates by electrodeposition on conductive substrate (Fig. 4.10) [100]. Initially, highly

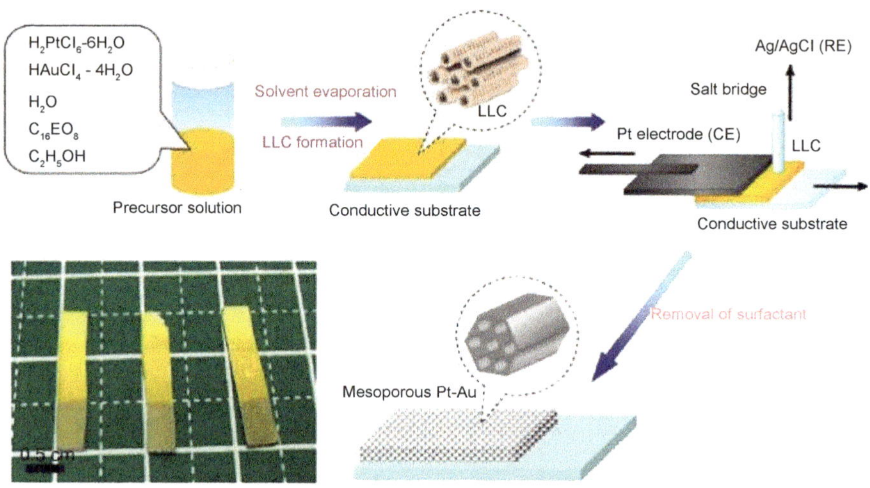

Figure 4.10. Scheme shows the experimental process used to prepare mesoporous Pt-Au films by electrochemical deposition through LLCs templates. Reprinted with permission from ref. [100]. Copyright 2012 American Chemical Society.

ordered and stable 2D hexagonally LLCs templates were prepared on the surface of conductive substrates by aqueous surfactant solution including water, a nonionic surfactant, ethanol and metal salts by drop-coating. The results reported that, the composition of the obtained Pt-Au was mainly determined by adjusting the initial precursor concentrations. Moreover, the LLCs templates were used for production mesoporous multi-layer film by dominating the charge of the deposition, which subsequently tune film thickness. Based on the same concept, Bartlett et al. obtained mesoporous Pd-Rh with multi-layers structure through control layer by layer deposition of Pd on Rh under specific deposition charges [99]. Similarly, mesoporous Pt-Pd with multi-layer structure was obtained by electrochemical layer-by-layer deposition [101].

2.2.2.2 Chemical reduction

The chemical reduction approach involves the reduction of metal precursor by using proper reducing agents in the presence of soft template such as LLCs for production of various porous metal NCs. Attard and coworkers synthesized for the first time hexagonal mesoporous Pt NCs by chemical reduction of Pt salts with less transition metals (e.g., Fe, Mg, Zn) in aqueous domains solution of LLC [102]. The formation mechanism is based on initial nucleation of Pt clusters in the confined pores among rods of LLCs templates due to higher reduction potential of Pt salt (Fig. 4.11a–b) [103]. Then formed nuclei grow to form spherical clusters (Fig. 4.11c). Finally, both nucleation and growth occur along with the mesospaces to form bumpy Pt frameworks (Fig. 4.11d–f). This is ascribed to the difference in the reduction potential among metal salts that facilitate initial nucleation and the high surface area of the formed clusters that require fast growth into spherical molecules in order to reduce their high total free energy. Various mesoporous Pt-based NCs such as Pt-Pd, Pt-Ru and Pt-Ni are obtained by chemical reduction of metal ions in the LLCs templates [96, 104, 105]. Interestingly, mesoporous multi-metallic (Fe, Co, Ni) with well controlled compositions were fabricated by adjusting the metal species ratio in the aqueous solution of non-ionic surfactant (Brij 56) [106, 107, 108]. The results clearly warrant that, the uniformity of mesoporous Ni-Co structure boosted on lessening the Co content which subsequently tune the magnetic properties of the obtained product. It should be noted that, using LLCs templates provide several benefits compared to hard templates owing to the high abundance of LLCs templates with different size, morphology and composition enabling a versatile approach for rational design of well-defined 3D porous structures with controlled shape and features. In spite of the great progress in using LLCs templates for preparing porous metals, it is not possible to produce some mesoporous metal such as Au.

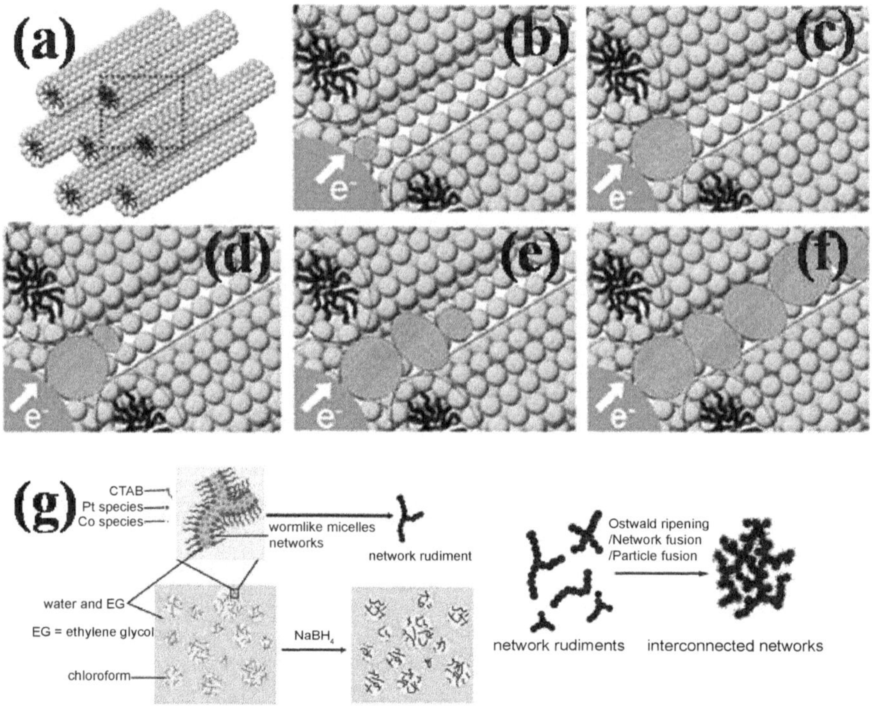

Figure 4.11. (a–f) Model represents the proposed mechanism for deposition of Pt in the aqueous domain of LLCs templates. Reprinted with permission from [103]. Copyright 2005 American Chemical Society. (g) Scheme shows the preparation of porous Pt-Co networks by using the micelles templates. Reprinted with permission from ref. [109]. Copyright 2012 Wiley-VCH.

To overcome these barriers, another soft template such as cetyltrimethylammonium bromide (CTAB) is used for preparing a mesoporous structure. This is based on the self-assembly property of CTAB in water/chloroform solvent into wormlike micelles as templates for fabrication of Pt nanowire networks. Zhang and coworker used CTAB for production of spongy-like Pt-Ni and Pt-Co by using ethylene glycol with the water/chloroform solvent (Fig. 4.11g) [109]. It is noteworthy that, the ionic surfactants, such as CTAB, cetyltrimethylammonium chloride (CTAC), and Sodium Dodecyl Sulfate (SDS) suffer from inevitable aggregation at a low temperature due to their inferior solubility at such temperatures which limits their utilization in the synthesis of 3D porous structure such as nanodendrites. This is can be avoided by increasing the solubility of ionic surfactant by heating to avoid surfactant aggregation. For instance, porous Pt NCs are created by reducing metal precursors in the presence of CTAB assisted by heating till 70°C.

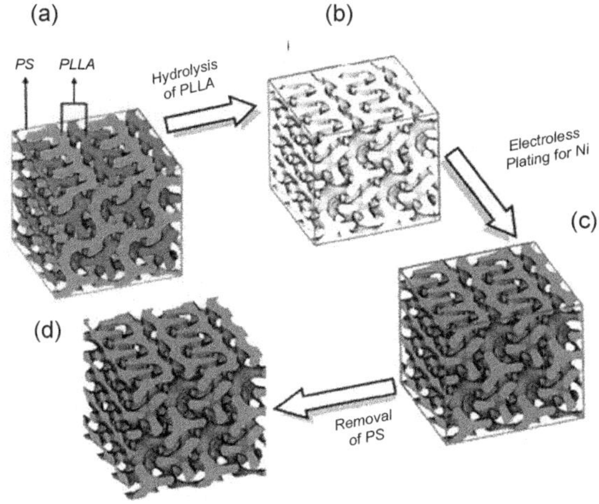

Figure 4.12. Scheme shows the fabrication process of a highly ordered mesoporous gyroid Ni by using BCP templates. Reprinted with permission from ref. [108]. Copyright 2011 Wiley-VCH.

Other highly interesting templates are bloc co-polymers that can self-assemble in a water-oil solvent to form unique micelles with highly ordered 3D interconnected porous structure. In this regard, polystyrene-b-poly (L-lactide) (PS-PLLA) has the ability to self-assemble into a 3D porous gyroide structure after selective erosion of the poly(L-lactide) network. Based on these phenomena, Hsueh et al. synthesized a highly ordered porous gyroid Ni structure with high crystanility by electroless plating of Ni species with PS-PLLA followed by etching the polystyrene matrix (Fig. 4.12) [108]. The electroless plating can be achieved under mild reaction conditions to create a mesoporous Ni gyroid nanostructure. Many mesoporous structures can be easily obtained by changing the metals salts in the electroless solution.

Interestingly, the electrochemical reduction of H^+ lead to release hydrogen bubbles which can be used as a dynamic template for fabrication of various porous metals [110]. The metal nanoparticles deposited simultaneously with the formed hydrogen bubbles in the form of porous structure without the help of additional steps for removing the template. This subsequently enables the fabrication of different types of porous monometallic such as Ni, Cu, Sn, Ag, Au, Pt and bi-metallic such as Cu_6Sn_5, PdCu.

2.2.3 Hybrid templates

The combination among different templates such as hard and soft templates is an effective and versatile route for synthesis of hierarchical porous NCs with different morphologies and compositions [11, 111, 112]. For instance, by using colloidal spheres with porous AAO templates, porous metal nanowires are obtained. Typically, the synthesis silica nanospheres are done first by templating into the confined 1D channel of AAO or porous polymer membranes. Afterwards, metal precursors can be infiltrated into AAO pores by electrodeposition. This enables synthesis of porous Au and Ni nanowires, and the results clearly reveal that this method can be extended for synthesis of other porous metallic NCs. The combination of multi-templates provides benefits in terms of structures and properties that are not available by using an individual template. In this regard, inorganic mesoporous materials such as silica are prepared by using dilute surfactant solutions along with a solvent evaporation. One of the most typical examples is using LLCs template together with AAO with assistance of solvent evaporation leading to the formation of a highly ordered and uniform structure that cannot be obtained by using single LLCs template [112]. This is due to the high viscosity of the LLCs templates solutions which cannot form a uniform phase to infiltrate in the pores of AAO.

Colloidal silica can be easily coated onto other metallic or metallic oxide NCs such as Ag, Au and Fe_2O_3 to form multi-model porous structures including rattle, cage, egg-yolk and core-satellite [11]. This can be easily achieved by initial coating of metallic NCs with silica by the sol-gel method followed by selective etching of silica and subsequent galvanic replacement of the metal in the core area by another active metal. This can be used for not only tuning pores structure morphology but also composition. In spite of the ease of fabrication and low cost of this method, it requires high dissimilar stability in the etching solution among core metal and silica shell, which can be avoided by using appropriate selective etching solution for core or metal. For instance, Zhao and coworkers synthesized Au-nanocage@ mesoporous SiO_2 by initial coating of Ag nanocubes with two silica layers followed by selective etching of the middle silica layer by Na_2CO_3 without affecting the Ag to produce Ag@mesoporous silica yolk-shell structure (Fig. 4.13a–c) [113]. Afterwards, Ag nanocubes are replaced by $HAuCl_4$ to form Au-nanocage@mesoporous SiO_2 (Fig. 4.13d–e). The results demonstrate that the amount of $HAuCl_4$ determine the pores distribution in the core area. Moreover, Ag NCs are not fully replaced by Au ions implying that the formed porous structure is Ag-Au nanocages@mesoporous silica. Therefore, this method is amenable to scaling to synthesize various multi-complex mesoporous structures through changing the templates type and reaction conditions.

2.3 Surfactant mediated synthesis

Surfactant-based synthesis is a robust way for fabrication spatial and complex porous structures with controlled morphology and

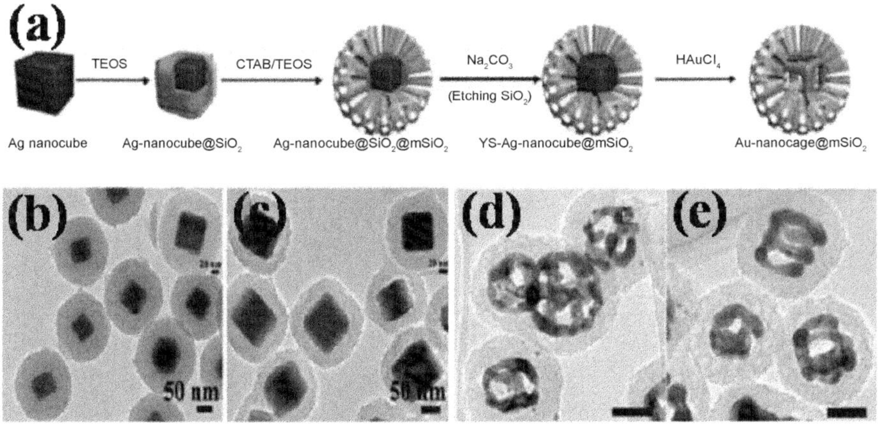

Figure 4.13. (a) Scheme display the synthesis of Au-nanocage@ Mesoporous Silica and its TEM images. (b) TEM images of Ag-nanocube@SiO_2 @$mSiO_2$, (c) yolk-shell Ag-nanocube@$mSiO_2$ structure (after etching the middle silica layer), (d) Au-nanocage@mesoporous SiO_2 after the spatially confined galvanic replacement reaction with 2.2 mL of 0.5 mM $HAuCl_4$ solution, and (e) after reacting with 4.0 mL of 0.5 mM $HAuCl_4$. The indicated scale bars in d and e are 50 nm. Reprinted with permission from ref. [113]. Copyright 2013 American Chemical Society.

composition [111, 114–117]. This approach is more adventitious over the traditional template synthetic method in the ease of preparation, low cost and obtaining a variety of products of ordered porous structures. Controlling of both composition and morphology of porous nanostructures are required for better utilization in various applications. In pursuit of this target, surfactant mediated synthesis stands out as facile and versatile way for fabricating well-ordered porous NCs with controllable morphology for various applications. In the surfactant-based synthesis, porous NCs are ubiquitously obtained by reduction of metal precursors in an aqueous solution of surfactant. This can be executed through either chemical reduction, thermal decomposition, self-assembly or seed mediated growth.

Shelnutt and coworkers used seeding and prompt autocatalytic growth for reduction metal precursors by Ascorbic Acid (AA) in an aqueous solution of surfactant for creation 2D and 3D porous Pt dendrite NCs. In the typical synthesis, Pt seeds are formed by reducing Pt salt by AA in the presence of sodium dodecyl sulfate (SDS) or polyoxyethylene (23) lauryl ether (Brij-35) combined with other surfactants (Fig. 4.14a–e) [111]. The results demonstrate that the morphology of the obtained particles depends greatly on the surfactant. Whereas, the concentration of the initial Pt seeds with assistance of a tin-porphyrin photocatalyst determine the final size of particles. Photocatalyst facilitates creation of massive catalytic growth sites. This subsequently leads to controllable growth into porous dendrites Pt sheets and solid foam-like structures by using liposome (Fig. 4.14a–d).

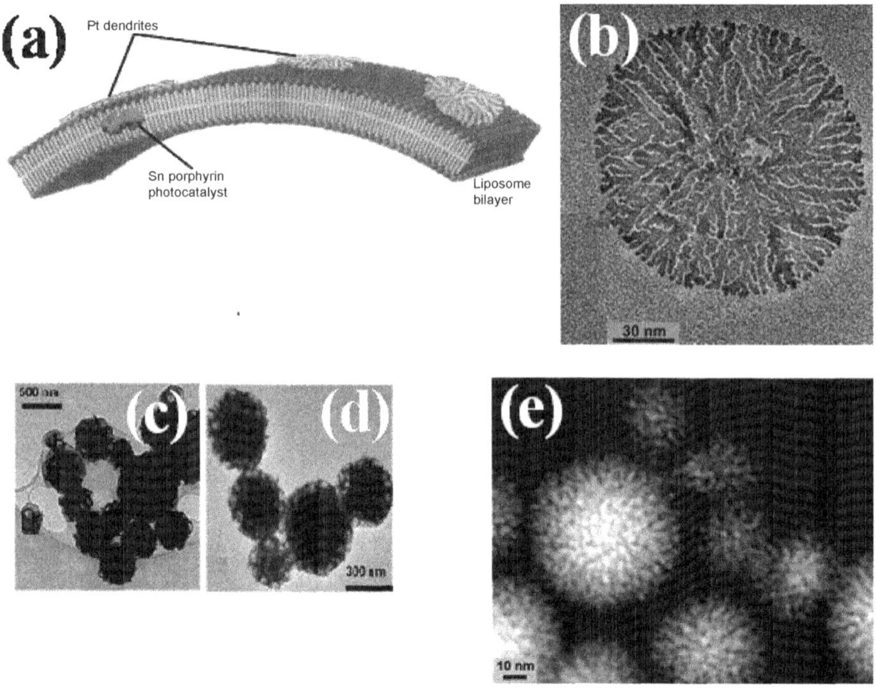

Figure 4.14. (a) Illustration model for the growth of dendritic Pt nanosheets on the liposomal surface and (b) its TEM image. (c and d) TEM images at different magnification of Foam-like platinum nanosheets. (e) HAADF image of the 3D platinum nanodendrites grown in the presence of Brij-35 micelles and the absence of photocatalyst. Reprinted with permission from ref. [111]. Copyright 2004 American Chemical Society.

3D Pt dendrites are obtained in the presence of Brij-35 micelles and the absence of photocatalyst (Fig. 4.14e). Block copolymers-mediated synthesis strategy was extended for syntheses of various multi-dimensions porous metallic NCs with different morphologies. For instance, dendritic Pt NCs with nanoporous in both exterior and interior are obtained by using Brij 700, Tetronic 1107, poly(vinyl pyrrolidone) (PVP), poly(1-vinylpyrrolidone-co-vinyl acetate) (PVP-co-VA) and hexadecylpyridinium chloride (HDPC) (Fig. 4.15a–f) respectively [118–120]. It is to be noted that ionic surfactant such as CTAB and SDS are not favored for synthesis of such porous nanodendrites structures due to their low solubility at room temperature. The reduction kinetic is a critical factor in such synthesis methods. Particularly, inferior reducing agents such as AA or formic acid were preferred for fabricating these porous dendritic metal NCs. Interestingly AA is used as both a structure directing agent and reducing agent for synthesis of porous dendritic PtNCs without any surfactant at room temperature. This

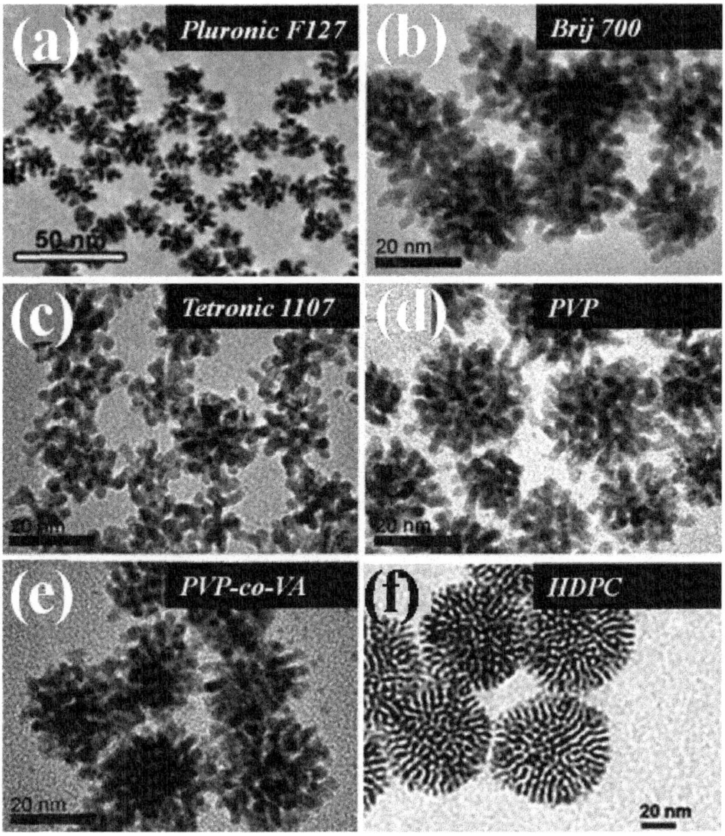

Figure 4.15. TEM images of porous branched NCs prepare with various surfactants (a) Pluronic F127. Reproduced with permission from ref. [118]. Copyright 2009 American Chemical Society. (b–e) Brij 700, Tetronic 1107, PVP and PVP-co-VA. Reproduced with permission from ref. [119]. Copyright 2010 American Chemical Society. (f) HDPC. Reproduced with permission from ref. [120]. Copyright 2013 Wiley-VCH.

is based on oxidation of AA into dehydroascorbic acid by platinum salt that is naturally hydrolyzed into 2, 3-diketo-1-gulonic acid [121]. This confirms the role of carboxylic group in the synthesis of Pt dendritic NCs. Similarly Pt dendritic NCs are obtained when formic acid is used as a reducing agent with or without surfactants. Conversely, aggregated Pt NCs particles are obtained when AA is replaced with $NaBH_4$ resulting from its high reducing power. These results warrant the role of carbonyl group with their reducing power in the synthesis of porous dendrite Pt NCs. This is shown in the utilization of colloidal silica templates together with AA for preparation of highly uniform nanoporous Pt NCs. The mechanism of formation dendritic Pt NCs is possibly ascribed to the mid reduction strength of AA which lead to homogeneous formation of initial Pt seeds followed by indiscriminate adsorption of surfactant on the different crystallographic planes to reduce their interfacial free energy and lead to spontaneous growth into porous dendritic structure. Interestingly, based on this approach various porous bi-metallic NCs are obtained by combining two metal precursors in the starting solutions in the presence of various nonionic surfactants such as Pluronic (F127). For instance, Pt-Pd, Au-Pt and Pt-Cu porous dendrites NCs are produced by reduction of bimetallic precursors in the presence of Pluronic F127 [122–124]. Impressively, this approach can be extended to synthesize tri-metallic core shell dendritic NCs including Au@Pd@Pt by reducing three metal precursors by AA in the presence of Pluronic F127 and PVP in one-step [125]. The mechanism was attributed to the natural isolation between the nucleation and growth step, because of the ability of AA to reduce three metal precursors in dissimilar rates based on their difference in the reduction potentials. Interestingly recently, we prepared porous PtPdRu dendritic NCs by simultaneously reducing three metal precursors in an aqueous solution of Pluronic F127 at room temperature [126]. The mediated autocatalytic process facilitated reduction of Ru salt, that is could not be reduced by the AA. This method is adventitious over traditional approaches such as seed-mediated growth and thermal decomposition due to the ease of preparation, absence of multi-steps, mixing metals at atomic scale instead of segregation. These features are highly favored in the electrocatalytic applications.

Xia and coworkers synthesized porous dendritic PdPt NCs with multi facets based on seed-mediated growth. This is includes initial using Pd seeds followed by growth using octahedral Pd seed in the presence of PVP as a structure direct agent [127]. Although the obtained dendritic NCs are highly uniform, their catalytic activity is unstable that is most probably owing to segregation of Pd with Pt. Recently, we synthesized highly uniform porous bimetallic PtCu NCs by reducing metal precursors by AA in the presence of PVP under ultrasonic treatment at room temperature (Fig. 4.16) [128]. The synthesized porous PtCu NCs have a high surface area in the range of 54.2 $m^2\ g^{-1}$ and with pore size around 2.5 nm. Ultrasonic treatment made the initial formation of Pt nuclei, which subsequently stimulate reduction of the Cu precursor by autocatalytic processes. This strategy opens a new frontier for fabrication of bimetallic Pt-based NCs including transition metals.

The ligand stabilized Pt nanoparticles in the presence of block copolymer along with organic solvent evaporation can be co-assembled to form mesoporous Pt nanostructures. Particularly, block copolymer N, N-di-2-propoxyethyl-N-3-mercaptopropyl-N-methylammonium chloride co-assembles with Pt NPs followed by annealing under an inert atmosphere to form mesoporous Pt-C composite. The formation of this composite is probably attributed to the catalytic properties of Pt. Ar-O plasma or acid-etch is used for removing C from Pt-C to produce highly ordered mesoporous

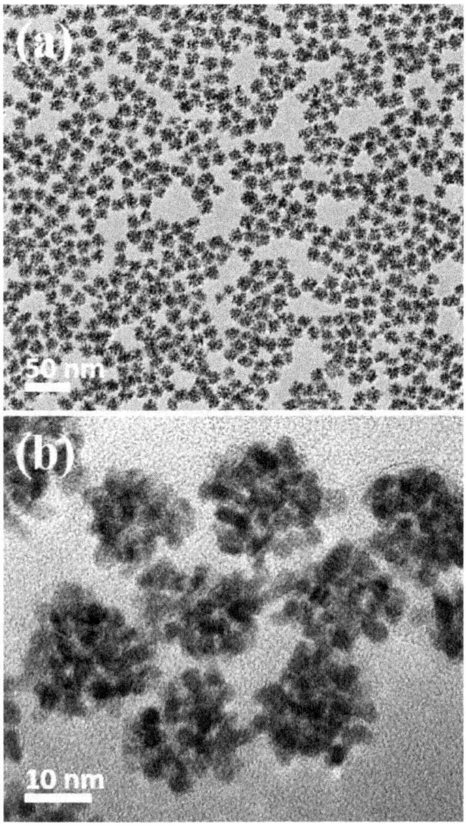

Figure 4.16. (a–b) TEM images of the porous PtCu NCs at different magnifications. Reprinted with permission from ref. [128]. Copyright 2015 The Royal Society of Chemistry.

Pt [129]. Nonetheless, in this approach various parameters should be considered including the solubility, size and volume fraction of the ligand stabilized metal nanoparticles.

2.4 Self-assembly

Self-assembly is a natural phenomenon including assembly of building blocks into well-defined porous structures at nanoscale. Particularly, amphiphilic molecules such as surfactants, phospholipids and ionic soaps self-assemble into micelles in dilute aqueous solution due to the hydrophobic interaction force. The self-assembly method attracted great attention because of its simplicity, flexibility and efficiency for tunable synthesis over both compositions and structure [12, 129–138]. The main key in the self-assembly method is inherent organization property of amphiphilic species into various morphologies driven by varying the reaction conditions. Novel properties such as electronic, magnetic and optical properties can be also originate from the interaction among inherent properties of the assembled components. Based on this, myriad multi-dimensional structures ranging from 0D to 3D are obtained [63, 129, 133, 134]. For example, a highly ordered mesoporous 3D Pt nanostructure is obtained on self-assembly of block copolymers followed by removal of C from the Pt-C composite [129]. A 1D nanotubes Bi is produced from layered Bi in solution under mild reaction conditions [135]. Colloidal crystal templates are utilized for synthesis of porous metal 2D films with different ordered nanosize and microsize arrays [85, 86]. Interestingly, these metals involve 0D core structures that can be spontaneously assembled to higher dimension structures. The ensembles of 0D NCs can predict the structurally related properties of such NCs [134]. Consequently, it is very important to sort out the self-assembly process to design new porous structures with desired features. Furthermore, it is crucial to understand the surfactant templates properties that play a paramount role in building porous nanostructure. In this regard, block copolymers can self-assemble to form various mesoporous structures such as spherical, cylindrical micelles, and vesicles (Fig. 4.17) [136]. In this case, two factors are required for the efficient self assembly process including high concentration of block co-polymer (higher than the Critical Micelle Concentration (CMC)) and hydrophobic force interactions.

Meanwhile, ionic surfactants have the ability to form micelles, which subsequently facilitate assembly of metal precursors to different morphologies. Moreover, the ionic surfactants can be easily adsorbed onto NCs surface to be charged. As an example, significant increment in both hydrophobic volume of the surfactant and pore size, can be tuned introducing hydrophobic organic species such as trimethylbenzene (TMB). Additionally, this alters structure morphology from a more curved to a less curved structure, and even a reverse micelle structure. Accordingly, Zhang and coworkers fabricated 3D porous Pt-based network-like NCs with well-controlled size and composition by using the self-assembly approach. This includes initial capping of Pt nuclei by Dodecyl Trimethyl Ammonium Bromide (DTAB) followed by Ostwald repining growth based on the electrostatic repulsion of inter-particle and heat-induced fusion during nucleation and growth (Fig. 4.18a) [139]. Similarly, Rosi and coworkers obtained porous spherical Pt-Co by using functionalized peptide molecules (Fig. 4.18b–c) [140]. This technique can be extended towards fabrication of other porous metallic NCs. The self-assembly method can be combined with other synthetic approaches such as template-based, seed-mediated, surfactant-mediated and dealloying for synthesis of multi-complex porous metal NCs (Table 4.1). The particle size and composition depend on reaction parameters and conditions such as reduction kinetic, pH,

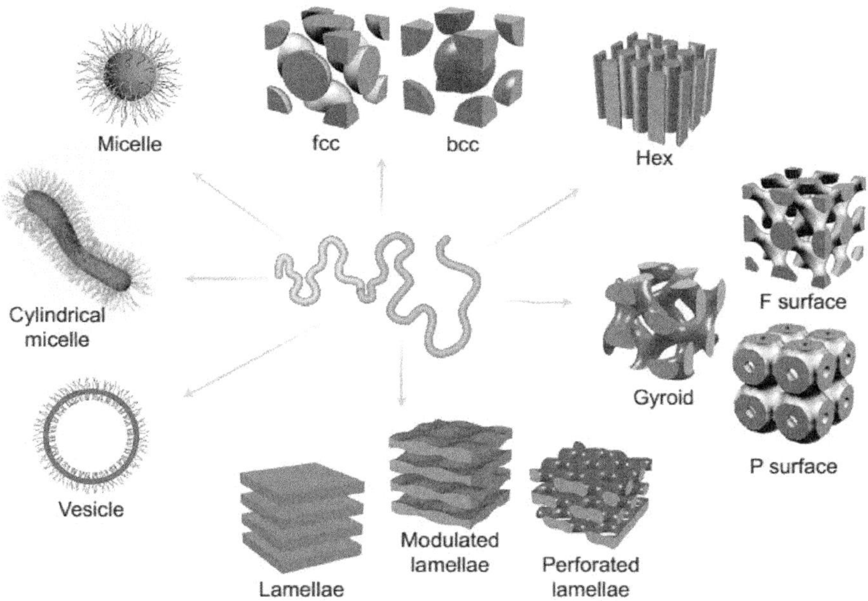

Figure 4.17. Scheme represents self-assembly of block copolymers into various mesoporous morphologies including spherical and cylindrical micelles, vesicles and spheres. Face-centered cubic (*fcc*), body-centered cubic (*bcc*), hexagonally packed cylinders, bicontinuous gyroid (gyroid, F surfaces, and P surface). Lamella, modulated lamellae and performed lamellae. Adapted with permission from ref. [136]. Copyright 2002 Wiley-VCH.

temperature, metal salt concentration and surfactant molecules [118–120]. Yamauchi and coworkers fabricated mesoporous monocrystalline Pt by controlling chemical reduction of Pt precursor by AA in the presence of KIT-6 or SBA-15 template followed by chemical etching of silica by HF (Fig. 4.19a–c) [90]. The particle size can be controlled by adjusting the reduction time. Similarly, they obtained novel 1D mesoporous Pt nanorods by using self-assembly combined with hard-template method [141]. These 1D mesoporous Pt nanorod formed the self-assembly of surfactant micelles in a confined space of a polycarbonate (PC) membrane by electrodeposition followed by chemical dealloying (Fig. 4.19d). Interestingly, the aspect ratio of mesoporous Pt nanorods can be easily controlled by modulating the electrodeposition times at constant potential of 0.2 V (vs. Ag/AgCl) at room temperature. Particularly, a Pt thin film is deposited on a side of the PC membrane and acts as a conductive layer. Both K_2PtCl_4 and Brij 58 are mixed in the electrolyte solution. The PC membranes and surfactant are dissolved after the electrochemical deposition by soaking in NaOH and ethanol solution to obtain 1D mesoporous Pt nanorods. Most recently we obtained tri-metallic PtPdRu nanocage with porous dendritic shell

Table 4.1. Combination of self-assembly method with other methods for preparing multi-complex porous nanostructure.

Method	Morphology	Composition	References
Self-assembly and direct surfactant-mediated synthesis	Branched nanoporous structures	Pt, Pd, PtPd, PtNi, AuPd, PtCo AuPdPt	[111, 118, 119, 121, 122, 143–150]
Self-assembly and hard template	Mesoporous monocrystalline nanoparticles	Pt	[90]
Self-assembly and dealloying	Wormhole nanostructure, nanocages with dendritic shells	PtNi, PtPd, PtPdRu	[55, 126, 151]
Self-assembly and self-etching	Nanocages, nanoskeletons	Co	[152]
Self-assembly and galvanic replacement	Nanocages, nanoboxes	PtCu, AuAg, PdAg, PtAg	[153–155]
Self-assembly and surfactant mediated	Mesoporous spherical, cylindrical, and lamellar domains	Au	[156]

through combining the self-assembly and chemical dealloying method (Fig. 4.20) [126]. This includes initial preparation of PtPdRu nanodendrites by chemical reduction of metal precursors by AA in the presence of PVP followed by selective chemical etching of Pd and Ru by HNO_3.

Recently, Zhao and coworkers prepared highly ordered mesoporous titania by combining self assembly with sol-gel chemistry (EISA) and soft-template and hard-template mediated (CASH) followed by calcinations at 450°C to obtain highly crystalline framework (Fig. 4.21) [142]. In particular, PEO-*b*-PS, titanium isopropoxide (TIPO) was used as a template and acetylacetone (AcAc) as a chelating agent. The AcAc facilitates assembly of templates into micelles with Ti during gel-formation based on delay of the hydrolysis and condensation of TIPO molecules (Fig. 4.21a). Interestingly PEO-*b*-PS micelles transformed to mesoporous carbon because of the catalytically activity of titania that acts as a solid supporter for preserving the mesostructure during the calcination process (Fig. 4.21b). This method can be scaled up for preparing other mesoporous metal oxides. Mesoporous metal Pt/C with 2D hexagonal structure is obtained by self assembly of a non-Pluronic block copolymer, PI-b-poly(dimethylaminoethyl methacrylate) (PI-*b*-PDMAEMA) with Pt-ligand stabilized throughout solvent evaporation process followed by calcinations in an inert atmosphere [129].

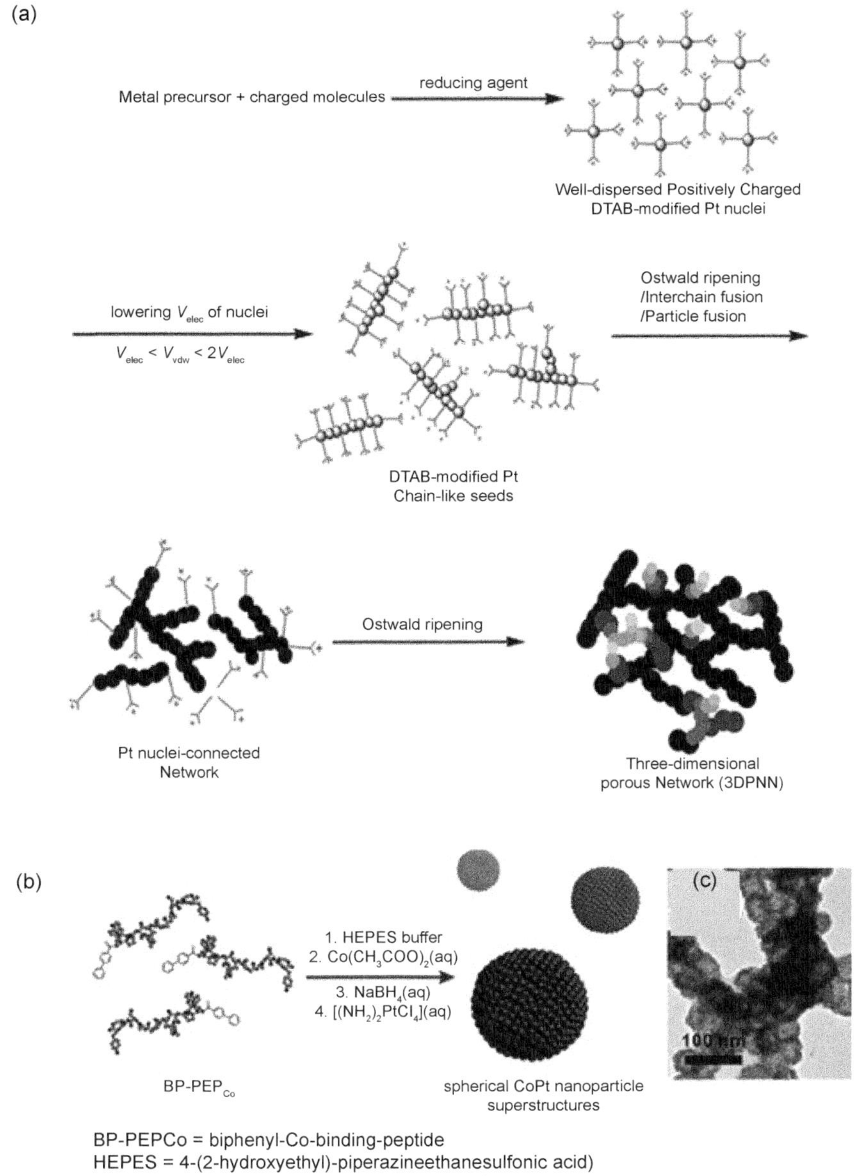

Figure 4.18. (a) Scheme shows fabrication of a 3D porous metal network based on the electrostatic repulsion of the charged nuclei with surfactant and heat-induced particle-fusion. Reprinted with permission from ref. [139]. Copyright 2012 The Royal Society of Chemistry. (b) Scheme depicts synthesis of hollow spherical-like Pt-Co NCs and (c) its TEM image. Reprinted with permission from ref. [140]. Copyright 2013 Wiley-VCH.

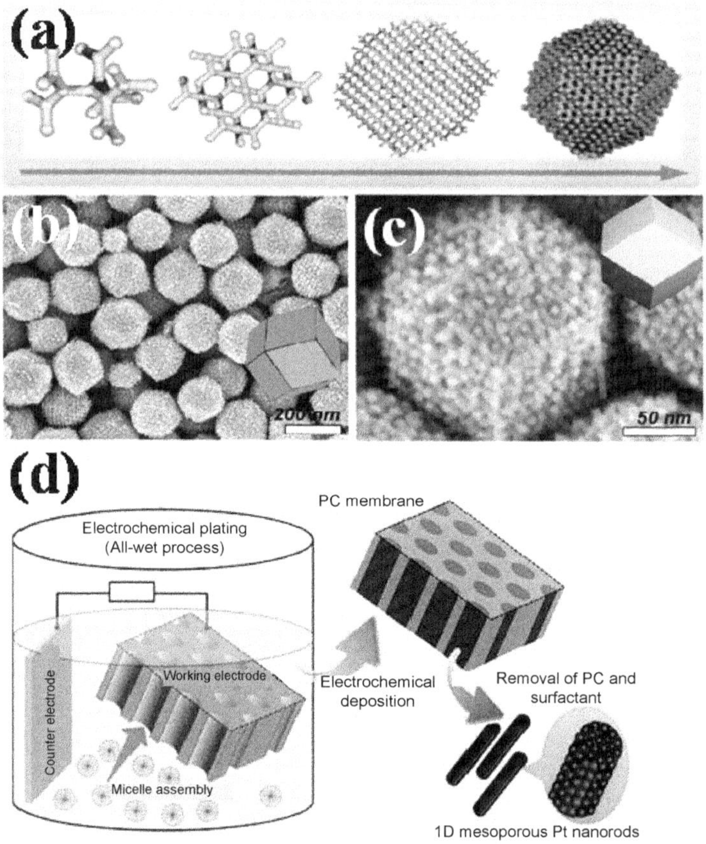

Figure 4.19. (a) scheme shows simulation for growth of mesoporous Pt NCs, based on formation of highly uniformed Pt seeds followed by their tracing inside the channels of silica KIT-6 template. (b) Low- and (c) high-magnification SEM images of the mesoporous Pt NCs. The insets in b and c represent the model of as prepared mesoporous PtNCs. Reprinted with permission from ref. [90]. Copyright 2011 American Chemical Society. (d) Scheme illustrates the preparation of 1D mesoporous Pt nanorods by using self-assembly combined with hard-template method followed by chemical dealloying. Reprinted with permission from ref. [141]. Copyright 2013 Wiley-VCH.

3. Electrochemical Applications

Mesoporous materials have unique physical and chemical properties, which enable their utilization in various electrocatalytic applications. These features include highly accessible surface area, pore structure, structural stability and high electrical conductivity that provide a superior pathway for tailoring their electrocatalytic activity towards different reactions. There are various mesoporous metallic NCs exhibiting catalytic activities

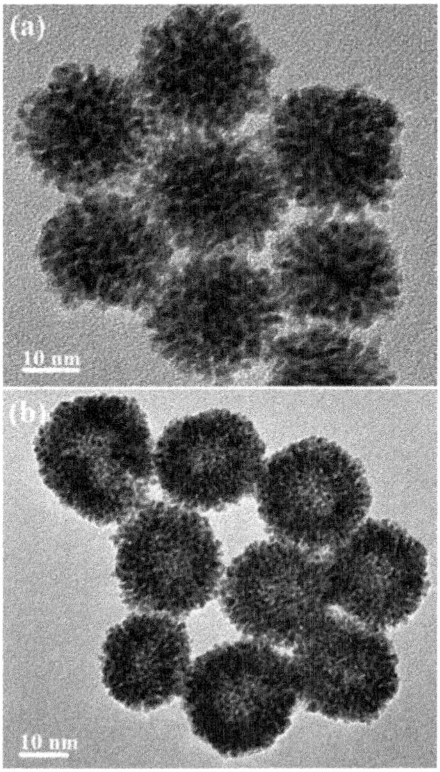

Figure 4.20. (a) TEM images of PtPdRu nanodendrites before chemical etching and (b) dendritic nanocages after chemical etching. Adapted with permission from ref. [126]. Copyright 2015 American Chemical Society.

utilized in either oxidation of organic molecules (e.g., methanol, ethanol, formic acid) or reduction of oxygen molecules which are the main reactions occurring in fuel cells [14, 35]. One of the most attractive and effective metals for these sorts of reactions are porous Pt-based nanostructures because of their dual electrocatalytic activity for the cathodic Oxygen Reduction Reaction (ORR) in membrane fuel cells (PEMFCs), and the anodic oxidation reactions in the direct methanol fuel cell (DMFCs) [8]. The porous geometry not only enhances the accessible surface area to the reactant molecules but also increases their diffusion rates that lead to effective mass transfer in the catalytic reactions. For instance, mesoporous Pt NCs is used as both a microelectrode and gas diffusion electrode for ORR. The reported mass activity is considerably superior to the commercial Pt/C, merely one quarter of the ultra-small pores are accessible to the electrolyte ions. Consequently, the surface areas, mostly of internal pores, are not included in the electrochemical reaction. Therefore, the electrochemical active surface

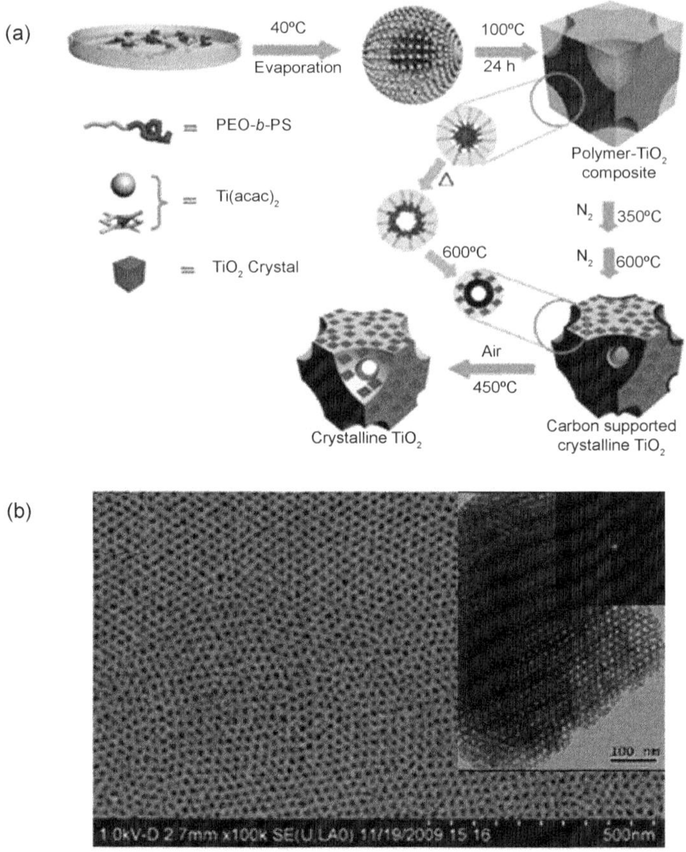

Figure 4.21. Scheme demonstrates synthesis of cubic mesoporous titania with a highly crystallized framework based on the ligand-assisted assembly method. (b) SEM of cubic mesoporous titania and its TEM images shown as inset. Adapted with permission from ref. [142]. Copyright 2011 Wiley-VCH.

area (ECSA) is used to represent the real surface area of porous Pt NCs. The ECSA of metallic NCs can be measured by different methods such as Cyclic Voltammetric (CV), -stripping and copper monolayer stripping. Particularly, the ECSA of porous Pt-based NCs can be measured by the equation (2)

$$Q = Q_o / q_o \tag{2}$$

Where Q_o is the surface charge obtained from the area under the CV trace of hydrogen desorption, while q_o is the charge required for electrochemical desorption of a monolayer of hydrogen on the metal surface. Meanwhile, the ECSA of other porous metals NCs especially Pd and Au can be estimated by the oxygen adsorption method. There are many studies that discuss further details about ECSA measurements.

Mesoporous Pt nanostructures with their high surface area can be integrated into the electro-chemical fuel cell devices as supported electrocatalysts removing the need for carbon supporter catalysts which is known to cause Ostwald ripening and subsequent dissolution of Pt [9]. In case of using Pt as supported electrocatalysts, both surface Roughness factor (Rf) and the geometric confinement effect of nanopores are crucial to determine their inherent electrocatalytic activity. Generally, the Rf is the ratio of the electrochemically active surface area to its geometric surface area. Thus Rf indicates the effect of internal nanopores on increment the surface area. Predominantly, the specific surface area, pore size distribution and pore volume are measured by the multipoint Brunauer-Emmett-Teller (BET) method on the basis of gas adsorption/desorption isotherms. The BET method is only applicable for samples in the powdered form whereas, electrochemical method is used for determination the surface area of solid electrode structures. Chung and coworkers thoroughly investigated the relation among film thickness and Rf in the electrochemical reactions. The results clearly implying that, nanoporous Pt thin films with small ($Rf < 40$) participate effectively in electrochemical reactions (Fig. 4.22a–b and e–f) [157]. Meanwhile, some diffusion-controlled reactions occur only at the surface of Pt when the thickness rises above the maximum penetration depth ($40 > Rf < 300$) (Fig. 4.22c). This is due to the sensitivity of electrochemical reactions to the film thickness. Subsequently, the faradaic current reaches a plateau that does not require enhancement in the surface area. This means that further increment in the surface area will not participate in the electrochemical reactions (Fig. 4.22d). This phenomenon is noticed for the faradaic reductions of O_2 and H_2O_2. In case of the slow electrochemical oxidation of glucose, the faradaic current boosts with increasing the thickness of porous film implying that, the system is still in the kinetic controlled regime. Although, the overall surface of porous electrode seems to contribute in the faradaic reactions (Fig. 4.22g) the sluggish reactions will reach the diffusion-controlled regime in the thicker films (Fig. 4.22h). Based on this investigation, thickness and pore structures of the nanoporous catalysts can be optimized for better improvement of catalytic performance. Also, it is of interest for selective detection of target species that suffer from inevitable slow or fast reactions.

One of the most observed effects of the Rf on the electrocatalytic reactions is the combined effect of both nanopores with surface area on enhancement the catalytic activities of mesoporous metallic NCs. The significant effect of nanopores is further observed by voltammetric and potentiometric responses. It is worthy noting that the increase in the electrocatalytic reactions of O_2 and H_2O_2 at nanoporous Pt electrodes is further noticed right after eliminating the Rf. This occurred on normalizing the Rf to the ECSA instead of the geometric surface resulting in observation

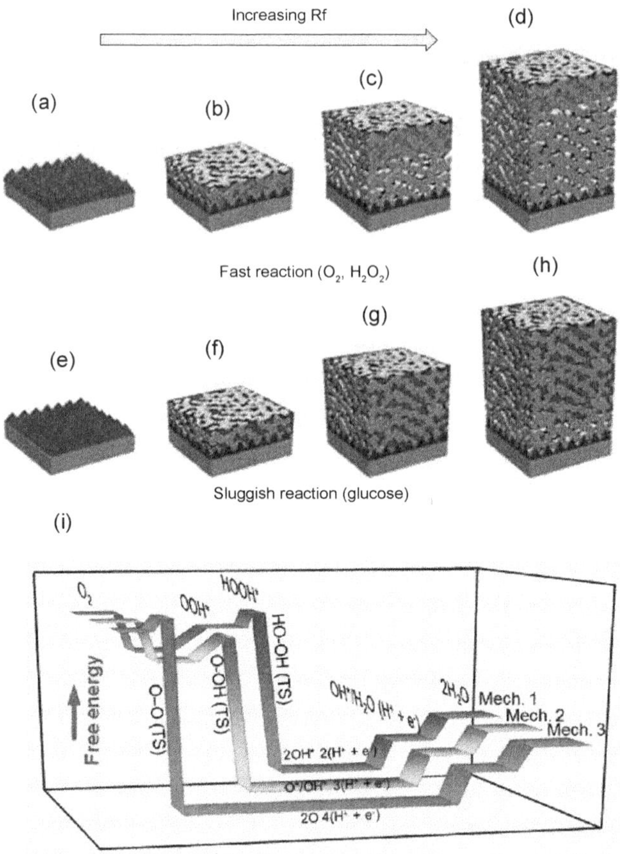

Figure 4.22 (a–h) Schematic represent the relation between porosity and Rf. (a and e) Solid Pt films without nanoporous structure Rf < 10, (b and f) nanoporous Pt films 10 < Rf < 40, (c and g) nanoporous Pt film 40 < Rf < 300, and (d and h) the highest Rf. Red and blue backgrounds show the electrode surface regions while fast (red) and sluggish (blue) reactions take place. Reprinted with permission from ref. [157]. Copyright 2010 Elsevier. (i) Schematic show the relative free energy levels of the ORR mechanism on Pt (111) surface at 0.8 V. Reprinted with permission from [159]. Copyright 2013 American Chemical Society. The drawing was based on the data reported in ref. [160].

of near Nernstian behavior. A nanoporous Pt electrode was found to exhibit faster response times and lower hysteresis relative to a flat Pt electrode with the same surface area. Considering this result, it is plausible that the confinement effect of nanoporous structure lead to increased residence time of reactant species.

Moreover, El-Sayed and coworkers revealed that, the electrocatalytic activity could be influenced by the nanopores because of the confinement effect. That led to increasing the steady-state concentration of the reactant molecules in the rate determining step of the reaction [158]. Subsequently, reactant species reside for a longer time inside the inner pores of the porous catalysts that allow reactants to collide with the electrode surface, which makes the porous catalysts more beneficial as they exhibit higher current density. Other factors, such as the stress and defect effect of mesoporous metallic NCs also play a main role in improving the electrocatalytic activity. Thus it would be of great value to fully understand the effect of geometric confinement for enhancing the electrocatalytic activity in porous metallic structures. Hence it is great of importance to tailor pores size of porous catalysts for improving the electrochemical reactions activity in the fuel cell devices (Fig. 4.23).

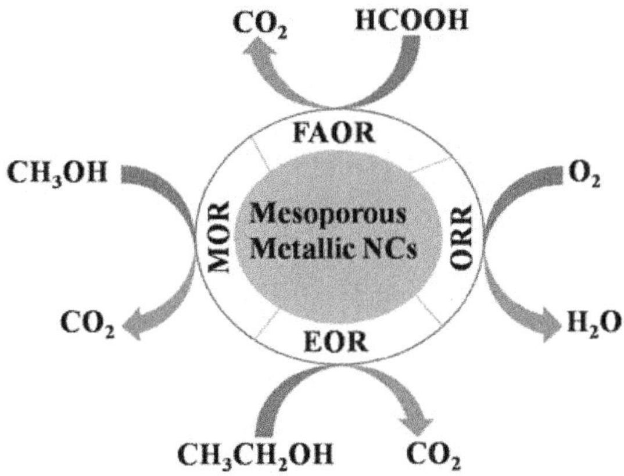

Figure 4.23. A scheme shows the utilization of mesoporous metallic NCs in various electrocatalytic applications including, oxygen reduction reaction (ORR), methanol oxidation reaction (MOR), ethanol oxidation reaction (EOR) and formic acid oxidation (FAOR).

3.1 Oxygen reduction reaction

Oxygen reduction reaction (ORR) is the main reaction in the PEMFCs, representing the new generation of environmental benign renewable energy sources. Various oxygenated intermediates species such as H_2O_2, O_2^- and O_2^{2-} are usually generated during ORR. Ubiquitously, oxygen is electrochemically reduced on the catalytic surfaces at the cathode by two pathways: four-electron reduction or two electron reduction [50]. In case of four-electron reduction pathways, O_2 is reduced electrochemically in the acidic medium to form water by combining oxygen and the two protons. This process typically takes place under a strong acidic environment for hydrogen fuel cells. The typical standard potential, for this reaction is 1.23 V.

The four electron pathway reactions are given in equation (3)

$$O_2 + 4H^+ + 4e^- = 2H_2O \quad (3)$$

In case of two electron pathway, the reactions are given in the equation (4)

$$O_2 + 2H^+ + 2e^- = 2H_2O_2 \; E^o = 0.67 \text{ eV} \quad (4)$$

$$H_2O_2 + 2H^+ + 2e^- = 2H_2O_2 \; E^o = 1.76 \text{ eV}$$

Although the two electron reaction pathway takes place at a lower potential, it is not preferred for the ORR. Various efforts are fuelled to understand the ORR mechanism for designing active and durable catalysts. A wide variety of factors such as potential bias and energy levels for ORR with different transition states are considered in the recent studies. In this regard, based on the free energy levels there are three main mechanisms in the acidic medium at potential of 0.8 V as shown in (Fig. 4.22i) [159, 160]. It is quite apparent that the three mechanisms are occurring according to the four electron reaction pathway that entail the adsorbed intermediated oxygen species, including OOH*, OH* and HOOH*.

In the PEMFCs, the oxygen molecules adsorb on the catalyst surface and follow by the splitting of O-O bond that lead to the formation of OH groups at potential of 0.8–0.9 V. Subsequently, the formed OH group reacts with a proton to form H_2O_2. Pt is the most active and durable catalyst for ORR due to its dual ability to hinder formation of adsorbed oxygenated molecules at potentials greater than 0.8 V and decreasing the energy pathway. However, the sluggish ORR kinetic on Pt catalyst surface in the acidic medium, Pt cost and Pt scarcity inhibit development of cost-effective PEMFCs. Added to that, the low accessible surface area of Pt catalysts is not vulnerable to be integrated into the fuel cell devices. Various efforts are devoted to overcome these barriers culminating in control of Pt catalyst morphology and composition [14].

Various porous Pt NCs are prepared by different methods such as seed-mediated growth, template-based and self-assembly for augmentation of the electrocatalytic activity toward ORR. For instance, mesoporous Pt with size around 50–300 nm synthesized by electro-deposition of Pt into a mesoporous Double Gyroid (DG) silica film (Fig. 4.24a–b) [161]. The prepared mesoporous Pt exhibited higher specific activity and durability towards ORR than the commercial Pt/C electro-catalyst. This is ascribed to the presence of the twisted network structure of the DG mesoporous Pt, which prevents the Ostwald ripening pathway causing subsequent agglomeration of Pt. However, the measured mass activity for mesoporous Pt is about 0.04 A/mg_{Pt} that is not enough to replace the commercial Pt/C catalysts which exhibit mass activity between 0.1–0.33 A/mg_{Pt}. Additionally, it is very important to explore new catalysts to rationalize use of highly expensive and scarce Pt metal.

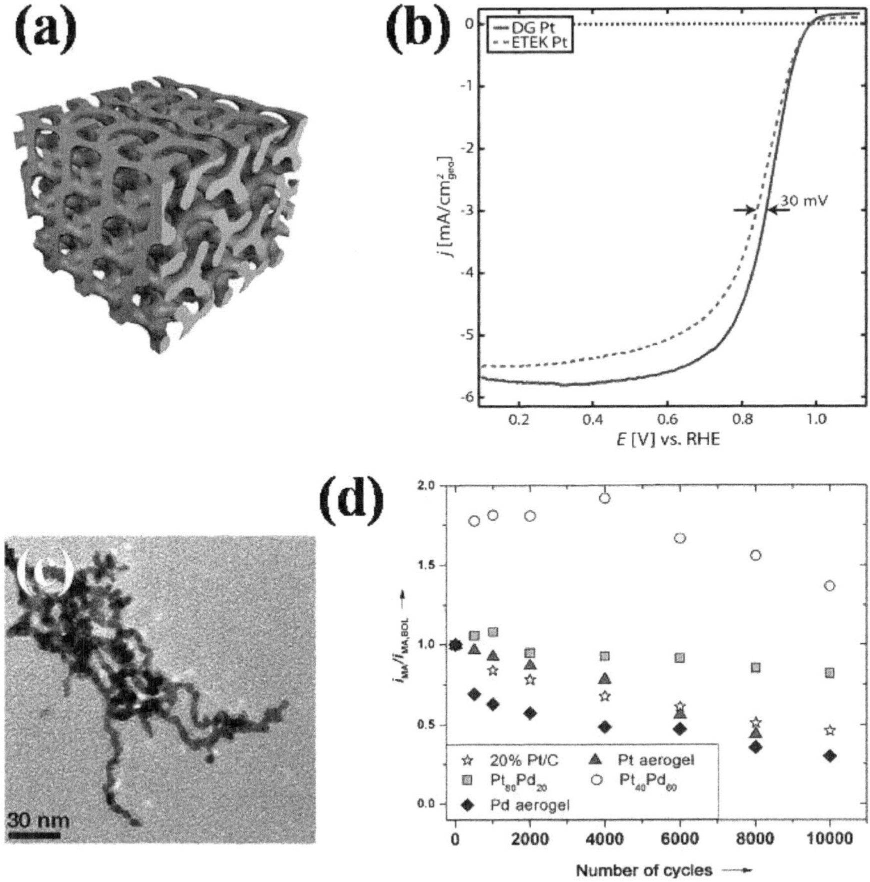

Figure 4.24. (a) Structural model for the mesoporous DG Pt and (b) ORR polarization curves for DG Pt and ETEK Pt/C measured in O_2-saturated $HClO_4$ 0.1 M at 23°C, sweep rate 5 mVs^{-1}, and 1600 rpm at 0.9 V. Reprinted with permission from ref. [161]. Copyright 2012 American Chemical Society. (c) TEM image of the $Pt_{50}Pd_{50}$ aerogel. (d) The comparison of mass activity of PtxPdy catalysts of different compositions as a function of the number of potential cycles (0.5 V to 1.0 V) measured in O_2-saturated $HClO_4$ 0.1 M, 10 mVs^{-1}, and 1600 rpm at room temperature after correction of ORR polarization curves to electrolyte resistances. Reprinted with permission from ref. [162]. Copyright 2013 Wiley-VCH.

Mixing Pt with less expensive metals in the form of random alloys, intermetallic alloys, near surface alloys and core shell nanostructure not only reduce the amount of Pt metal used but also improve its catalytic activity and durability. This is based on modification of the electronic state of Pt and strain effects which optimize the binding energies of intermediate oxygenated species onto the catalyst surface.

The synthesized porous bi-metallic Pt-based NCs consume low Pt mass and yield high accessible surface area that lead to better utilization of Pt metal for ORR. Additionally, the porous Pt alloy can be effectively used as self-supporting catalysts reducing the use of carbon supporter which suffers from degradation resulting from corrosion.

The stability of the second metals in the electrolyte medium and their electronic structure should be considered before mixing with Pt. This consequently leads to further understanding the ORR mechanism of porous Pt-based alloys. In this manner, the surface composition of porous Pt-based alloys is easily changed reversibly in gas phase environments such as CO, H_2 and O_2. Such changes can be easily measured by X-ray Photoelectron Spectroscopy (XPS). For example, the surface composition of PtCu can be controlled by reducing gases such as CO. Specifically, CO is adsorbed onto PtCu surface and Cu pulled to the surface. This also facilitates understanding the relation among modulating the oxygenated species and the surface composition. Considering these factors various bi-metallic porous Pt-based NCs such as PtFe, PtAg, PtPd, PtCu, PtCo and PtNi are fabricated for ORR [41, 57]. Therefore, multilayered Pt-M surfaces (M=Co, Cu and Ni) are prepared with an acid solution, followed by an annealing treatment showing enhancement in the specific activity of ORR over 2.5 mA/cm^2_{Pt} [57].

Self-supported nanoporous PtFe nanowires prepared by combination between the electro-spinning techniques exhibiting a specific activity of two fold greater than that of commercial Pt/C catalysts [41]. This is attributed to the strain effect resulting from massive difference in the lattice mismatching between Pt and Fe. Bi-metallic Pt_xPd_y aerogels (x = 80, 50 and 20 while y = 20, 50 and 80) with controlled composition is prepared by chemical reduction of Pt and Pd precursors by $NaBH_4$ and that showed a drastic activity and durability towards ORR. The $Pt_{80}Pd_{20}$ showed mass activity of five fold greater than Pt/C ascribed to their massive surface area 168 $m^2\ g^{-1}$ and 3D porous structure (Fig. 4.24c–d) [162]. In particular, the significant increment in the ORR activity and durability is ascribed to the strong Pt-Pd legend effect resulting from synergy between Pt and Pd metals. This increases the adsorption rate of oxygen molecules on the catalyst surface, facilitating splitting of O-O, and increasing oxidation potential of Pt.

Based on the volcano plot, Pt_3Ni is the most active and durable catalyst for the ORR due to their unique ability to counterbalance among strong adsorption rate for the oxygenated molecules and retarding adsorption of the intermediated oxygenated species onto catalyst surfaces [148]. Recently, porous Pt_3Ni nanoframe achieved a factor of 36 and 22 enhancement in mass activity and specific activity, respectively, toward ORR compared with Pt/C [95]. This is attributed to the porous interior and exterior surfaces which provide 3D surface molecular accessibility and optimal use of both metals effectively.

Although great advances in the synthesis of porous bi-metallic NCs have taken place, their catalytic activity is still far from replacing the commercial Pt/C. Another effective strategy for improving the ORR activity is synthesis tri-metallic porous Pt-based NCs due to their unique composition effect. There synthesis of porous ternary metallic NCs have not been studied significantly relative to bi-metallic NCs. There are few presented examples of porous tri-metallic such as PtPdCu and PtCuCoNi [41]. For example, PtCuCoNi nanotubes prepared by the templating-based technique by using the AAO method followed by dealloying which revealed outstanding ORR activity and durability. PtPdBi Porous NCs are prepared by the thermal decomposition method followed by acid pickling and that showed excellent ORR activity and durability relative to Pt/C. Particularly, porous PtPdBi nanowire exhibits mass activity of 1.73 times of PtPdBi nanoparticles and 6.82 of Pt/C [163]. Most recently, we synthesized porous PtPdRu nanocages by selective chemical etching of PtPdRu dendrites at room temperature. The mass activity of as prepared PtPdRu nanocages (2.61 mA μg^{-1}) is 14.1 times higher than that of Pt/C (0.185 mA μg^{-1}) whereas the specific activity (2.9 mA cm^{-2}) is 8.78 times higher than Pt/C (0.33 mA cm^{-2}) (Fig. 4.25a–b) [126]. Moreover, PtPdRu nanocages preserved most of their initial ECSA (95.5%) and showed only a 3.7 mV degradation in the half-wave potential ($E_{1/2}$) due to both composition and structure effect (Fig. 4.25c–d). Porous dendritic surface with multi corners provides massive active catalytic sites for O_2 molecules and facilitates their diffusion rate to the inner core resulting in better utilization of buried metals. Also the synergy between Pt and Pd increases the adsorption of O_2 molecules, while the strain effect between Pt and Ru retard adsorption of intermediate oxygenated species. This leads to increase of both ORR activity and durability. This newly developed strategy may open a new window to synthesize tri-metallic Pt-based NCs.

There are several studies demonstrating the catalytic activity of different Pt-based catalysts in the real PEMFCs. For example, Du and coworkers demonstrated the impact of the matrix materials on the Pt nanowires (Pt-NWs) and the PEMFCs performance [164]. They prepared two types of matrixes with various Pt-NW loadings as cathode electrocatalyst layers in PEMFCs. The results clearly showed that, the optimum Pt-NW loadings of 0.30 mg cm^{-2} in the Carbon Matrix (CM) and 0.20 mg cm^{-2} for the Pt/C Matrix (PM) were obtained. The Pt-NWs grown in the CM exhibited superior catalyst activity than those in the PM (Fig. 4.26) [164]. Particularly, the power density on the polarization curves increased from 0.16 Wcm^{-2} at a Pt-NW loading of 0.10 mg cm^{-2} to 0.46 Wcm^{-2} at 0.30 mg cm^{-2} (Fig. 4.26a). Meanwhile, power density decreased with increasing Pt-NW loading.

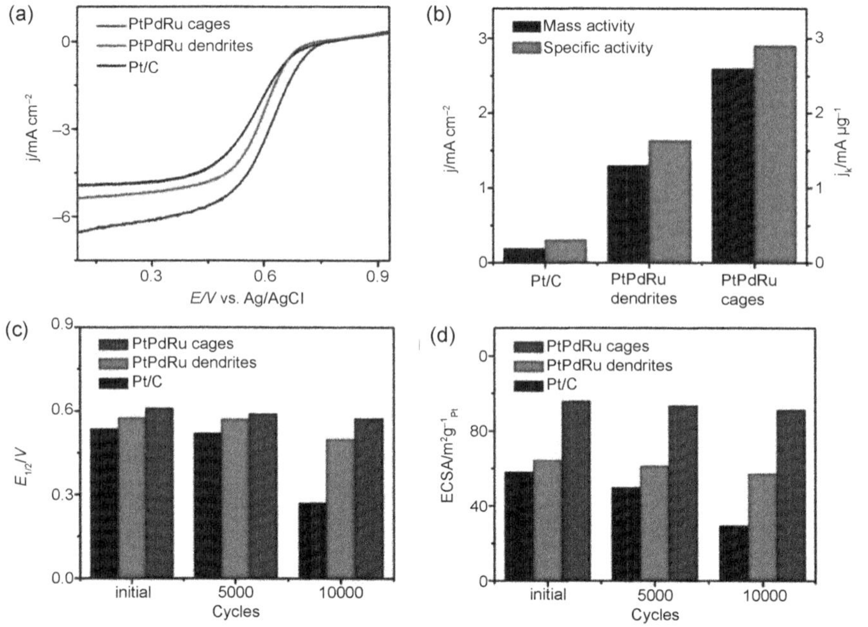

Figure 4.25. (a) ORR polarization curves of PtPdRu nanocages, PtPdRu nanodendrites and Pt/C benchmarked in O_2-saturated $HClO_4$ 0.1 M solution, at scan rate of 10 mVs^{-1}, and 1600 rpm at room temperature (b) Comparisons of the specific and mass activities at 0.6 V, (c) comparison of ECSA, and (d) comparison of degradation of $E_{1/2}$. Reprinted with permission from ref. [126]. Copyright 2015 American Chemical Society.

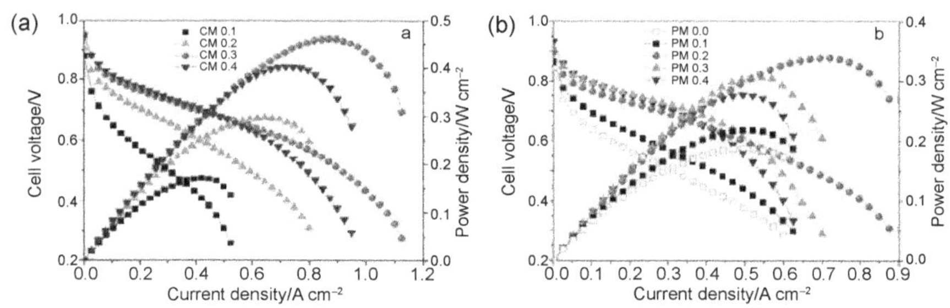

Figure 4.26. (a) Polarization curves for MEAs at various Pt-NW loadings (unit: mg cm^{-2}) in the CM and (b) in the PM. Reprinted with permission from ref. [164]. Copyright 2015 Wiley-VCH.

Interestingly by using the same Pt-NW loadings of 0.10 mg cm^{-2} the power density of 0.22 Wcm^{-2} with the PM was greater than that of 0.16 Wcm^{-2} with the CM due to the presence of Pt nanoparticles in the matrix (Fig. 4.26b).

Strasser and coworkers reported a comparative study of dealloyed binary PtM_3 (M = Co, Cu, Ni) and ternary $PtNi_3M$ (M = Co, Cu, Fe, Cr) electrocatalysts for PEMFCs relative to Pt/C [165]. The results revealed that, the catalytic activity of the dealloyed binary PtM_3 was superior to Pt/C (Fig. 4.27a). Particularly, the mass activity of dealloyed $PtCo_3$, $PtCu_3$ catalysts were 3.4 fold higher than those of Pt/C, whereas $PtNi_3$ was 2.5 fold of Pt/C (Fig. 4.27b). Meanwhile, all ternary catalysts, showed unambiguous enhancement in the ORR performance compared to $PtNi_3$ and Pt/C in real single Membrane Electrode Assemblies (MEAs) (Fig. 4.27c). The mass activity of $PtNi_3Co$ and $PtNi_3Cu$ exhibited more than fourfold of Pt/C, whereas $PtNi_3Cr$ and $PtNi_3Fe$ depicted more than threefold and twofold respectively (Fig. 4.27d). Further work is urgently required for exploring new methods for fabrication multi-metallic Pt-based for improvement the ORR activity in the real PEMFCs.

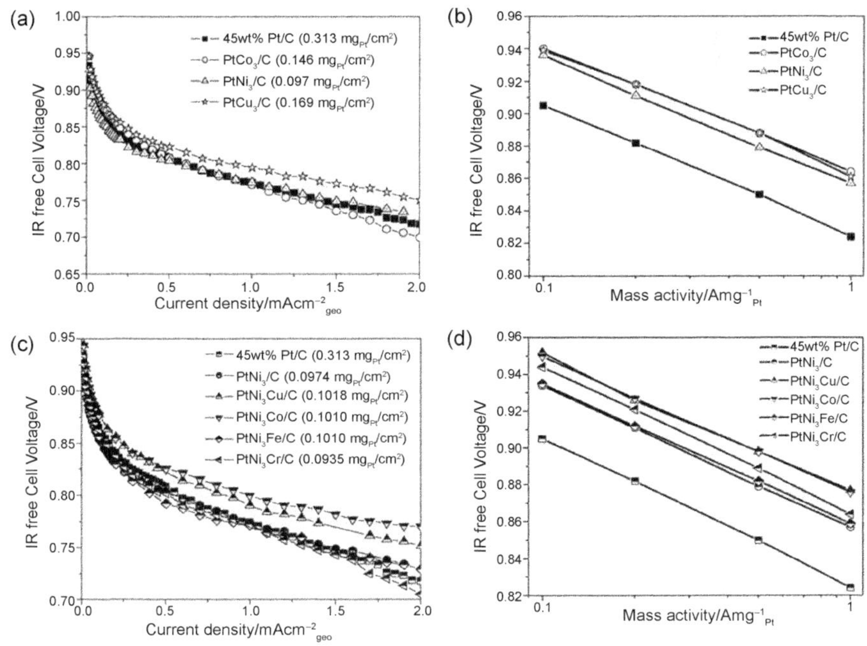

Figure 4.27. (a) Polarization curves of single cell MEAs prepared using carbon-supported PtM_3 (M = Co, Cu, Ni) on the cathode and (b) comparison of the mass activity. (c) Polarization curves of MEAs prepared using $PtNi_3M$ (M = Co, Cu, Fe, Cr) alloy precursors on the cathode and (d) comparison of the mass activity. All measurements were done after *in situ* voltammetric dealloying of the cathode catalysts. Reprinted with permission from ref. [165]. Copyright 2011 Elsevier.

3.2 Methanol oxidation reaction

DMFCs stand out as one of the most promising energy conversion technology due to its high energy yield, liquid state and low environmental impact. In the DMFCs methanol is directly oxidized electrochemically at the anode throughout multi-step reaction process including the adsorption and subsequent oxidation of the methanol molecules on the catalyst [50, 166]. The mechanism of Methanol Oxidation Reaction (MOR) is given in equation (5)

$$CH_3OH = (CH_3OH)_{ads} \text{ (methanol adsorption)} \quad (5)$$

$$(CH_3OH)_{ads} = (CO)_{ads} + 4H^+ + 4e^- \text{ (C–H bond activation)}$$

$$(CO)_{ads} + H_2O -= CO_2 + 2H^+ + 2e^- \text{ (CO oxidation)}$$

Mechanistically, methanol is oxidized to CO_2 via a CO or $HCOO^-$ reactive intermediates. The two pathways need a drastic catalyst that has the ability to absorb methanol molecules strongly, cleaves the C-H bond and facilitates reaction of the resulting species CO or $HCOO^-$ with oxygenated species to form CO_2 or HCOOH. This is not only to ease MOR but also circumvent the poisoning catalyst by adsorption of intermediated CO species. Although, Pt is the most active catalyst for MOR, its high cost, insufficient activity, low availability and instability are the main reasons that hinder DMFC commercialization. Various efforts are devoted to overcome these barriers including control size, morphology and composition. Porous morphologies with their unique features including high accessible surface area to volume ratio, low density and structural geometry reveal enhancement in the MOR activity and durability. Owing to that, porous structures provide active catalytic sites for methanol molecules and ease their transfer. This consequently decreases the current cost through decreasing Pt loading as well as increases the energy yield and improve the Pt durability.

In view of this, various porous Pt-based NCs are synthesized for MOR benefiting from their uncommon geometry that provides a pathway for tailoring both physical and chemical properties [166]. For instance, mesoporous Pt NCs synthesized by reduction of Pt precursor by AA in presence of Brij 58 as a structure directing agent show current density of 4.2 times higher than that of the Pt black (Fig. 4.28a) [167]. Also, mesoporous Pt NCs shows superior durability toward MOR relative to Pt black. This is due to the effect of both mesoporous dendrite structures and rich atomic steps, which provide massive accessible catalytic sites for methanol species. Mesoporous Pt nanotube prepared by galvanic replacement utilizing Ag nanowires as a sacrificial template show a specific activity 2.01 times higher than Pt/C and 1.26 times greater than bulk polycrystalline (BP-Pt) (Fig. 4.28b) [168].

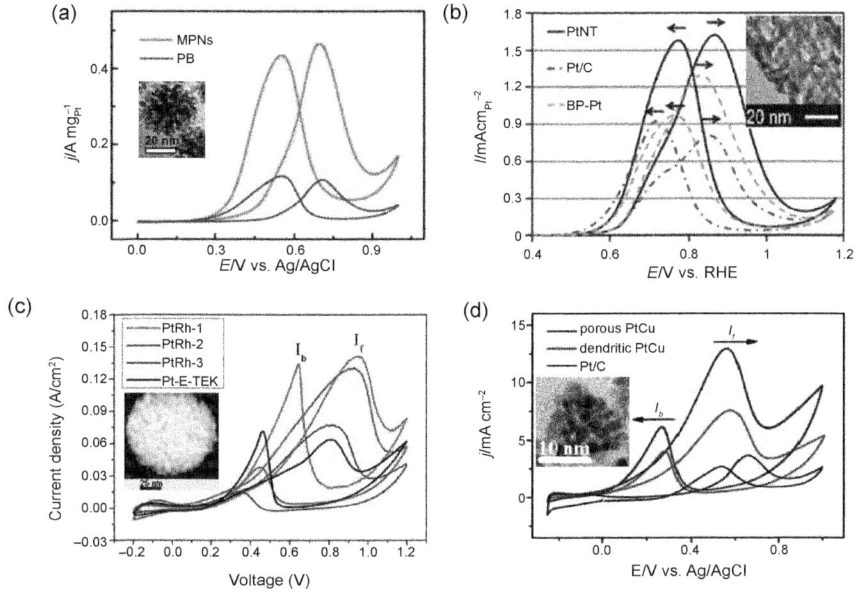

Figure 4.28. (a) CVS of mesoporous Pt (MPNs) and Pt black (PB) measured in an aqueous solution of 0.5 M H_2SO_4 + 1 M CH_3OH at a scan rate of 50 mVs^{-1}. The inset in (a) shows the TEM image of MPNCs. Reprinted with permission from ref. [167]. Copyright 2011 Wiley-VCH. (b) CV of PtNTs, Pt/C and BP-Pt in an argon-saturated 0.5 M H_2SO_4 + 1 M CH_3OH at a scan rate of 50 mV s^{-1}. The inset in (b) reveals the TEM image of Pt NT. Reprinted with permission from ref. [168]. Copyright 2010 Wiley-VCH. (c) CVs of PtRh-3 ($Pt_{80.1}6Rh_{19.84}$), PtRh-2 ($Pt_{84.42}$ $Rh_{15.58}$), and PtRh-1 ($Pt_{88.53}Rh_{11.47}$) measured in a solution of 0.5 M H_2SO_4 + 1 M CH_3OH at a scan rate of 50 mV s^{-1}. Reprinted with permission from ref. [169]. Copyright 2012 Wiley-VCH. (d) CVs of porous PtCu NCs, PtCu nanodendrites, and Pt/C measured in 0.1 M $HClO_4$ + 1 M CH_3OH at a scan rate of 50 mV s^{-1}. Reprinted with permission from ref. [128]. Copyright 2015 American Chemical Society.

These porous Pt NCs displays better electrocatalytic performance and better durability towards MOR relative to Pt/C or Pt black catalysts, because of their outstanding ability to enhance transfer rate of reactant molecules to catalytic active sites. However, the resulted MOR activity and durability is not sufficient to replace the commercial Pt/C or PtB catalyst due to the inevitable poisoning of Pt by CO species.

An effective thrust to defeat these barriers is modification of the electronic structure of Pt by coupling with other metals such as Pd, Ru, Ag, Rh, Au, Co, Ni, Pb, Ir, Sn, Mo and W. The synergy effect among two metals increases the adsorption rate of methanol molecules, while their strain effect retards adsorption of CO species. Furthermore, alloying Pt with other metals down shift the d-band center activates water to provide the needed oxygen for the complete oxidation of methanol to CO_2 at lower potential.

This triggered scientists to design wide varieties of bi-metallic Pt-based NCs with tunable morphology and composition. Various factors are critical for selecting the second metal alloyed with Pt such as stability in the electrolyte medium acid or base, synergy, cost and ability for tolerance of poisoning species. The ratio of I_f/I_b is usually used to demonstrate the ability of catalyst to retard adsorption of the intermediates species generated during the MOR. A higher I_f/I_b ratio means the MOR occur smoothly with minimum adsorption of carbonaceous intermediates on the catalyst surface during the anodic forward scan. For instance, PtRh with 3D porous nanostructures synthesized by controlled aggregation of nanoparticles in oleylamine by using a modified template-free self-assembly method. The atomic composition of PtRh is tuned by changing the concentration of initial Pt and Rh precursor resulting in significant increment in the catalytic activity and durability for MOR relative to PtRh nanoparticles and commercial Pt E-TEK catalysts. Interestingly the MOR performance depends on the particle composition and higher Rh content does not improve poisoning tolerance. Particularly the anodic peak forward (I_f) to the reverse current (I_b) of PtRh-3 is about 5.31, that is noticeably larger than that of PtRh-1 (I_f/I_b = 1.05) and Pt catalyst (Pt E-TEK) (I_f/I_b = 0.88) with the same loading amount of Pt (Fig. 4.28c) [169].

Interestingly, Yamauchi and coworker synthesized Pt-Pd nanocage with a dendritic shell by selective chemical etching of Pd core through HNO_3. The prepared Pt-Pd nanocages exhibit mass activity 5.2 greater than Pt black and 1, 4 times of PdPt nanodendrites [55]. The stability of Pt-Pd nanocages was also distinctly higher than those of Pt-Pd nanodendrites and Pt black. This is due to the spatial porous nanocages which provide huge active catalytic sites at the inner and outer surface. Recently, we prepared porous PtCu NCs by ultrasonic treatment. The synthesized porous PtCu NCs revealed mass activity 3.9 times higher than that of dendritic PtCu NCs and 10.5 times higher than that of Pt/C (Fig. 4.28d) [128]. The I_f/I_b ratio of porous PtCu NCs (2.43) was clearly greater than those of dendritic PtCu NCs (1.7) and Pt/C (1.21). Furthermore, porous PtCu NCs reserved around 95% of its initial current density after 2,000 durability cycles due to its stable active surface catalytic sites. Also, the Pt-Cu strain effect facilitated methanol oxidation at lower potential and stabilized Pt against oxidation.

Another effective strategy for improving the MOR activity and durability is by synthesis ternary Pt-based NCs [128]. Different from monometallic and bi-metallic catalyst, tri-metallic NCs shows better improvement in the MOR performance due to their ability to facilitate methanol oxidation and subsequently prevent accumulation of carbonious species onto catalyst surfaces. For instance, porous tri-metallic PtNiP nanotube enhances activity toward MOR with better CO tolerance than binary PtNi catalysts with the same shape.

Most recently, we synthesized porous PtPdRu nanodendrites in one-step by reduction of three metal precursor by AA in presence of Pluronic F127 as a structure direct agent which exhibit mass activity of (1.82 mA μg^{-1}) that is 3.0 and 11.4 times higher than those of Pt-Pd nanoflowers (0.61 mA μg^{-1}) and Pt/C (0.16 mA μg^{-1}), respectively with same Pt loading amount (Fig. 4.29a–b) [170]. Moreover, Pt-Pd-Ru nanodendrites reserve around 88% of its initial ECSA after 2000 cycles, compared with Pt/C 54% under the same conditions. This is attributed to the synergy effect among Pt and Pd which enhances adsorption of methanol molecules onto multi corner surface, while strains amongst Pt and Ru give better tolerance for CO poisoning. Tri-metallic PtNiFe dandelion-like prepared by using Ni nanoparticle as seed followed by galvanic replacement of Pt and oxide etching by Fe^{3+} showed high electrocatalytic activity and durability toward MOR that depended on the particle composition. Particularly,

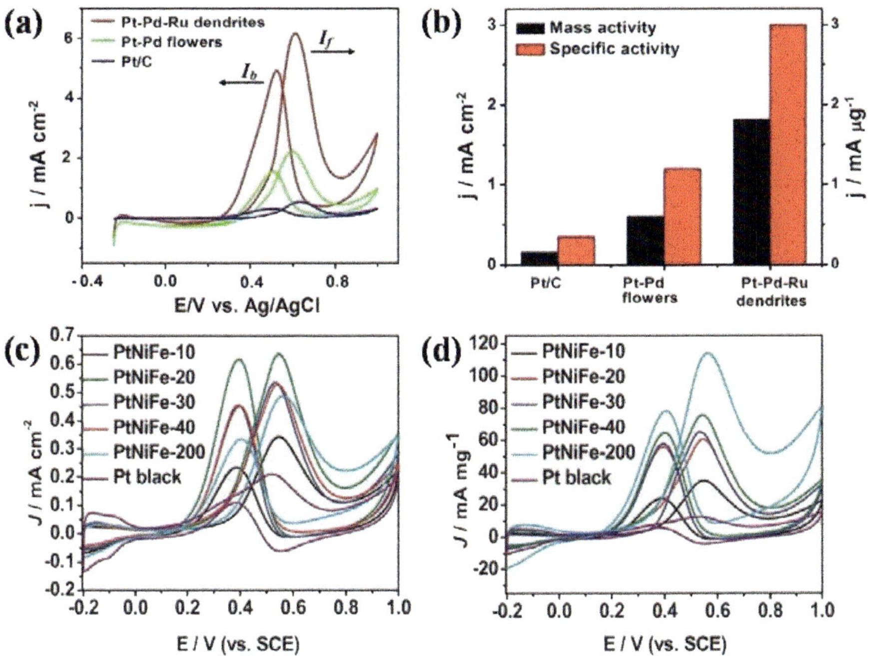

Figure 4.29. (a) CVs of trimetallic Pt-Pd-Ru nanodendrites, bimetallic Pt-Pd nanoflowers and Pt/C tested in 0.1 M $HClO_4$ + 1 M CH_3OH at a scan rate of 50 mV s^{-1}. (b) The comparisons of the mass activities and specific activities of the three materials at 0.6 V. Reprinted with permission from ref. [170]. Copyright 2015 The Royal Society of Chemistry. (c and d) Specific activity and mass activity on PtNiFe-10, -20, -30, -40, -200 and Pt black catalysts measured at 0.55 V in 0.1 M $HClO_4$ + 0.1 M CH_3OH at a scan rate of 50 mV s^{-1}. The initial mole Pt/Fe ratios are represented by 10–200. Reprinted with permission from ref. [171]. Copyright 2013 The Royal Society of Chemistry.

PtNiFe-200 displays the highest mass activity, 8.7 fold greater than Pt black (Fig. 4.29c–d) [171]. The enhancement in the MOR activity resulted from the shape and composition effect. It is expected that porous tri-metallic Pt-based NCs are the best candidates for improving MOR activity and durability. However, the synthesis of porous tri-metallic NCs for MOR is rarely reported. More attention should be devoted for the tailoring synthesis of porous ternary metallic NCs in better understanding their electrocatalytic properties for DMFCs.

3.3 Formic acid oxidation reaction

Recently, Direct Formic Acid Fuel Cells (DFAFCs) attracted much attention as a promising clean power source. Formic acid is a highly promising clean fuel due to its handling flexibility (e.g., transportation and storage), low fuel crossover through the membrane and yield of potential [50, 172]. Mechanistically, the formic acid oxidation (FAOR) takes place on porous Pt-based electrodes proceeds via two pathways namely dehydrogenation and dehydration. In particular the dehydrogenation path includes oxidation to produce CO_2 whereas, the dehydration path involves formation CO_{ads} followed by its subsequent oxidation to CO_2 at high potentials as shown in the following equations [173].

$$HCOOH \rightarrow CO_2 + 2H^+ + 2e^- \text{ dehydrogenation path equation} \quad (6)$$

$$HCOOH \rightarrow CO_{ads} + H_2O \rightarrow CO_2 + 2H^+ + 2e^- \text{ dehydration path equation} \quad (7)$$

Interestingly Osawa et al. revealed that both direct and indirect pathways involve $HCOO_{ads}$ as a reactive intermediate. This is executed by using a systematic Attenuated Total Reflectance-Surface-Enhanced Infrared Absorption Spectroscopy (ATR-SEIRAS) [174]. It is noteworthy that Pt is the most popular electrocatalyst for FAOR although their catalytic activity is lower compared with Pd catalyst. This is due to the fact that Pt is more stable than Pd in the acidic medium. Nevertheless, Pt is prone to be poisoned by CO species and loses its catalytic activity. One of the most efficient techniques to solve this problem is by controlling Pt shape, size and composition. Porous Pt and Pd are prepared with various shapes such as flowers, tubes and cubes show higher electro-oxidation for formic acid [8, 173, 175]. Regarding the composition effect, porous Pt-based NCs in forms of alloy demonstrates significant enhancement in the catalytic activity of FAOR relative to mono Pt NCs. Inspired by this, various binary Pt-based NCs such as Pt-Pd, Pt-Au and Pt Cu are synthesized by different methods for improving the catalytic activity towards FAOR.

For instance, Wang et al. synthesized nanoporous Pt, PtRu, PtIr, PtPd and PtPb by a hydrothermal deposition approach. The results clearly displayed that the catalytic activity of as synthesized binary Pt-based NCs catalysts were superior to mono Pt catalyst towards FAOR (Fig. 4.30a) [176]. These results also warrant that Pt-Pb is superior to Pt-Pd, Pt-Ir, Pt-Ru, nanoporous Pt and polycrystalline Pt electrode. This is attributed to the ability of Pt-Pb to oxidize formic acid directly to CO_2 through the dehydrogenation path which prevented adsorption of CO species. Porous Au-Pd core-shell NCs prepared with different composition by porous Au spheres by an electrodeposition method show higher catalytic activity towards formic acid than Pt black.

Recently, Zheng and coworkers prepared Excavated Rhombic Dodecahedral (ERD) $PtCu_3$ by a co-reduction method in the presence of n-butylamine. The prepared ERD $PtCu_3$ shows mass activity higher than Pt black with oxidation current density about 9.2, 1.7 and 1.46 times higher than those of Pt black and octahedral (OCT), edge-concave octahedral (EC-OCT) and ERD $PtCu_3$ (Fig. 4.30b–c) [177]. This is due to the high accessible surface

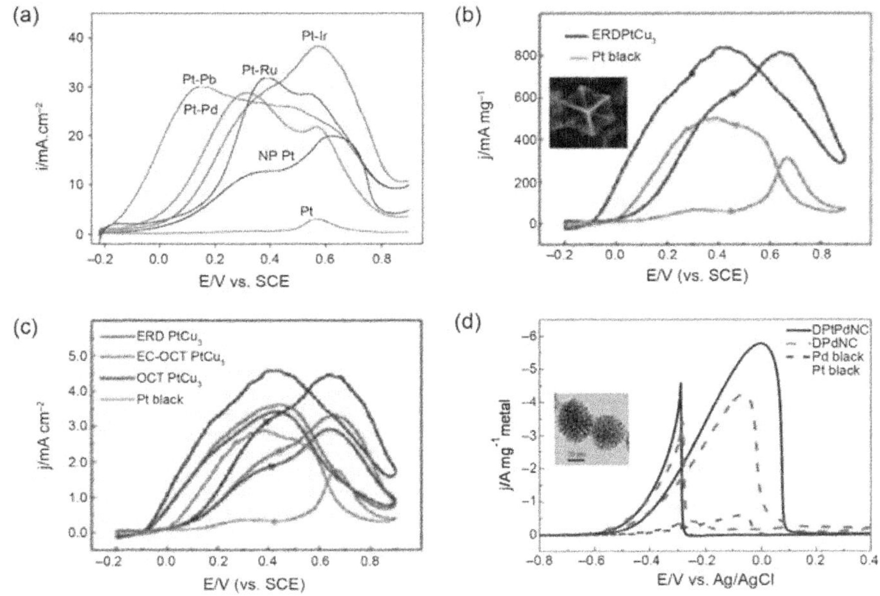

Figure 4.30. (a) CVs of Pt-Ir, Pt-Ru, Pt-Pd, Pt-Pb, nanoporous Pt and bulk polycrystalline Pt measured in 0.1 M HCOOH + 0.1 M H_2SO_4 at a scan rate of 20 mVs^{-1}. Reprinted with permission from ref. [176]. Copyright 2008 Elsevier (b) CVs of ERD $PtCu_3$ NCs and commercial Pt black tested in a N_2-saturated 0.5 M H_2SO_4 + 0.25 M HCOOH solution at a scan rate of 50 mV s^{-1}. (c) CVs of $PtCu_3$ NCs with different shapes relative to commercial Pt black measured in a N_2-saturated 0.5 M H_2SO_4 + 0.25 M HCOOH solution at a scan rate of 50 mV s^{-1}. Reprinted with permission from ref. [177]. Copyright 2014 American Chemical Society. (d) CVs of dandelion-like PtPd nanoclusters (DPtPdNC), dandelion-like Pd nanoclusters DPdNC, Pd black, Pt black catalyst measured in 0.5 M NaOH + 1.0 M ethanol at s scan rate of 50 mV s^{-1}. Reprinted with permission from ref. [182]. Copyright 2015 Elsevier.

area and synergetic effect among Pt and Cu. The I_f/I_b ratio of ERD $PtCu_3$ is higher than EC-OCT, OCT and Pt black implying the better anti-poisoning ability of EDR compared with other catalysts attributed to its excavated structure and large accessible surface area. However, more attention should be fueled to design new porous multi-metallic NCs for FAOR.

3.4 Electrochemical oxidation of ethanol

Ethanol is a highly effective eco-friendly fuel that is oxidized at the anode of Direct Alcoholic Fuel Cells (DAFCs). Different from methanol, ethanol offers many benefits such as low toxicity, higher power yield, ease of transportation/storage and low permeability when it across Proton Exchange Membrane (PEM) [178]. Highly active catalyst is required for the Ethanol Oxidation Reaction (EOR), which is attributed to the structure complexity of ethanol. In particular aside from the molecular formula, ethanol is the smallest alcohol involving a strong sigma C-C bond. This bond is very stable and needs high energy and drastic oxidative catalyst to dissociate [8]. Therefore, it is of great importance to develop an active catalyst able to cleavage the C-C bond and oxidize ethanol to CO_2 under low potentials. Uptil now, Pt-based catalysts are still the most active catalysts and highly promising for substituting the electrochemical oxidation approach for ethanol in the anode of DAFCs. Similar to other electrocatalytic applications porous Pt NCs are superior to their solid counterparts in the EOR [179]. For instance, Webley and coworkers inferred that porous Pt nanowires were more active and durable catalyst than solid Pt nanowore towards EOR with less Pt loading [180]. This means better optimization for the highly expensive Pt metal. The enhancement in the electrocatalytic activity and durability originates from massive accessible active catalytic sites and fast electrons transfer rate.

Recently, Mourdikoudis and coworkers synthesized porous Pt nanodendrites by using dimethylformamide (DMF)-mediated synthesis. The prepared Pt nanodendrites display outstanding catalytic performance and durability in various electrocatalytic applications such as sensing, selectivity and cycle-ability. Similar to the MOR Pt is also prone to poison by CO species during EOR that cause Pt aggregation and loss of its catalytic activity. Alloying Pt with other metals is also an efficient strategy not only to overcome this problem but also to improve Pt catalytic activity because of the electronic, composition and strain affect [179]. Triggered by this various binary porous Pt-based NCs such as PtW, PtPd, PtRu, PtSn and PtIr are prepared and used as self supported catalysts [8]. For example, porous PtPd network with 10 nm size is synthesized by the co-reduction method at room temperature in one step which depict higher catalytic activity than pure Pt and commercial Pt black toward EOR [181]. This is attributed to the synergy effect among Pt and Pd which facilitates the breaking of C-C at lower potential. Most recently, dandelion-like PtPd (DPtPdNC) was fabricated via one-step hydrothermal method in the presence of hexadecylpyridinium chloride monohydrate as a structure directing agent. The synthesized DPtPdNC showed mass activity of 1.2, 14 and 8.8 times higher than that of dandelion Pd, Pt black and Pd black, respectively under the same reaction conditions (Fig. 4.30d) [182]. The durability also is greater than those of the dandelion Pd, Pd black and Pt black catalysts because of the alloy and shape effect. This opens a new frontier to design other multi-metallic Pt-based NCs for improvement their catalytic activity towards EOR.

It should be noted that, the high cost, low abundance and instability of Pt-based catalysts hinder their practical application for Direct Ethanol Fuel Cells (DEFCs). Therefore, it is still a great challenge to develop highly active, durable and inexpensive Pt-free catalysts for EOR [183]. Alternatively, Pd-based catalysts are the most suitable candidate for Pt-catalyst not only because of the low-cost and high abundance of Pd but also for its high activity and lower degree of corrosion relative to Pt catalysis in the alkaline medium [184]. The catalytic activity and durability of Pd-based catalysts depend on the particle size, morphology and composition. Consequently, various porous Pd-based catalysts such as nanowires, raspberry-like, nanoplates, nanoboxes and snowflake-like were synthesized for various electrocatalytic applications [185–189]. For example, Pd aerogels modified by α-, β- or γ-cyclodextrins (PdCD) was synthesized by reduction of Pd precursor by sodium borohydride ($NaBH_4$) in the presence of α-, β- or γ-CD, the PdCD hydrogels [185]. The obtained PdCD aerogels depicted higher catalytic activity and durability than commercial Pd/C towards EOR. This is due to the high surface area and high stability of self-supporting ultrafine Pd network. Furthermore, the results clearly warranted that, Pd aerogels can be enlarged for the industrial scale due to its high catalytic activity in addition to ease of preparation. Porous raspberry-like PdA NCs prepared by galvanic replacement show superior electrocatalytic activity and stability than commercial Pd/C in alkaline medium [188]. This is attributed to the alloying and structure effect, which facilitate the oxidation reaction and inhibit poisoning of the active Pd sites. This mechanism is similar to various reported bi-metallic Pd-based electrocatalysts such as hollow Au@Pd, Au@Pt and PdPt nanosponges [190, 191]. PdCo nanotube prepared by using ZnO nanorod arrays as a template followed by supporting on Carbon-Fiber Cloth (CFC) after removing the template show higher catalytic activity and durability than Pd/CFC nanotube and Pd/C [192]. Additionally, the electrochemical performance of PdCo/CFC nanotube toward EOR can be kept without any significant change at different reaction states such as normal, bending and twisting. Although the great progress in this filed the catalytic activity and durability of Pt-based and Pd-based catalysts cannot replace the commercial Pt/C and Pd/C catalysts. Development of various Pt-free catalysts is greatly required for practical applications of (DEFCs).

4. Mesoporous Carbon Materials

Mesoporous Carbon Materials (MCM) are highly desired in various electrocatalytic applications, such as catalysis, fuel cells and energy storage due to their unique properties such as high accessible surface area, electrical conductivity and thermal stability [193, 194]. Generally, template-based synthesis, including hard- and soft-templating is the main method used for controlling synthesis of MCM [195]. This is usually achieved by the replica approach, including impregnation of the proper carbon source into the template structures followed by carbonization and template removal. The hard-templating method is a time-consuming, high-costly, unsuitable for large-scale synthesis and the hazardous in nature relative to soft-templating [12, 196, 197]. The soft-templating method mainly depends on an organic-organic self-assembly of amphiphilic molecules and a carbon precursor [12, 198, 199]. The polymerization, pyrolysis and carbonization are the main steps in the soft-templating approach. Particularly, the carbon precursor and soft template should co-assemble together into an ordered macrostructure composite. Then formed composite must be stable during polymerization of the carbon source. Meanwhile, the template should decompose during pyrolysis producing a carbon framework preserving the mesostructured morphology during the carbonization.

Wide varieties of MCM with different morphologies were prepared by the template-based synthesis [12, 193–199]. For instance, Ryoo and coworkers synthesized ordered MCM with tunable pore size by using sucrose as a carbon source and mesoporous silicates, including MCM-48, SBA-1 and SBA-15 as hard templates (Fig. 4.31a) [200, 201]. Right after this pioneering work, MCM with different pore size, structure and morphologies were prepared by using different hard templates under different reaction conditions [193–199, 202]. Tang and Xie fabricated MC by using ZnO nanoparticles as a template and polyacrylamide as a carbon source followed by carbonization in an inert atmosphere [203]. Ariga and coworkers synthesized MCM with novel nanocages structure by the replica method in the presence of mesoporous silica materials KIT-5 as a template under different pyrolysis temperature (Fig. 4.31b) [204].

Highly ordered mesoporous carbons (MC) with tunable pores size of 10 nm were prepared by using PEO-*b*-PPO-*b*-PEO as a soft template [198, 205]. The pore size can be increased up to 27 nm by using a pore expander [199]. MCM with 2D hexagonally aligned mesopores was synthesized by solvent Evaporation Induced Self-Assembly (EISA) combined with solvent annealing in presence of PS-*b*-P4VP template and

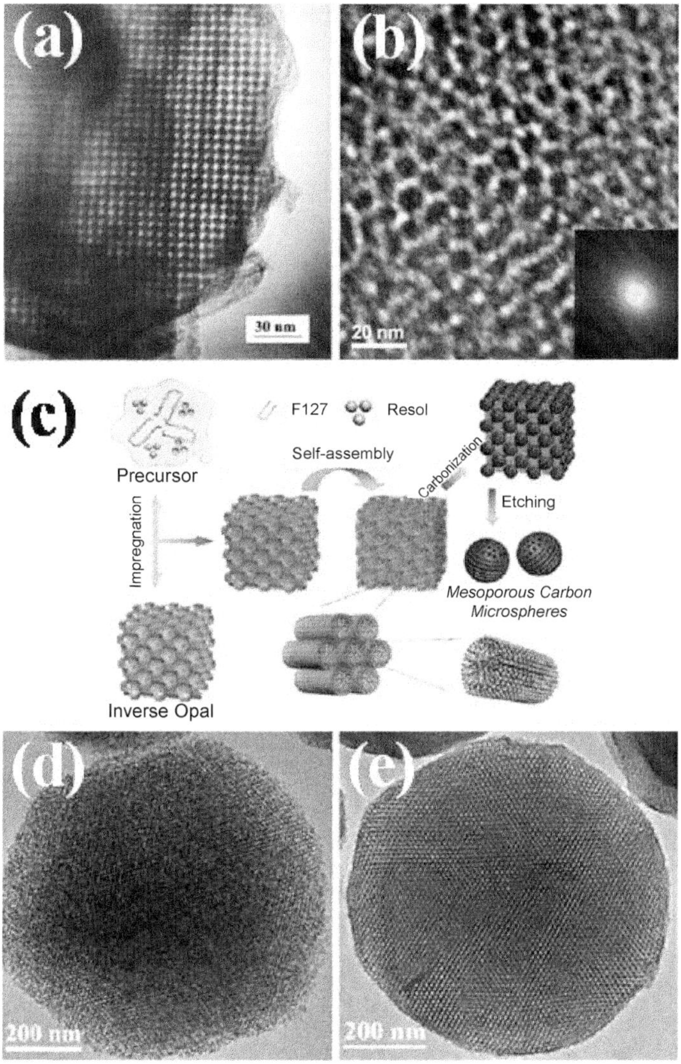

Figure 4.31. (a) TEM image of ordered MC, prepared by carbonization of sucrose in presence of mesoporous silica molecular sieve MCM-48 template and sulfuric acid catalyst followed by removal of silica template by aqueous solution of NaOH and ethanol. Reprinted with permission from ref. [200]. Copyright 1999 American Chemical Society. (b) HRTEM images of MCM nanocage structure prepared by using sucrose as a carbon source, mesoporous silica KIT-5 as a template, and sulfuric acid as a catalyst. The inset reveals its corresponding Fourier transformed image. Reprint with permission from ref. [204]. Copyright 2005 The Royal Society of Chemistry. (c) Scheme shows the preparation process of ordered MC microspheres with variable mesostructures upon self assembly strategy. (d and e) TEM images of hexagonal mesoporous carbon MC-MSs and crushed MC-MSs respectively. Reprinted with permission from ref. [217]. Copyright 2013 American Chemical Society.

Resorcinol-Formaldehyde (RF) carbon precursor [206]. The hydrogen bonding between P4VP segments and the RF facilitates the co-assembly process before polymerization into a highly stable mesostructured composite. Then the PS-*b*-P4VP decomposes selectively after carbonization yielding a mesoporous structure with pore size up to 36.0 nm and wall thickness of 10.0 nm. MC films were prepared by the spin-coating method, followed by thermosetting at 180°C and calcinations at 460°C in the presence of PS-*b*-P4VP template and carbohydrate precursors (turanose, raffinose, glucose, etc.) as carbon sources [207]. Noteworthy, the obtained MCM were disordered due to the low cross-linking degree between carbon sources and template. This result suggested that, highly interaction degree among the carbon source and template is necessary to obtain ordered MCM. Alternatively, a highly ordered MC with pore size of 22.6 nm was obtained by EISA approach in the presence of PEO-*b*-PS and resol [208]. The phenol hydroxyl groups in the resol strongly interact with PEO segments of the PEO-*b*-PS by hydrogen bonding and drive the formation of ordered mesoporous structures. Deng et al. prepared MC with adjustable pore size by the EISA approach in the presence of PEO-*b*-PS template, resol carbon source, homopolystyrene (*h*-PS_{49}) pore expander and tetrahydrofuran (THF) solvent [209]. The pore size was tuned by the changing the ratio of *h*-PS_{49} to PEO_{125}-*b*-PS.

The pore size can also be determined by adjusting the length of the hydrophobic chain in the template. For instance, using PEO-*b*-PS template with different PS chain's lengths, including EO_{125}-*b*-St_{120}, EO_{125}-*b*-St_{230} and EO_{125}-*b*-St_{305} led to the formation of ordered MC with excellent surface area of 1000 $m^2\,g^{-1}$. Interestingly, the obtained products have a fcc mesostructured shape with pore diameter of 12–33 nm and pore wall thickness of 5–11 nm [210]. Controlling the pore size and pore wall thickness of MCM are important for functionalizing the pore wall with various metals and metal oxides as well as formation of graphitized framework materials. This can be achieved by modifying the chemical structure of the soft template with a specific surfactant or polymer. For example, ordered MC with fcc structure, pore size of 10.5 nm and thick pore wall of 12 nm was obtained by using PEO-*b*-PMMA and resol [211]. This is attributed to both hydrophilic and hydrophobic nature of the PEO-b-PMMA templates, which drive the formation of MC with thick pore walls. Inspired by this idea, ABC types triblock copolymers PEO-*b*-PMMA-*b*-PS templates were designed and successfully used for fabrication of MC with adjustable wall thickness [212]. Particularly, MC with pore diameter of 20.0 nm and wall thickness range from 10 to 19 nm was produced by adjusting the resol/template ratio. The variation in the wall thickness was ascribed to the gradient hydrophilicity of PEO-*b*-PMMA-*b*-PS templates. Particularly, resol molecules bind strongly with the PEO chains and partially associate with the PMMA segments.

Another efficient way to obtain MC with tunable wall thickness is through controlling the pore diameter and wall thickness of the mother template by hiring proper reaction parameters and conditions such as synthesis temperature 356. The prepared SBA-15 was used as a template to control synthesis of MC and MC nanopipes with adjustable wall thickness and pore size [213, 214]. Yamauchi and co-workers prepared MC with different content of fullerene cage by using a fullerenol-based precursor solution [215]. The pore wall thickness of MC was controlled by adjusting the filling rate of mesoporous silica template with the carbon sucrose [216]. In particular, increasing the weight ratio of sucrose to silica led to increase in the pore size and pores wall thicknesses. Zhao and coworkers fabricated various morphologies of ordered MCM with controllable pore size and wall thickness by sol-gel and self-assembly methods [208–212]. Interestingly, they prepared highly ordered MC microspheres by using 3-D-ordered macroporous silica and resol in the presence of pluronic F127 (Fig. 4.31c–e) [217]. The formed MC have variable symmetry (hexagonal *p6*mm or cubic *Im3m*) of mesostructures, highly accessible surface areas (500–1100 m^2g^{-1}) and pore size (7–10.3 nm). The results clearly demonstrated that, the morphology of MC can be tuned by using various shapes of mesoporous silica templates.

MCM are usually used as a supporter for Pt-based catalyst in the fuel cell applications due to their superior surface area and thermal stability. Interestingly MC doped with various traces such as N, S and Fe can be used as efficient catalysts instead of Pt [218]. For example, Xu and coworkers fabricated MC doped with Fe and N which exhibit superior catalytic activity towards ORR relative to Pt/C in an alkaline medium. In fact the amount of Fe traces had an unambiguous effect on the catalytic activity [219]. MC doped with nitrogen (MC-N) is among the most intriguing catalysts for various electrocatalytic reactions such as ORR, MOR and EOR. N enhances adsorption rate of oxygen and facilitates oxygen reduction on the carbon atoms next to N heteroatom or on the pyridinic-N species [220]. There are two main methods used for preparing MC-N the first one is *in situ* doping, including direct integration of N into the carbon matrix during the preparation process [221, 222]. This is usually achieved through physical methods such as low temperature CVD, high-temperature arc-discharge and laser ablation in addition to chemical methods such as solvothermal and sol-gel [223–225]. These methods were successfully used for preparing various MC-N such as nanotubes, nanofibers and nanospheres. Furthermore the chemical methods are preferred more than physical methods not only for fabrication MC-N with high N content but also for controlling the doping sites [225, 226].

The second method used for fabrication of MC-N is by treatment of prepared MC with N-containing gas such as N_2, NH_3 and CH_3CN or by harsh oxidation in an HNO_3 under refluxing. This method is favored for production of MC-N with specific N-containing groups [227]. Nevertheless, due to the high temperature (900–1200°C) used during the synthesis process, the obtained MC-N products suffer from inevitable morphological defects and structural degradation in addition to formation graphitic-N in the carbon matrix [228]. Therefore, the *in situ* doping methods at a low-temperature are preferred for production MC-N. However, much greater efforts should be dedicated to increase the N content and to control the N-doping sites in the synthesized MC-N. Also, the utilization of these MC-NS as catalysts in the fuel cells applications are not highlighted enough relative to Pt-based NCs catalyst. MC nanocage doped with 10 wt% of N formed by using MgO template and pyridine as a precursor at various growth temperatures ranged from 600 to 900°C, exhibited a drastic electrocatalytic activity towards ORR in the alkaline solution [229]. Antonietti and coworkers fabricated MC-N by carbonization of nucleobases dissolved in ionic liquid in the presence of silica template (Fig. 4.32a) [230]. The prepared materials show excellent accessible surface area up to 1500 $m^2 g^{-1}$ and high N content up to 12 wt%. Furthermore, the obtained mesoporous carbon revealed higher catalytic activity toward ORR with over potential of 197 V, than Pt/C in the alkaline mediums with high tolerance to methanol (Fig. 4.32b–c). Recently, Liang and coworkers developed the vaporization-capillary condensation method for preparing ordered MC-N (Fig. 4.32d) [231]. This method yielded a highly ordered MC-N with N content (2.79–11.43 wt%) and high surface area of 661 $m^2 g^{-1}$ which exhibited superior catalytic activity towards ORR relative to Pt catalyst (Fig. 4.32e). Interestingly the catalytic activity was correlated to the N-activated carbon atoms (Fig. 4.32f). Also, the prepared MC-N depicted a highly catalytic activity towards methanol fuel cell test (Fig. 4.32g).

5. Conclusion and Future Outlook

Mesoporous structured materials with their unique structures and catalytic properties enable their effective use in fuel cell applications. Great efforts have been devoted to develop synthetic methods for mesoporous electrocatalysts with controlled pore sizes, morphologies and compositions in which dealloying, templated routes and self-assemble approaches are the general routes. Based on these methods, various mesoporous electrocatalysts are prepared for electrode reactions in fuel cells. Fuel cells are very important clean energy sources. Reducing their high cost at the industrial scale still remains a great challenge. Control over mesoporous structures and compositions are very important for designing new electrocatalysts with enhanced performance (e.g., activity and durability), which will greatly facilitate the commercialization of fuel cells.

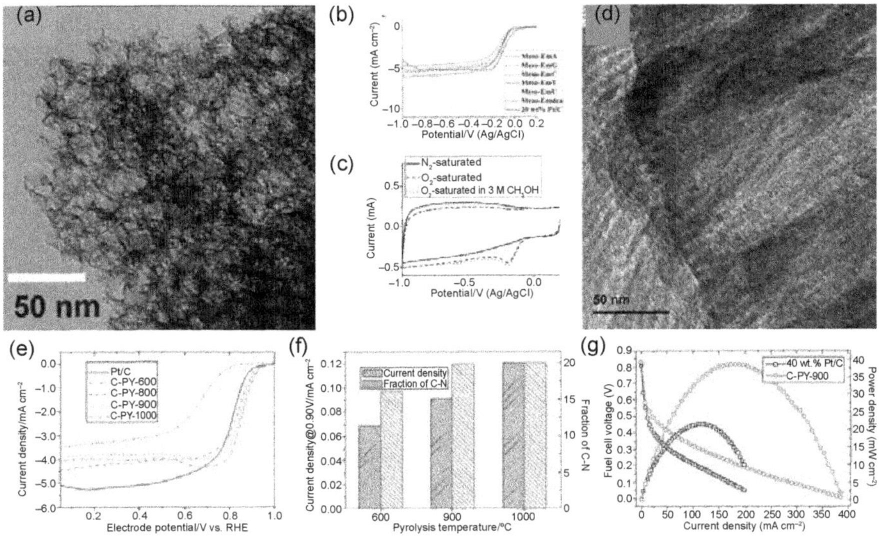

Figure 4.32. (a) TEM image of MC-N synthesized by carbonization of nucleobases dissolved in ionic liquid (1-ethyl-3-methylimidazolium dicyanamide) in presence of silica nanoparticles template. (b) ORR polarization curves for MC-N (meso-Em, meso-EmA, meso-EmG, mesoEmC, meso-EmT and meso-EmU prepared from adenine, guanine, cytosine, thymine and uracil solutions respectively) compared with Pt/C measured in O_2-saturated 0.1 M KOH at a scan rate of 10 mV s^{-1} and rotation speed of 1600 rpm. (c) CVs of meso-EmG measured in N_2-saturated 0.1 M KOH, O_2-saturated 0.1 M KOH, and O_2-saturated 0.1 M KOH and 3 M CH_3OH solutions. Reprinted with permission from ref. [230]. Copyright 2011 American Chemical Society. (d) TEM images of MC-N namely C-PY-600 formed by vaporization-capillary condensation method. (e) ORR polarization curves of the nitrogen-doped mesoporous carbon catalyst prepared at different temperatures (C-PY-X is for polypyrrole while X refers to the pyrolysis temperature 600, 800, 900 and 1000°C. (f) The fraction of the activated carbon-nitrogen (C-N) and the current density at 0.90 V of different catalysts. (g) Direct methanol fuel cell test of C-PY-900 relative to Pt/C catalysts. Reprinted with permission from ref. [231]. Copyright 2015 Elsevier.

Acknowledgement

This work is supported by the National Natural Science Foundation of China (No. 21273218 and 21601154).

References

[1] Zhang, J. and Li, C.M. 2012. Nanoporous metals: fabrication strategies and advanced electrochemical applications in catalysis, sensing and energy systems. *Chem. Soc. Rev.*, 41: 7016–7031.

[2] Yamauchi, Y. and Kuroda, K. 2008. Rational design of mesoporous metals and related nanomaterials by a soft-template approach. *Chem, Asian J.*, 3: 664–676.

[3] Liang, J., Du, X., Gibson, C. et al. 2013. N-Doped graphene natively grown on hierarchical ordered porous carbon for enhanced oxygen reduction. *Adv. Mater.*, 25: 6226–6231.
[4] Kränzlin, N. and Niederberger, M. 2015. Controlled fabrication of porous metals from the nanometer to the macroscopic scale. *Mater. Horiz.*, 2: 359–377.
[5] Bae, J.H., Han, J.-H. and Chung, T.D. 2012. Electrochemistry at nanoporous interfaces: new opportunity for electrocatalysis. *Phys. Chem. Chem. Phys.*, 14: 448–463.
[6] Lang, X., Hirata, A., Fujita, T. et al. 2011. Nanoporous metal/oxide hybrid electrodes for electrochemical supercapacitors. *Nat. Nanotechnol.*, 6: 232–236.
[7] Zheng, X.T. and Li, C.M. 2012. Single cell analysis at the nanoscale. *Chem. Soc. Rev.*, 41: 2061–2071.
[8] Xu, Y. and Zhang, B. 2014. Recent advances in porous Pt-based nanostructures: synthesis and electrochemical applications. *Chem. Soc. Rev.*, 43: 2439–2450.
[9] Kloke, A., von Stetten, F., Zengerle, R. et al. 2011. Strategies for the fabrication of porous platinum electrodes. *Adv. Mater.*, 23: 4976–5008.
[10] Erlebacher, J., Aziz, M.J., Karma, A. et al. 2001. Evolution of nanoporosity in dealloying. *Nature*, 410: 450–453.
[11] Liu, Y., Goebl, J. and Yin, Y. 2013. Templated synthesis of nanostructured materials. *Chem. Soc. Rev.*, 42: 2610–2653.
[12] Deng, Y., Wei, J., Sun, Z. et al. 2013. Large-pore ordered mesoporous materials templated from non-Pluronic amphiphilic block copolymers. *Chem. Soc. Rev.*, 42: 4054–4070.
[13] Velev, O., Tessier, P., Lenhoff, A. et al. 1999. Materials: a class of porous metallic nanostructures. *Nature*, 401: 548.
[14] Zhu, C., Du, D., Eychmüller, A. et al. 2015. Engineering ordered and nonordered porous noble metal nanostructures: synthesis, assembly, and their applications in electrochemistry. *Chem. Rev.*, 115: 8896–8943.
[15] Raney, M. 1927. Method of producing finely-divided nickel. *US Pat.*, 1: 628–190.
[16] Swann, P. and Pickering, H. 1963. Implications of the stress aging yield phenomenon with regard to stress corrosion cracking. *Corrosion*, 19: 369t–372t.
[17] Forty, A. 1979. Corrosion micromorphology of noble metal alloys and depletion gilding. *Nature*, 282: 597–598.
[18] Erlebacher, J. and Seshadri, R. 2009. Hard materials with tunable porosity. *Mrs Bulletin*, 34: 561–568.
[19] Wittstock, A., Biener, J. and Bäumer, M. 2010. Nanoporous gold: a new material for catalytic and sensor applications. *Phys. Chem. Chem. Phys.*, 12: 12919–12930.
[20] Ding, Y. and Chen, M. 2009. Nanoporous metals for catalytic and optical applications. *Mrs Bulletin*, 34: 569–576.
[21] Kertis, F., Snyder, J., Govada, L. et al. 2010. Structure/processing relationships in the fabrication of nanoporous gold. *JOM*, 62: 50–56.
[22] Huang, J.F. and Sun, I.W. 2005. Fabrication and surface functionalization of nanoporous gold by electrochemical alloying/dealloying of Au-Zn in an ionic liquid, and the self-assembly of L-Cysteine monolayers. *Adv. Funct. Mater.*, 15: 989–994.
[23] Parida, S., Kramer, D., Volkert, C. et al. 2006. Volume change during the formation of nanoporous gold by dealloying. *Phys. Rev. Lett.*, 97: 035504–035508.
[24] Kraehnert, R., Ortel, E., Paul, B. et al. 2015. Electrochemically dealloyed platinum with hierarchical pore structure as highly active catalytic coating. *Catal. Sci. Technol.*, 5: 206–216.
[25] Wang, D., Yu, Y., Zhu, J. et al. 2015. Morphology and activity tuning of Cu_3Pt/C ordered intermetallic nanoparticles by selective electrochemical dealloying. *Nano Lett.*, 15: 1343–1348.
[26] Erlebacher, J. 2004. An atomistic description of dealloying porosity evolution, the critical potential, and rate-limiting behavior. *J. Electrochem. Soc.*, 151: C614–C626.

[27] Stratmann, M. and Rohwerder, M. 2001. Materials science: A pore view of corrosion. *Nature*, 410: 420–423.

[28] Pareek, A., Borodin, S., Bashir, A. et al. 2011. Initiation and inhibition of dealloying of single crystalline Cu_3Au (111) surfaces. *J. Am. Chem. Soc.*, 133: 18264–18271.

[29] Koh, S. and Strasser, P. 2007. Electrocatalysis on bimetallic surfaces: modifying catalytic reactivity for oxygen reduction by voltammetric surface dealloying. *J. Am. Chem. Soc.*, 129: 12624–12625.

[30] Snyder, J., Asanithi, P., Dalton, A.B. et al. 2008. Stabilized nanoporous metals by dealloying ternary alloy precursors. *Adv. Mater.*, 20: 4883–4886.

[31] Yu, J., Ding, Y., Xu, C. et al. 2008. Nanoporous metals by dealloying multicomponent metallic glasses. *Chem. Mater.*, 20: 4548–4550.

[32] Yeh, F.-H., Tai, C.-C., Huang, J.-F. et al. 2006. Formation of porous silver by electrochemical alloying/dealloying in a water-insensitive zinc chloride-1-ethyl-3-methyl imidazolium chloride ionic liquid. *J. Phys. Chem. B*, 110: 5215–5222.

[33] Li, C., Malgras, V., Aldalbahi, A. et al. 2015. Dealloying of mesoporous PtCu alloy film for the synthesis of mesoporous Pt films with high electrocatalytic activity. *Chem. Asian J.*, 10: 316–320.

[34] Rudi, S., Gan, L., Cui, C. et al. 2015. Electrochemical dealloying of bimetallic ORR nanoparticle catalysts at constant electrode potentials. *J. Electrochem. Soc.*, 162: F403–F409.

[35] Chen, A., Wang, J., Wang, Y. et al. 2015. Effects of pore size and residual Ag on electrocatalytic properties of nanoporous gold films prepared by pulse electrochemical dealloying. *Electrochim. Acta*, 153: 552–558.

[36] Xu, C., Wang, R., Chen, M. et al. 2010. Dealloying to nanoporous Au/Pt alloys and their structure sensitive electrocatalytic properties. *Phys. Chem. Chem. Phys.*, 12: 239–246.

[37] Liu, Z., Koh, S., Yu, C. et al. 2007. Synthesis, dealloying, and ORR electrocatalysis of PDDA-stabilized Cu-rich Pt alloy nanoparticles. *J. Electrochem. Soc.*, 154: B1192–B1199.

[38] Luo, X., Li, R., Huang, L. et al. 2013. Nucleation and growth of nanoporous copper ligaments during electrochemical dealloying of Mg-based metallic glasses. *Corros. Sci.*, 67: 100–108.

[39] Sun, L., Chien, C.-L. and Searson, P.C. 2004. Fabrication of nanoporous nickel by electrochemical dealloying. *Chem. Mater.*, 16: 3125–3129.

[40] Oezaslan, M., Heggen, M. and Strasser, P. 2011. Size-dependent morphology of dealloyed bimetallic catalysts: linking the nano to the macro scale. *J. Am. Chem. Soc.*, 134: 514–524.

[41] Shui, J.l., Chen, C. and Li, J. 2011. Evolution of nanoporous Pt-Fe alloy nanowires by dealloying and their catalytic property for oxygen reduction reaction. *Adv. Funct. Mater.*, 21: 3357–3362.

[42] Li, X., Chen, Q., McCue, I. et al. 2014. Dealloying of noble-metal alloy nanoparticles. *Nano Lett.*, 14: 2569–2577.

[43] Liu, W., Chen, L., Yan, J. et al. 2015. Dealloying solution dependence of fabrication, microstructure and porosity of hierarchical structured nanoporous copper ribbons. *Corros. Sci.*, 94: 114–121.

[44] Zhao, C., Qi, Z., Wang, X. et al. 2009. Fabrication and characterization of monolithic nanoporous copper through chemical dealloying of Mg-Cu alloys. *Corros. Sci.*, 51: 2120–2125.

[45] Lechtman, H. 1984. Andean value systems and the development of prehistoric metallurgy. *Technology and Culture*, 25: 1–36.

[46] Zhang, Z., Wang, Y., Qi, Z. et al. 2009. Generalized fabrication of nanoporous metals (Au, Pd, Pt, Ag, and Cu) through chemical dealloying. *J. Phys. Chem. C*, 113: 12629–12636.

[47] Zhang, Z., Wang, Y., Qi, Z. et al. 2009. Nanoporous gold ribbons with bimodal channel size distributions by chemical dealloying of Al-Au alloys. *J. Phys. Chem. C*, 113: 1308–1314.
[48] Wang, X., Qi, Z., Zhao, C. et al. 2009. Influence of alloy composition and dealloying solution on the formation and microstructure of monolithic nanoporous silver through chemical dealloying of Al-Ag Alloys. *J. Phys. Chem. C*, 113: 13139–13150.
[49] Qi, Z., Zhao, C., Wang, X. et al. 2009. Formation and characterization of monolithic nanoporous copper by chemical dealloying of Al-Cu alloys. *J. Phys. Chem. C*, 113: 6694–6698.
[50] Rahman, M.A., Zhu, X. and Wen, C. 2015. Fabrication of nanoporous Ni by chemical dealloying Al from Ni-Al alloys for lithium-ion batteries. *Int. J. Electrochem. Sci.*, 10: 3767–3783.
[51] Qiu, H.-J., Peng, L., Li, X. et al. 2015. Using corrosion to fabricate various nanoporous metal structures. *Corros. Sci.*, 92: 16–31.
[52] Wu, T., Wang, X., Huang, J. et al. 2015. Characterization and functional applications of nanoporous Ag foams prepared by chemical dealloying. *Metall. Mater. Trans. B*, 46: 2296–2304.
[53] Qiu, H.-J., Xu, H., Li, X. et al. 2015. Core-shell-structured nanoporous PtCu with high Cu content and enhanced catalytic performance. *J. Mater. Chem. A*, 3: 7939–7944.
[54] Ding, Y. and Erlebacher, J. 2003. Nanoporous metals with controlled multimodal pore size distribution. *J. Am. Chem. Soc.*, 125: 7772–7773.
[55] Wang, L. and Yamauchi, Y. 2013. Metallic nanocages: synthesis of bimetallic Pt-Pd hollow nanoparticles with dendritic shells by selective chemical etching. *J. Am. Chem. Soc.*, 135: 16762–16765.
[56] Qiu, H.-J., Shen, X., Wang, J. et al. 2015. Aligned nanoporous Pt-Cu bimetallic microwires with high catalytic activity towards methanol electro-oxidation. *ACS Catalysis*, 5: 3779–3785.
[57] Wang, R., Xu, C., Bi, X. et al. 2012. Nanoporous surface alloys as highly active and durable oxygen reduction reaction electrocatalysts. *Energy Environ. Sci.*, 5: 5281–5286.
[58] Fujita, T., Kanoko, Y., Ito, Y. et al. 2015. Nanoporous metal papers for scalable hierarchical electrode. *Adv. Sci.*, 2: 1500086.
[59] Liu, H., Qu, J., Chen, Y. et al. 2012. Hollow and cage-bell structured nanomaterials of noble metals. *J. Am. Chem. Soc.*, 134: 11602–11610.
[60] White, R., Weber, J.N. and White, E. 1972. Replamineform: a new process for preparing porous ceramic, metal, and polymer prosthetic materials. *Science*, 176: 922–924.
[61] Zhao, X., Su, F., Yan, Q. et al. 2006. Templating methods for preparation of porous structures. *J. Mater. Chem.*, 16: 637–648.
[62] Kulinowski, K.M., Jiang, P., Vaswani, H. et al. 2000. Porous metals from colloidal templates. *Adv. Mater.*, 12: 833–838.
[63] Stein, A., Li, F. and Denny, N.R. 2007. Morphological control in colloidal crystal templating of inverse opals, hierarchical structures, and shaped particles. *Chem. Mater.*, 20: 649–666.
[64] Masuda, H. and Fukuda, K. 1995. Ordered metal nanohole arrays made by a two-step replication of honeycomb structures of anodic alumina. *Science*, 268: 1466–1468.
[65] Braun, P.V. and Wiltzius, P. 1999. Microporous materials: Electrochemically grown photonic crystals. *Nature*, 402: 603–604.
[66] Petkov, N., Platschek, B., Morris, M.A. et al. 2007. Oriented growth of metal and semiconductor nanostructures within aligned mesoporous channels. *Chem. Mater.*, 19: 1376–1381.
[67] Walcarius, A. 2010. Template-directed porous electrodes in electroanalysis. *Anal. Bioanal. Chem.*, 396: 261–272.
[68] Lai, M. and Riley, D.J. 2008. Templated electrosynthesis of nanomaterials and porous structures. *J. Colloid Interface Sci.*, 323: 203–212.

[69] Guliants, V., Carreon, M. and Lin, Y. 2004. Ordered mesoporous and macroporous inorganic films and membranes. *J. Membr. Sci.*, 235: 53–72.

[70] Meldrum, F.C. and Seshadri, R. 2000. Porous gold structures through templating by echinoid skeletal plates. *Chem Commun.*, 29–30.

[71] Lakshmi, B.B., Patrissi, C.J. and Martin, C.R. 1997. Sol-gel template synthesis of semiconductor oxide micro- and nanostructures. *Chem. Mater.*, 9: 2544–2550.

[72] Limmer, S.J., Seraji, S., Wu, Y. et al. 2002. Template-based growth of various oxide nanorods by sol-gel electrophoresis. *Adv. Funct. Mater.*, 12: 59–64.

[73] Martin, B.R., Dermody, D.J., Reiss, B.D. et al. 1999. Orthogonal self-assembly on colloidal gold-platinum nanorods. *Adv. Mater.*, 11: 1021–1025.

[74] Hurst, S.J., Payne, E.K., Qin, L. et al. 2006. Multisegmented one-dimensional nanorods prepared by hard-template synthetic methods. *Angew. Chem. Int. Ed.*, 45: 2672–2692.

[75] Wang, J.-G., Tian, M.-L., Kumar, N. et al. 2005. Controllable template synthesis of superconducting Zn nanowires with different microstructures by electrochemical deposition. *Nano Lett.*, 5: 1247–1253.

[76] Klein, J.D., Herrick, R.D., Palmer, D. et al. 1993. Electrochemical fabrication of cadmium chalcogenide microdiode arrays. *Chem. Mater.*, 5: 902–904.

[77] Wang, W., Li, N., Li, X. et al. 2006. Synthesis of metallic nanotube arrays in porous anodic aluminum oxide template through electroless deposition. *Mater. Res. Bull.*, 41: 1417–1423.

[78] Liu, L., Pippel, E., Scholz, R. et al. 2009. Nanoporous Pt-Co alloy nanowires: fabrication, characterization, and electrocatalytic properties. *Nano Lett.*, 9: 4352–4358.

[79] Liang, Z., Susha, A.S., Yu, A. et al. 2003. Nanotubes prepared by layer-by-layer coating of porous membrane templates. *Adv. Mater.*, 15: 1849–1853.

[80] Hou, S., Wang, J. and Martin, C.R. 2005. Template-synthesized protein nanotubes. *Nano Lett.*, 5: 231–234.

[81] Hou, S., Wang, J. and Martin, C.R. 2005. Template-synthesized DNA nanotubes. *J. Am. Chem. Soc.*, 127: 8586–8587.

[82] Qin, Y., Pan, A., Liu, L. et al. 2011. Atomic layer deposition assisted template approach for electrochemical synthesis of Au crescent-shaped half-nanotubes. *ACS Nano.*, 5: 788–794.

[83] Dickey, M.D., Weiss, E.A., Smythe, E.J. et al. 2008. Fabrication of arrays of metal and metal oxide nanotubes by shadow evaporation. *ACS Nano.*, 2: 800–808.

[84] Shimizu, T., Xie, T., Nishikawa, J. et al. 2007. Synthesis of vertical high-density epitaxial Si (100) nanowire arrays on a Si (100) substrate using an anodic aluminum oxide template. *Adv. Mater.*, 19: 917–920.

[85] Bartlett, P.N., Birkin, P.R. and Ghanem, M.A. 2000. Electrochemical deposition of macroporous platinum, palladium and cobalt films using polystyrene latex sphere templates. *Chem. Commun.*, 1671–1672.

[86] Bartlett, P., Baumberg, J., Birkin, P.R. et al. 2002. Highly ordered macroporous gold and platinum films formed by electrochemical deposition through templates assembled from submicron diameter monodisperse polystyrene spheres. *Chem. Mater.*, 14: 2199–2208.

[87] Jiang, P., Cizeron, J., Bertone, J.F. et al. 1999. Preparation of macroporous metal films from colloidal crystals. *J. Am. Chem. Soc.*, 121: 7957–7958.

[88] Shin, H.J., Ryoo, R., Liu, Z. et al. 2001. Template synthesis of asymmetrically mesostructured platinum networks. *J. Am. Chem. Soc.*, 123: 1246–1247.

[89] Shi, Y., Hua, C., Li, B. et al. 2013. Highly ordered mesoporous crystalline $MoSe_2$ material with efficient visible-light-driven photocatalytic activity and enhanced lithium storage performance. *Adv. Funct. Mater.*, 23: 1832–1838.

[90] Wang, H., Jeong, H.Y., Imura, M. et al. 2011. Shape- and size-controlled synthesis in hard templates: sophisticated chemical reduction for mesoporous monocrystalline platinum nanoparticles. *J. Am. Chem. Soc.*, 133: 14526–14529.

[91] Wang, D., Luo, H., Kou, R. et al. 2004. A general route to macroscopic hierarchical 3D nanowire networks. *Angew. Chem. Int. Ed.*, 43: 6169–6173.
[92] Sun, Y., Mayers, B. and Xia, Y. 2003. Metal nanostructures with hollow interiors. *Adv. Mater.*, 15: 641–646.
[93] González, E., Arbiol, J. and Puntes, V.F. 2011. Carving at the nanoscale: Sequential galvanic exchange and Kirkendall growth at room temperature. *Science*, 334: 1377–1380.
[94] Oh, M.H., Yu, T., Yu, S.-H. et al. 2013. Galvanic replacement reactions in metal oxide nanocrystals. *Science*, 340: 964–968.
[95] Chen, C., Kang, Y., Huo, Z. et al. 2014. Highly crystalline multimetallic nanoframes with three-dimensional electrocatalytic surfaces. *Science*, 343: 1339–1343.
[96] Yamauchi, Y., Suzuki, N., Radhakrishnan, L. et al. 2009. Breakthrough and future: nanoscale controls of compositions, morphologies, and mesochannel orientations toward advanced mesoporous materials. *Chem. Rec.*, 9: 321–339.
[97] Attard, G.S., Bartlett, P.N., Coleman, N.R. et al. 1997. Mesoporous platinum films from lyotropic liquid crystalline phases. *Science*, 278: 838–840.
[98] Wang, H., Wang, L., Sato, T. et al. 2012. Synthesis of mesoporous Pt films with tunable pore sizes from aqueous surfactant solutions. *Chem. Mater.*, 24: 1591–1598.
[99] Bartlett, P. and Marwan, J. 2003. Electrochemical deposition of nanostructured (H1-e) layers of two metals in which pores within the two layers interconnect. *Chem. Mater.*, 15: 2962–2968.
[100] Yamauchi, Y., Tonegawa, A., Komatsu, M. et al. 2012. Electrochemical synthesis of mesoporous Pt-Au binary alloys with tunable compositions for enhancement of electrochemical performance. *J. Am. Chem. Soc.*, 134: 5100–5109.
[101] Wang, H., Ishihara, S., Ariga, K. et al. 2012. All-metal layer-by-layer films: bimetallic alternate layers with accessible mesopores for enhanced electrocatalysis. *J. Am. Chem. Soc.*, 134: 10819–10821.
[102] Attard, G.S., Corker, J.M., Göltner, C.G. et al. 1997. Liquid-crystal templates for nanostructured metals. *Angew. Chem. Int. Ed.*, 36: 1315–1317.
[103] Yamauchi, Y., Momma, T., Fuziwara, M. et al. 2005. Unique microstructure of mesoporous Pt (HI-Pt) prepared via direct physical casting in lyotropic liquid crystalline media. *Chem. Mater.*, 17: 6342–6348.
[104] Yamauchi, Y., Ohsuna, T. and Kuroda, K. 2007. Synthesis and structural characterization of a highly ordered mesoporous Pt-Ru alloy via "evaporation-mediated direct templating". *Chem. Mater.*, 19: 1335–1342.
[105] Takai, A., Saida, T., Sugimoto, W. et al. 2009. Preparation of mesoporous Pt-Ru alloy fibers with tunable compositions via evaporation-mediated direct templating (EDIT) method utilizing porous anodic alumina membranes. *Chem. Mater.*, 21: 3414–3423.
[106] Yamauchi, Y., Komatsu, M., Fuziwara, M. et al. 2009. Ferromagnetic mesostructured alloys: design of ordered mesostructured alloys with multicomponent metals from lyotropic liquid crystals. *Angew. Chem. Int. Ed.*, 48: 7792–7797.
[107] Yamauchi, Y., Yokoshima, T., Momma, T. et al. 2004. Fabrication of magnetic mesostructured nickel-cobalt alloys from lyotropic liquid crystalline media by electroless deposition. *J. Mater. Chem.*, 14: 2935–2940.
[108] Hsueh, H.Y., Huang, Y.C., Ho, R.M. et al. 2011. Nanoporous gyroid nickel from block copolymer templates via electroless plating. *Adv. Mater.*, 23: 3041–3046.
[109] Xu, Y., Yuan, Y., Ma, A. et al. 2012. Composition-tunable Pt-Co alloy nanoparticle networks: facile room-temperature synthesis and supportless electrocatalytic applications. *ChemPhysChem.*, 13: 2601–2609.
[110] Shin, H.-C., Dong, J. and Liu, M. 2003. Nanoporous structures prepared by an electrochemical deposition process. *Adv. Mater.*, 15: 1610–1614.
[111] Song, Y., Yang, Y., Medforth, C.J. et al. 2004. Controlled synthesis of 2-D and 3-D dendritic platinum nanostructures. *J. Am. Chem. Soc.*, 126: 635–645.

[112] Yamauchi, Y., Takai, A., Nagaura, T. et al. 2008. Pt fibers with stacked donut-like mesospace by assembling Pt nanoparticles: guided deposition in physically confined self-assembly of surfactants. *J. Am. Chem. Soc.*, 130: 5426–5427.

[113] Yang, J., Shen, D., Zhou, L. et al. 2013. Spatially confined fabrication of core-shell gold nanocages@mesoporous silica for near-infrared controlled photothermal drug release. *Chem. Mater.*, 25: 3030–3037.

[114] Yang, P., Zhao, D., Margolese, D.I. et al. 1998. Generalized syntheses of large-pore mesoporous metal oxides with semicrystalline frameworks. *Nature*, 396: 152–155.

[115] Blin, J.L., Léonard, A., Yuan, Z.Y. et al. 2003. Hierarchically mesoporous/macroporous metal oxides templated from polyethylene oxide surfactant assemblies. *Angew. Chem. Int. Ed.*, 42: 2872–2875.

[116] Huang, L. and Kruk, M. 2015. Versatile surfactant/swelling-agent template for synthesis of large-pore ordered mesoporous silicas and related hollow nanoparticles. *Chem. Mater.*, 27: 679–689.

[117] Song, Y., Garcia, R.M., Dorin, R.M. et al. 2007. Synthesis of platinum nanowire networks using a soft template. *Nano Lett.*, 7: 3650–3655.

[118] Wang, L. and Yamauchi, Y. 2009. Block copolymer mediated synthesis of dendritic platinum nanoparticles. *J. Am. Chem. Soc.*, 131: 9152–9153.

[119] Wang, L., Wang, H., Nemoto, Y. et al. 2010. Rapid and efficient synthesis of platinum nanodendrites with high surface area by chemical reduction with formic acid. *Chem. Mater.*, 22: 2835–2841.

[120] Huang, X., Li, Y., Chen, Y. et al. 2013. Palladium-based nanostructures with highly porous features and perpendicular pore channels as enhanced organic catalysts. *Angew. Chem. Int. Ed.*, 125: 2520–2524.

[121] Sun, L., Wang, H., Eid, K. et al. 2016. Shape-controlled synthesis of porous AuPt nanoparticles and their superior electrocatalytic activity for oxygen reduction reaction. *Sci. Tech. Adv. Mater.*, 17: 58–62.

[122] Wang, L., Nemoto, Y. and Yamauchi, Y. 2011. Direct synthesis of spatially-controlled Pt-on-Pd bimetallic nanodendrites with superior electrocatalytic activity. *J. Am. Chem. Soc.*, 133: 9674–9677.

[123] Ataee-Esfahani, H., Wang, L. and Yamauchi, Y. 2010. Block copolymer assisted synthesis of bimetallic colloids with Au core and nanodendritic Pt shell. *Chem. Commun.*, 46: 3684–3686.

[124] Zhang, J., Ma, J., Wan, Y. et al. 2012. Dendritic Pt-Cu bimetallic nanocrystals with a high electrocatalytic activity toward methanol oxidation. *Mater. Chem. Phys.*, 132: 244–247.

[125] Wang, L. and Yamauchi, Y. 2011. Strategic synthesis of trimetallic Au@Pd@Pt core-shell nanoparticles from poly(vinylpyrrolidone)-based aqueous solution toward highly active electrocatalysts. *Chem. Mater.*, 23: 2457–2465.

[126] Eid, K., Wang, H., Malgras, V. et al. 2015. Trimetallic PtPdRu dendritic nanocages with three-dimensional electrocatalytic surfaces. *J. Phys. Chem. C*, 119: 19947–19953.

[127] Lim, B., Jiang, M., Camargo, P.H. et al. 2009. Pd-Pt bimetallic nanodendrites with high activity for oxygen reduction. *Science*, 324: 1302–1305.

[128] Eid, K., Wang, H., He, P. et al. 2015. One-step synthesis of porous bimetallic PtCu nanocrystals with high electrocatalytic activity for methanol oxidation reaction. *Nanoscale*, 7: 16860–16866.

[129] Warren, S.C., Messina, L.C., Slaughter, L.S. et al. 2008. Ordered mesoporous materials from metal nanoparticle-block copolymer self-assembly. *Science*, 320: 1748–1752.

[130] Philp, D. and Stoddart, J.F. 1996. Self-assembly in natural and unnatural systems. *Angew. Chem. Int. Ed.*, 35: 1154–1196.

[131] Whitesides, G.M. and Grzybowski, B. 2002. Self-assembly at all scales. *Science*, 295: 2418–2421.

[132] Nie, Z., Petukhova, A. and Kumacheva, E. 2010. Properties and emerging applications of self-assembled structures made from inorganic nanoparticles. *Nat. Nanotechnol.*, 5: 15–25.
[133] Li, Y., Wang, J., Deng, Z. et al. 2001. Bismuth nanotubes: a rational low-temperature synthetic route. *J. Am. Chem. Soc.*, 123: 9904–9905.
[134] Ying, J., Yang, X.-Y., Tian, G. et al. 2014. Self-assembly: an option to nanoporous metal nanocrystals. *Nanoscale*, 6: 13370–13382.
[135] Hillmyer, M.A., Batesm, F.S., Almdal, K. et al. 1996. Complex phase behavior in solvent-free nonionic surfactants. *Science*, 271: 976.
[136] Förster, S. and Plantenberg, T. 2002. From self-organizing polymers to nanohybrid and biomaterials. *Angew. Chem. Int. Ed.*, 41: 688–714.
[137] Förster, S. and Konrad, M. 2003. From self-organizing polymers to nano-and biomaterials. *J. Mater. Chem.*, 13: 2671–2688.
[138] Bucknall, D.G. and Anderson, H.L. 2003. Polymers get organized. *Science*, 302: 1904–1905.
[139] Cui, J., Zhang, H., Yu, Y. et al. 2012. Synergism of interparticle electrostatic repulsion modulation and heat-induced fusion: a generalized one-step approach to porous network-like noble metals and their alloy nanostructures. *J. Mater. Chem.*, 22: 349–354.
[140] Song, C., Wang, Y. and Rosi, N.L. 2013. Peptide-directed synthesis and assembly of hollow spherical CoPt nanoparticle superstructures. *Angew. Chem. Int. Ed.*, 52: 3993–3997.
[141] Li, C., Sato, T. and Yamauchi, Y. 2013. Electrochemical synthesis of one-dimensional mesoporous Pt nanorods using the assembly of surfactant micelles in confined space. *Angew. Chem. Int. Ed.*, 52: 8050–8053.
[142] Zhang, J., Deng, Y., Gu, D. et al. 2011. Ligand-assisted assembly approach to synthesize large-pore ordered mesoporous titania with thermally stable and crystalline framework. *Adv. Energy Mater.*, 1: 241–248.
[143] Eid, K., Wang, H., Malgras, V. et al. 2015. One-step solution-phase synthesis of bimetallic PtCo nanodendrites with high electrocatalytic activity for oxygen reduction reaction. *J. Electroanal. Chem.*, doi:10.1016/j.jelechem.2015.10.035.
[144] Wang, L. and Yamauchi, Y. 2009. Facile synthesis of three-dimensional dendritic platinum nanoelectrocatalyst. *Chem. Mater.*, 21: 3562–3569.
[145] Wang, L., Imura, M. and Yamauchi, Y. 2012. Tailored design of architecturally controlled Pt nanoparticles with huge surface areas toward superior unsupported Pt electrocatalysts. *ACS Appl. Mater. Interfaces*, 4: 2865–2869.
[146] Lu, S., Eid, K., Li, W. et al. 2016. Gaseous NH_3 confers porous Pt nanodendrites assisted by halides. *Sci. Rep.*, 6: 26196.
[147] Eid, K., Wang, H., Malgras, V. et al. 2016. Facile synthesis of porous dendritic bimetallic PtNi nanocrystals as efficient catalysts for oxygen reduction reaction. *Chem. Asian J.*, 11: 1388–1393.
[148] Huang, X., Zhu, E., Chen, Y. et al. 2013. A facile strategy to Pt_3Ni nanocrystals with highly porous features as an enhanced oxygen reduction reaction catalyst. *Adv. Mater.*, 25: 2974–2979.
[149] Shi, L., Wang, A., Zhang, T. et al. 2013. One-step synthesis of Au-Pd alloy nanodendrites and their catalytic activity. *J. Phys. Chem. C*, 117: 12526–12536.
[150] Wang, L. and Yamauchi, Y. 2010. Autoprogrammed synthesis of triple-layered Au@Pd@Pt core-shell nanoparticles consisting of a Au@Pd bimetallic core and nanoporous Pt shell. *J. Am. Chem. Soc.*, 132: 13636–13638.
[151] Snyder, J., McCue, I., Livi, K. et al. 2012. Structure/processing/properties relationships in nanoporous nanoparticles as applied to catalysis of the cathodic oxygen reduction reaction. *J. Am. Chem. Soc.*, 134: 8633–8645.
[152] Wang, X., Fu, H., Peng, A. et al. 2009. One-pot solution synthesis of cubic cobalt nanoskeletons. *Adv. Mater.*, 21: 1636–1640.

[153] Xia, B.Y., Wu, H.B., Wang, X. et al. 2012. One-pot synthesis of cubic $PtCu_3$ nanocages with enhanced electrocatalytic activity for the methanol oxidation reaction. *J. Am. Chem. Soc.*, 134: 13934–13937.
[154] Chen, J., McLellan, J.M., Siekkinen, A. et al. 2006. Facile synthesis of gold-silver nanocages with controllable pores on the surface. *J. Am. Chem. Soc.*, 128: 14776–14777.
[155] Chen, J., Wiley, B., McLellan, J. et al. 2005. Optical properties of Pd-Ag and Pt-Ag nanoboxes synthesized via galvanic replacement reactions. *Nano Lett.*, 5: 2058–2062.
[156] Lin, Y., Daga, V.K., Anderson, E.R. et al. 2011. Nanoparticle-driven assembly of block copolymers: a simple route to ordered hybrid materials. *J. Am. Chem. Soc.*, 133: 6513–6516.
[157] Park, S., Song, Y.J., Han, J.-H. et al. 2010. Structural and electrochemical features of 3D nanoporous platinum electrodes. *Electrochim Acta*, 55: 2029–2035.
[158] Mahmoud, M.A., Narayanan, R. and El-Sayed, M.A. 2013. Enhancing colloidal metallic nanocatalysis: sharp edges and corners for solid nanoparticles and cage effect for hollow ones. *Acc. Chem. Res.*, 46: 1795–1805.
[159] Wu, J. and Yang, H. 2013. Platinum-based oxygen reduction electrocatalysts. *Acc. Chem. Res.*, 46: 1848–1857.
[160] Nilekar, A.U. and Mavrikakis, M. 2008. Improved oxygen reduction reactivity of platinum monolayers on transition metal surfaces. *Surf. Sci.*, 602: L89–L94.
[161] Kibsgaard, J., Gorlin, Y., Chen, Z. et al. 2012. Meso-structured platinum thin films: Active and stable electrocatalysts for the oxygen reduction reaction. *J. Am. Chem. Soc.*, 134: 7758–7765.
[162] Liu, W., Rodriguez, P., Borchardt, L. et al. 2013. Bimetallic aerogels: high-performance electrocatalysts for the oxygen reduction reaction. *Angew. Chem. Int. Ed.*, 52: 9849–9852.
[163] Liao, H. and Hou, Y. 2013. Liquid-phase templateless synthesis of Pt-on-$Pd_{0.85}Bi_{0.15}$ nanowires and PtPdBi porous nanoparticles with superior electrocatalytic activity. *Chem. Mater.*, 25: 457–465.
[164] Su, K., Yao, X., Sui, S. et al. 2015. Matrix material study for *in situ* grown Pt nanowire electrocatalyst layer in proton exchange membrane fuel cells (PEMFCs). *Fuel Cells*, 15: 449–455.
[165] Mani, P., Srivastava, R. and Strasser, P. 2011. Dealloyed binary PtM_3 (M = Cu, Co, Ni) and ternary $PtNi_3M$ (M = Cu, Co, Fe, Cr) electrocatalysts for the oxygen reduction reaction: performance in polymer electrolyte membrane fuel cells. *J. Power Sources*, 196: 666–673.
[166] Rauber, M., Alber, I., Müller, S. et al. 2011. Highly-ordered supportless three-dimensional nanowire networks with tunable complexity and interwire connectivity for device integration. *Nano Lett.*, 11: 2304–2310.
[167] Wang, L. and Yamauchi, Y. 2011. Synthesis of mesoporous Pt nanoparticles with uniform particle size from aqueous surfactant solutions toward highly active electrocatalysts. *Chem. Eur. J.*, 17: 8810–8815.
[168] Alia, S.M., Zhang, G., Kisailus, D. et al. 2010. Porous platinum nanotubes for oxygen reduction and methanol oxidation reactions. *Adv. Funct. Mater.*, 20: 3742–3746.
[169] Zhang, Y., Janyasupab, M., Liu, C.W. et al. 2012. Three dimensional PtRh alloy porous nanostructures: tuning the atomic composition and controlling the morphology for the application of direct methanol fuel cells. *Adv. Funct. Mater.*, 22: 3570–3575.
[170] Eid, K., Malgras, V., He, P. et al. 2015. One-step synthesis of trimetallic Pt-Pd-Ru nanodendrites as highly active electrocatalysts. *RSC Adv.*, 5: 31147–31152.
[171] Guo, Z., Dai, X., Yang, Y. et al. 2013. Highly stable and active PtNiFe dandelion-like alloys for methanol electrooxidation. *J. Mater. Chem. A*, 1: 13252–13260.
[172] Koper, M.T., Lai, S. and Herrero, E. 2009. *Fuel cell catalysis. A surface science approach.* John Wiley & Sons. 1: 159–207.
[173] Yu, X. and Pickup, P.G. 2008. Recent advances in direct formic acid fuel cells (DFAFC). *J. Power Sources*, 182: 124–132.

[174] Cuesta, A., Cabello, G., Gutiérrez, C. et al. 2011. Adsorbed formate: the key intermediate in the oxidation of formic acid on platinum electrodes. *Phys. Chem. Chem. Phys.*, 13: 20091–20095.
[175] Zhu, Y., Khan, Z. and Masel, R. 2005. The behavior of palladium catalysts in direct formic acid fuel cells. *J. Power Sources*, 139: 15–20.
[176] Wang, J., Holt-Hindle, P. and MacDonald, D. 2008. Thomas DF, Chen A. Synthesis and electrochemical study of Pt-based nanoporous materials. *Electrochim. Acta*, 53: 6944–6952.
[177] Jia, Y., Jiang, Y., Zhang, J. et al. 2014. Unique excavated rhombic dodecahedral PtCu3 alloy nanocrystals constructed with ultrathin nanosheets of high-energy {110} facets. *J. Am. Chem. Soc.*, 136: 3748–3751.
[178] Antolini, E. 2007. Catalysts for direct ethanol fuel cells. *J. Power Sources*, 170: 1–12.
[179] Lee, Y.W., Kim, M., Kim, Y. et al. 2010. Synthesis and electrocatalytic activity of Au–Pd alloy nanodendrites for ethanol oxidation. *J. Phys. Chem. C*, 114: 7689–7693.
[180] Zhang, X., Lu, W., Da, J. et al. 2009. Porous platinum nanowire arrays for direct ethanol fuel cell applications. *Chem. Commun.*, 195–197.
[181] Hou, S., Xu, Y., Liu, Y. et al. 2012. Room-temperature fast synthesis of composition-adjustable Pt-Pd Alloy Sub-10-nm nanoparticle networks with improved electrocatalytic activities. *Chem. Lett.*, 41: 546–548.
[182] Pan, Y., Guo, X., Li, M. et al. 2015. Construction of dandelion-like clusters by PtPd nanoseeds for elevating ethanol eletrocatalytic oxidation. *Electrochim. Acta*, 159: 40–45.
[183] Lee, C.L. and Tseng, C.M. 2008. Ag-Pt Nanoplates: galvanic displacement preparation and their applications as electrocatalysts. *J. Phys. Chem. C*, 112: 13342–13345.
[184] Wu, H., Li, H., Zhai, Y. et al. 2012. Facile synthesis of free standing Pd-based nanomembranes with enhanced catalytic Performance for methanol/ethanol oxidation. *Adv. Mater.*, 24: 1594–1597.
[185] Liu, W., Herrmann, A.-K., Geiger, D. et al. 2012. High-performance electrocatalysis on palladium aerogels. *Angew. Chem. Int. Ed.*, 51: 5743–5747.
[186] Lu, Y.Z. and Chen, W. 2012. PdAg alloy nanowires: facile one-step synthesis and high electrocatalytic activity for formic acid oxidation. *ACS Catal.*, 2: 84–90.
[187] Sun, Y.G., Mayers, B.T. and Xi, Y. 2002. Template-engaged replacement reaction: a one-step approach to the large-scale synthesis of metal nanostructures with hollow interiors. *Nano Lett.*, 2: 481–485.
[188] Peng, C., Hu, Y., Liu, M. et al. 2015. Hollow raspberry-like PdAg alloy nanospheres: high electrocatalytic activity for ethanol oxidation in alkaline media. *J. Power Sources*, 278: 69–75.
[189] Sun, Y.G., Tao, Z.L., Chen, J. et al. 2004. Ag nanowires coated with Ag/Pd alloy sheaths and their use as substrates for reversible absorption and desorption of hydrogen. *J. Am. Chem. Soc.*, 126: 5940–5941.
[190] Zhu, C., Guo, S. and Dong, S. 2013. Rapid, general synthesis of PdPt bimetallic alloy nanosponges and their enhanced catalytic performance for ethanol/methanol electrooxidation in an alkaline medium. *Chem. Eur. J.*, 19: 1104–1111.
[191] Song, H.M., Anjum, D.H., Sougrat, R. et al. 2012. Hollow Au@Pd and Au@Pt core-shell nanoparticles as electrocatalysts for ethanol oxidation reactions. *J. Mater. Chem.*, 22: 25003–25010.
[192] Wang, A.-L., He, X.-J., Lu, X.-F. et al. 2015. Palladium-Cobalt nanotube arrays supported on carbon fiber cloth as high-performance flexible electrocatalysts for ethanol oxidation. *Angew. Chem. Int. Ed.*, 54: 3669–3673.
[193] Lu, A., Schmidt, W., Matoussevitch, N. et al. 2004. Nanoengineering of a magnetically separable hydrogenation catalyst. *Angew. Chem. Int. Ed.*, 43: 4303–4306.
[194] Zhai, Y.P., Dou, Y.Q., Zhao, D.Y. et al. 2011. Carbon materials for chemical capacitive energy storage. *Adv. Mater.*, 23: 4828–4850.

[195] Ohkubo, T., Miyawaki, J. and Kaneko, K. 2002. Adsorption properties of templated mesoporous Carbon (CMK-1) for nitrogen and supercritical methane experiment and GCMC Simulation. *J. Phys. Chem. B*, 106: 6523–6528.
[196] Go´rka, J. and Jaroniec, M. 2011. Hierarchically porous phenolic resin-based carbons obtained by block copolymer-colloidal silica templating and post-synthesis activation with carbon dioxide and water vapor. *Carbon*, 49: 154–160.
[197] Liang, C.D., Li, Z.J. and Dai, S. 2008. Mesoporous carbon materials: synthesis and modification. *Angew. Chem. Int. Ed.*, 47: 3696–3717.
[198] Liang, C.D. and Dai, S. 2006. Synthesis of mesoporous carbon materials via enhanced hydrogen-bonding interaction. *J. Am. Chem. Soc.*, 128: 5316–5317.
[199] Wickramaratne, N.P. and Jaroniec, M. 2013. Phenolic resin-based carbons with ultra-large mesopores prepared in the presence of poly(ethylene oxide)-poly(butylene oxide)-poly(ethylene oxide) triblock copolymer and trimethyl benzene. *Carbon*, 51: 45–51.
[200] Ryoo, R., Joo, S.H. and Jun, S. 1999. Synthesis of highly ordered carbon molecular sieves via template-mediated structural transformation. *J. Phys. Chem. B*, 103: 7743–7746.
[201] Jun, S., Joo, S.H., Ryoo, R. et al. 2000. Synthesis of new, nanoporous carbon with hexagonally ordered mesostructure. *J. Am. Chem. Soc.*, 122: 10712–10713.
[202] Gierszal, K.P. and Jaroniec, M. 2006. Carbons with extremely large volume of uniform mesopores synthesized by carbonization of phenolic resin film formed on colloidal silica template. *J. Am. Chem. Soc.*, 128: 10026–10027.
[203] Tang, W. and Xie, J. 2011. A novel template route for synthesizing mesoporous carbon materials. *Chem. Lett.*, 40: 116–117.
[204] Vinu, A., Miyahara, M., Sivamurugan, V. et al. 2005. Large pore cage type mesoporous carbon, carbon nanocage: a superior adsorbent for biomaterials. *J. Mater. Chem.*, 15: 5122–5127.
[205] Tanaka, S., Nishiyama, N., Egashira, Y. et al. 2005. Synthesis of ordered mesoporous carbons with channel structure from an organic-organic nanocomposite. *Chem. Commun.*, 2125–2127.
[206] Liang, C., Hong, K., Guiochon, G.A. et al. 2004. Synthesis of a large-scale highly ordered porous carbon film by self-assembly of block copolymers. *Angew. Chem. Int. Ed.*, 43: 5785–5789.
[207] Rodriguez, A.T., Li, X., Wang, J. et al. 2007. Facile synthesis of nanostructured carbon through self-assembly between block copolymers and carbohydrates. *Adv. Funct. Mater.*, 17: 2710– 2716.
[208] Deng, Y., Yu, T., Wan, Y. et al. 2007. Ordered mesoporous silicas and carbons with large accessible pores templated from amphiphilic diblock copolymer poly(ethylene oxide)-b-polystyrene. *J. Am. Chem. Soc.*, 129: 1690–1697.
[209] Deng, Y., Liu, J., Liu, C. et al. 2008. Ultra-large-pore mesoporous carbons templated from poly(ethylene oxide)-b-polystyrene diblock copolymer by adding polystyrene homopolymer as a pore expander. *Chem. Mater.*, 20: 7281–7286.
[210] Deng, Y., Cai, Y., Sun, Z. et al. 2010. Controlled synthesis and functionalization of ordered large-pore mesoporous carbons. *Adv. Funct. Mater.*, 20: 3658–3665.
[211] Deng, Y., Liu, C., Gu, D. et al. 2008. Thick wall mesoporous carbons with a large pore structure templated from a weakly hydrophobic PEO-PMMA diblock copolymer. *J. Mater. Chem.*, 18: 91–97.
[212] Zhang, J., Deng, Y., Wei, J. et al. 2009. Design of amphiphilic ABC triblock copolymer for templating synthesis of large-pore ordered mesoporous carbons with tunable pore wall thickness. *Chem. Mater.*, 21: 3996–4005.
[213] Vinu, A., Streb, C., Murugesan, V. et al. 2003. Adsorption of cytochrome c on new mesoporous carbon molecular sieves. *J. Phys. Chem. B*, 107: 8297–8299.
[214] Kruk, M., Jaroniec, M., Kim, T.-W. et al. 2003. Synthesis and characterization of hexagonally ordered carbon nanopipes. *Chem. Mater.*, 15: 2815–2823.

[215] Doi, Y., Takai, A., Makino, S. et al. 2010. Synthesis of mesoporous carbon using a fullerenol-based precursor solution via nanocasting with SBA-15. *Chem. Lett.*, 39: 777–779.
[216] Vinu, A., Srinivasu, P., Takahashi, M. et al. 2007. Controlling the textural parameters of mesoporous carbon materials. *Microporous Mesoporous Mater.*, 100: 20–26.
[217] Sun, Z., Liu, Y., Li, B. et al. 2013. General synthesis of discrete mesoporous carbon microspheres through a confined self-assembly process in inverse opals. *ACS Nano.*, 7: 8706–8714.
[218] Zheng, Y., Jiao, Y., Jaroniec, M. et al. 2012. Nanostructured metal-free electrochemical catalysts for highly efficient oxygen reduction. *Small*, 8: 3550–3566.
[219] Liu, J., Sun, X., Song, P. et al. 2013. High-performance oxygen reduction electrocatalysts based on cheap carbon black, nitrogen, and trace iron. *Adv. Mater.*, 25: 6879–6883.
[220] Woods, M., Biddinger, E., Matter, P. et al. 2010. Correlation between oxygen reduction reaction and oxidative dehydrogenation activities over nanostructured carbon catalysts. *Catal. Lett.*, 136: 1–8.
[221] Ewels, C.P. and Glerup, M. 2005. Nitrogen doping in carbon nanotubes. *J. Nanosci. Nanotech.*, 5: 1345–1363.
[222] Shao, Y., Sui, J., Yin, G. et al. 2008. Nitrogen-doped carbon nanostructures and their composites as catalytic materials for proton exchange membrane fuel cell. *Appl. Catal. B*, 79: 89–99.
[223] Glerup, M., Castignolles, M., Holzinger, M. et al. 2003. Synthesis of highly nitrogen-doped multi-walled carbon nanotubes. *Chem. Commun.*, 2542–2543.
[224] Glerup, M., Steinmetz, J., Samaille, D. et al. 2004. Synthesis of N-doped SWNT using the arc-discharge procedure. *Chem. Phys. Lett.*, 387: 193–197.
[225] Cao, C., Huang, F., Cao, C. et al. 2004. Synthesis of carbon nitride nanotubes via a catalytic-assembly solvothermal route. *Chem. Mater.*, 16: 5213–5215.
[226] Feng, S., Li, W., Shi, Q. et al. 2014. Synthesis of nitrogen-doped hollow carbon nanospheres for CO_2 capture. *Chem. Commun.*, 50: 329–331.
[227] Wang, X., Lee, J.S., Zhu, Q. et al. 2010. Ammonia-treated ordered mesoporous carbons as catalytic materials for oxygen reduction reaction. *Chem. Mater.*, 22: 2178–2180.
[228] Jaouen, F., Lefèvre, M., Dodelet, J.-P. et al. 2006. Heat-treated Fe/N/C catalysts for O_2 electroreduction: are active sites hosted in micropores? *J. Phys. Chem. B*, 110: 5553–5558.
[229] Chen, S., Bi, J., Zhao, Y. et al. 2012. Nitrogen-doped carbon nanocages as efficient metal-free electrocatalysts for oxygen reduction reaction. *Adv. Mater.*, 24: 5593–5597.
[230] Yang, W., Fellinge, T.-P. and Antonietti, M. 2011. Efficient metal-free oxygen reduction in alkaline medium on high-surface-area mesoporous nitrogen-doped carbons made from ionic liquids and nucleobases. *J. Am. Chem. Soc.*, 133: 206–209.
[231] Wan, K., Long, G.-F., Liu, M.-Y. et al. 2015. Nitrogen-doped ordered mesoporous carbon: synthesis and active sites for electrocatalysis of oxygen reduction reaction. *Applied Catalysis B: Environmental*, 165: 566–571.

CHAPTER 5

Mesoporous Materials for Batteries

Kuang-Hsu (Tim) Wu,[1] *Hamid Arandiyan*,[2] *Ju Sun*,[2] *Cui Yanglansen*[2] and *Da-Wei Wang*[2,*]

ABSTRACT

This chapter will review recent advances in utilizing mesostructured materials, mainly carbons and metal oxides, for lithium storage in Li/Na-ion batteries, Li-S batteries and Li-Air batteries. It will elaborate why mesoporous materials are advantageous for battery applications, and how the cell performance is improved through tuning the porous texture and compositions.

Keywords: Mesoporous materials, lithium ion batteries, sodium ion batteries, lithium-sulfur batteries, lithium-air batteries

1. Introduction

Electricity plays a significant role in daily life. Electrical energy supplies the power for industrial processes and drives the communication technologies as well as the future electrified transportations. Electrical energy is primarily converted from mechanical energy or thermal energy at the coal or hydroelectric power plant. The mobility of electrical energy, however, is realized through the (electro) chemical conversion by using either primary or secondary batteries.

[1] Institute of Metal Research, Chinese Academy of Sciences, Shenyang, China.
[2] School of Chemical Engineering, University of New South Wales, Sydney, NSW, Australia.
* Corresponding author

Rechargeable batteries (also termed secondary batteries) utilize reversible electrochemical processes to store and transport electrical energy, and are being applied universally in various fields. Nowadays, the urgent demand on high energy and high power rechargeable batteries has put Li-ion batteries at the forefront of battery technologies. Li-ion batteries can provide much higher energy densities (> 100 Wh/kg) than other secondary batteries (20 ~ 50 Wh/kg), such as Lead-Acid and Ni-MH. Li-ion batteries utilize an intercalation anode, which is usually graphitic materials, and an insertion cathode, which is based on lithium metal oxides. Upon discharge, the electrons are forced to flow from the anode to the cathode through an external circuit, where the work is generated. The Li-ions diffuse synchronously from the anode to the cathode through the electrolyte. The electrode reactions on both electrodes rely on the bulk diffusion of Li ions, which determines both the power and the energy performance of Li-ion batteries. Precisely controlled electrode porosity has been developed to facilitate the bulk diffusion of Li-ions and the other relevant reactions to increase the overall device performance. Among various porous materials, mesoporous solids have demonstrated a wide range of tailorable texture, composition, morphology and other properties. These rich features of mesoporous materials have drawn remarkable attention to their application as advanced electrode materials in Li-ion batteries. Very recently, the emergence of Li-S batteries and Li-air batteries has opened up a new avenue for the battery application of mesoporous materials. In this chapter, the recent progress on the application of mesoporous materials in Li-ion, Li-S, and Li-air batteries is discussed.

2. Mesoporous Materials for Li/Na-ion Storage

2.1 Mesoporous transition metal oxides

The progress of advanced electrode materials for Li-Ion Batteries (LIBs) has continuously been an important field for engineering and science researchers. An advanced electrode materials for LIBs might be studied on a big category of nanostructures which combined through suitable electrochemical properties, such as CoO, Co_3O_4, Fe_3O_4, Fe_2O_3, NiO, MnO and Mn_3O_4 [1]. Meanwhile, a large number of different morphologies of Transition-Metal Oxides (TMOs) studied and expended as anode materials for LIBs showing high performance [2–5]. Due to this, they give reversible capacities up to 2–3 times greater than that the theoretical capacity of graphite, as well as high resistance to corrosion, superior rate capability and low processing cost. As electrode materials for LIBs, iron mesoporous TMOs (mainly Fe_2O_3 and Fe_3O_4), tin dioxide (SnO_2) provide much greater capacity than the graphite anodes used in commercial batteries

(372 mAh/g), non-toxicity, enhanced cycling performance and improved safety. The ideal capacity of Fe_3O_4 and Fe_2O_3 is 926 and 1005 mAh/g, respectively. Fe_2O_3 has types of α, β, γ and ε-Fe_2O_3 [6]. Each of them has its own exceptional characteristics. For instance, α-Fe_2O_3 has very suitable electrochemical activity and excellent thermodynamic stability, γ-Fe_2O_3 and Fe_3O_4 are spinel architecture demonstrating very effective electrical conductivity. Additionally, the density of Fe_2O_3 (ca. 5.0 g/cm^3) is two times greater than that of conventional graphite (ca. 2.2 g/cm^3), offering a better volumetric specific capacity [6]. The redox property reaction of iron-oxide-based anodes plays a vital rule on Li storage mechanism, which the iron-oxide-based anodes are reduced to metallic phase and distributed in a Li_2O matrix due to lithiation, additionally are then reversibly resumed to early oxidation phases of matrix through delithiation. The reaction mechanism of Fe_2O_3 might be explained as follows: $Fe_2O_3 + 6Li^+ + 6e^- \leftrightarrow 3Li_2O + 2Fe$ [6]. Nevertheless, they frequently perform low coulombic performance, weak level capability and quick capacity fading, due to huge volume transformation through cycling [5]. Recently, these problems have been solved using the construction of mesoporous nanostructure frameworks, by which one can control alignment and improve electrochemical reactions [2–4]. An alternative approach is to introduce carbonaceous components into the mixed-oxides to synthesis nanocomposites, which is also extraordinarily significant in the performance of TMOs [7]. By focusing on mixed oxides SnO_2 and Fe_2O_3, researchers try to find a rational consideration of the connection among appropriate mesoporous and improved physicochemical properties of the effective negative electrode anode material in LIBs; expectantly uncovering new promises to harvest advanced nanostructures for next generation in LIBs. Hence, quicker charge/discharge capability and better specific capacity might be estimated for LIBs due to promising mesoporous transition metal oxides electrode materials [7]. To increase the benefits of nanomaterials, the alignment, component, permeability, and surface chemistry characteristics require to be enhanced. In a related study, Cho et al. stated a simple synthesis of mesoporous spindle-like porous α-Fe_2O_3 via calcining of an iron-based structure pattern [8]. As illustrated in Fig. 5.1A–C, the morphology of surface texture changes considerably through the etching period. The microparticles with different porosities show separate ramping stability since assessed as negative electrode materials for LIBs (Fig. 5.1D). Particularly, in spite of the comparable preliminary capacity, a much greater reversible capacity of 662 mAh/g might be maintained by the porous materials at 100 charge/discharge cycles, however the non-etched sample can just provide a capacity of 341 mAh/g. Wang et al. described a facile hydrothermal method in the combination of deionized water, isopropanol and aminopolycarboxylic acid [9]. The high resolution FESEM image illustrates that the nanowires keep a

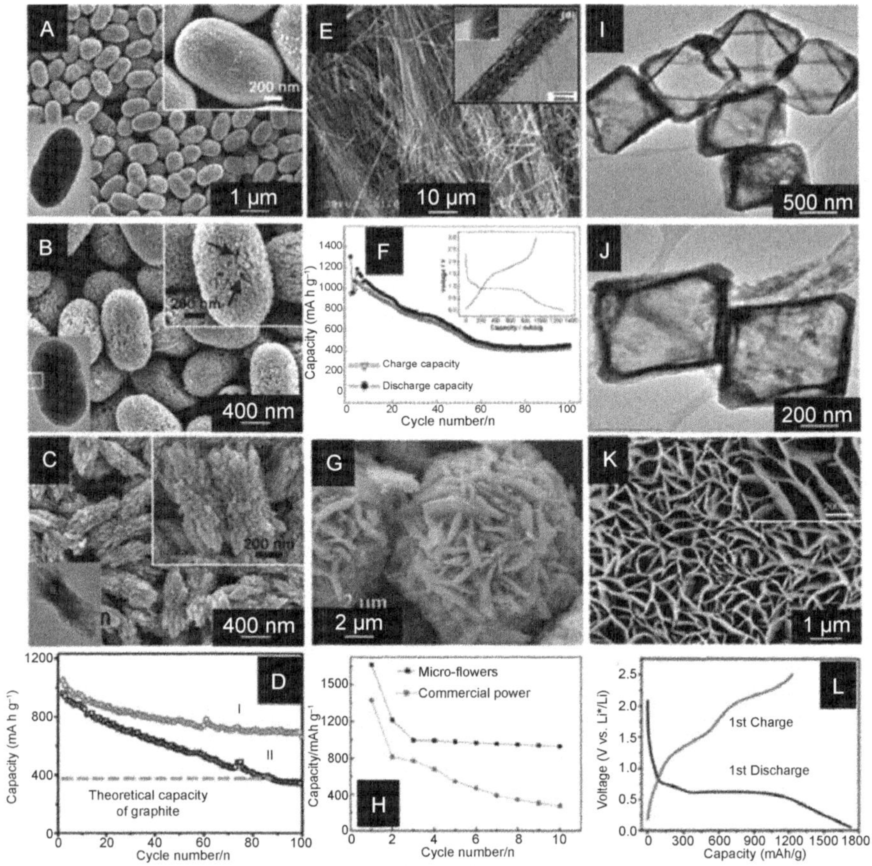

Figure 5.1. FESEM and TEM (inset) images of the melon-like α-Fe_2O_3 microparticles obtained with different oxalic acid etching durations: (A) 0 hour, (B) 36 hours, and (C) 42 hours. (D) Comparative cycling performance of the α-Fe_2O_3 microparticles shown in a (I) and b (II) [9]. (E) FESEM and TEM (inset) images and (F) cycling performance of iron oxide nanowires [9] (G) FESEM image and (H) cycling performance of α-Fe_2O_3 flowers constructed by nanosheets [10] (I) TEM images of $Fe(OH)_x$ nanoboxes, (G) octahedral α-Fe_2O_3 nanocages [11]. (K) FESEM image and (L) cycling performance of α-Fe_2O_3 nanosheets grown on Cu foil [12].

size as long as 100 μm (DI = 200 nm) (Fig. 5.1E). The high resolution-TEM image shows the polycrystalline characteristic of an individual nanowire that is created with little α-Fe_2O_3 nanoparticles (inset of Fig. 5.1E). When used as anode materials at 0.1 C, these α-Fe_2O_3 nanowires perform great initial discharge capacity of 1303 mAh/g. The discharge capacity of 456 mAh/g is kept during 100 cycles (Fig. 5.1F). Wang et al. fabricated flower-like hematite α-Fe_2O_3 synthesized by heat-treatment from the iron(III)-oxyhydroxide precursor [10]. The flower-like hematite frame works with a length between

7–10 μm are created from micro-flowers nanopetals with a thickness of about 20 nm (Fig. 5.1G) with high BET (115 m^2/g). The flower-like hematite (α-Fe_2O_3) electrode demonstrates high initial capacity with current density of 100 mA/g after 10 cycles at 929 mAh/g as illustrated in Fig. 5.1H. Following, a hydrothermal method scheme was investigated to fabricate up and down ranged, single phase α-Fe_2O_3 nanosheets (NSs) raised on nickel bubble [11]. The high resolution FESEM image illustrates that the α-Fe_2O_3 NSs have a wide open setup architecture constructed via interrelated NSs (Fig. 5.1I, J). The FESEM image illustrates that the open area between nanowalls (NWs) is comparatively considerable and the NWs are about thickness 20 nm with 2 μm of height (Fig. 5.1J). Due to the 2D characteristic and the close interaction with the current collector, the α-Fe_2O_3 nanosheets perform secure cycling operation with reversible capacity after 50 cycles in the potential between 0.05–2.5 V of 518 mAh/g at 0.1C. Additionally, Chowdari et al. studied a remarkably simple procedure to fabricate nanosheets of single-crystalline α-Fe_2O_3 by heating iron coated substrates in air [12]. The high resolution-FESEM image of the α-Fe_2O_3NSs illustrates that the nanoscale flakes are largely directed to the copper support (Fig. 5.1K). As expected, such an integrated electrode composed of α-Fe_2O_3 shows stable capacity of 700 mAh/g with no obvious capacity fading on first charge and discharge (Fig. 5.1L), when cycled in the potential between 0.005–3.0 V at 65 mA/g.

The reason why 1D materials of SnO_2 are investigated better than the 2D counterpart is perhaps due to the complexity in synthesizing single crystal phase of SnO_2 NSs. Jiang et al. studied a solvothermal route to synthesize very thin SnO_2 NSs [13]. The NSs samples display high Li storage properties contrasted to SnO_2 hollow NPs and nanospheres, with a reversible capacity after 50 charge-discharge cycles of 534 mAh/g. Recently, Lou's research group established a comparable hydrothermal system to fabricate flower-like 3D frameworks assembled from SnO_2 NSs [14]. The high-resolution SEM images (Fig. 5.2A) illustrate the homogeneous particles with a diameter around 2μm. The sample shows excellent cycling activity contrasted to commercial SnO_2 nanomaterials after 50 charge-discharge cycles with an extraordinary reversible capacity of 516 mAh/g (Fig. 5.2B). The growth of SnO_2 on α-Fe_2O_3 nanostructures has also been illustrated in a series of particles [15]. Zhou et al. have investigated growing α-Fe_2O_3 nanorods on SnO_2 nanowires (Fig. 5.2C) [16]. The achievement of succeeding such advanced heterogeneous structure keeps in the slight lattice difference between the (101) that the SnO_2 nanorods raise on (110) facets of the α-Fe_2O_3 nanotubes. The enhanced Li storage performing of the hybrid material as related to the pure phase counterpart (Fig. 5.2D) might additionally be due to the potential synergy effect among elements [17]. Ju-Song Chen et al. also studied to load the hollow SnO_2 nanospheres (NSPs) fabricated from the template-free procedure via hydrothermal treatment

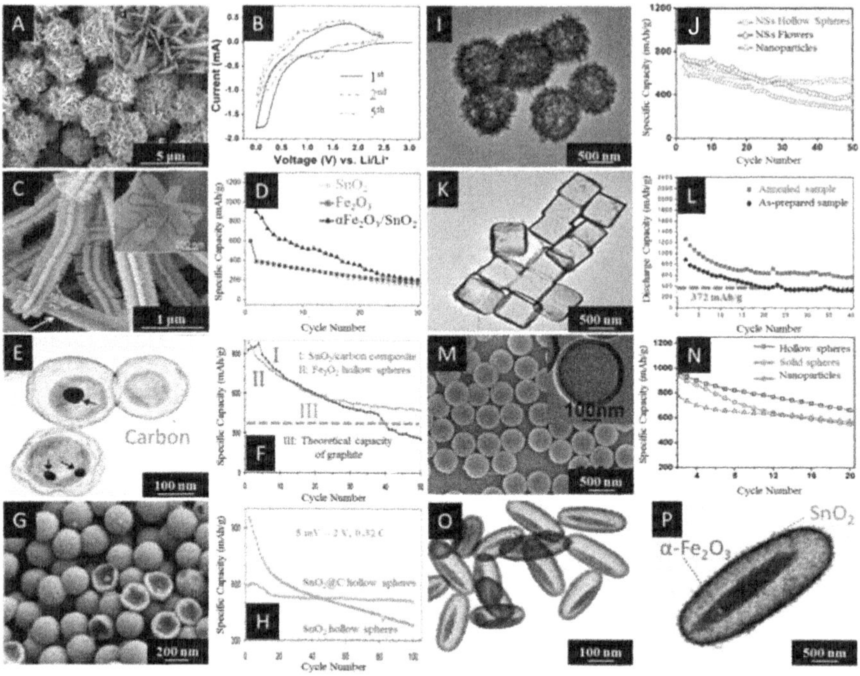

Figure 5.2. (A) Electron micrograph and (B) the comparative cycling performance of the SnO_2 hierarchical spheres [14].; SEM image (C) and comparative cycling performance (D) of the α-Fe_2O_3/SnO_2 structure [16]; Electron micrograph (E) and comparative cycling performance (F) of the carbon-coated SnO_2 hollow nanospheres [18]; Electron micrograph (G) and cycling performance (H) of the SnO_2@ carbon coaxial hollow nanospheres [22]; (I) Electron micrograph of SnO_2NSs hollow spheres and (J) the comparative cycling performance of NSs hollow spheres [19]; TEM image (K) and the cycling performance (L) of SnO_2 nanoboxes [20]; TEM image (M) and the comparable cycling performance (N) of the SnO_2 hollow nanospheres. The inset in (M) shows the SnO_2 shell formed inside the SiO_2 hollow nanoreactor [23]; TEM images of cocoon-like ellipsoidal SnO_2 hollow nanostructures with double (P) and single (O) shells [21].

(Fig. 5.2E) [18]. The obtained SnO_2@C NSPs establish enhanced cyclic capacity retention contrasted to the pure SnO_2 NSPs (Fig. 5.2F). Moreover, the amorphous carbon is coated on the top of SiO_2@SnO_2 core-shell NSPs [14]. Following carbonization, the SiO_2 core is gone in a sodium hydroxide solution, to obtain SnO_2@carbon hollow NSPs. As illustrated in Fig. 5.2G, there are a number of spheres looking like punctured balloons with no framework failure, revealing that the mix NSPs are definitely very flexible. Consequently, the material performs exceptional cyclic capacity at high current densities for many cycles (Fig. 5.2H). Lou et al. have investigated a templating route to fabricate hollow spheres assembled from SnO_2 NSs [19] whose surface is protected with negatively charged $-SO_3-$ functional groups (Fig. 5.2I). Hence, the Sn^{2+} ions can certainly relate with these patterns. The nanomaterial illustrates a enhanced cyclic capacity retention contrasted to SnO_2 nanoflowers fabricated from SnO_2 NSs without the hollow interior, in addition to the SnO_2 nanoparticles (Fig. 5.2J), clearly displaying the positive effect of exceptional nanostructuring. Besides the usual hard-patterns aforementioned, however there are several types of soft-patterns. For instance Ju-Song Chen et al. have investigated a soft template-employed approach to fabricate SnO_2 nanoboxes [20]. As the reaction continues, a hardy SnO_2 shell is shaped then the Cu_2O core is slowly removed till its full consumption, directing to the single hollow looking like nanobox (NB) architecture (Fig. 5.2K). After calcining, the as-prepared SnO_2 NBs display high Li storage properties after 40 cycles with an excellent reversible capacity of 570 mAh/g (Fig. 5.2L). The SiO_2 microreactors are completely divided via dilute hydrofluoric acid, due to well-structured SnO_2 NSPs (Fig. 5.2M). As concerned with the anode for LIBs, the SnO_2 hollow NSPs perform some progress contrasted to NPs and SnO_2 solid spheres (Fig. 5.2N). Using potassium tin oxide (K_2SnO_3), a homogeneous layer of inadequately SnO_2 might be easily added on the surface of the SiO_2 through hydrothermal procedure. After separating the SiO_2 template by hydrofluoric acid, SnO_2 NSPs with double or single shells can be achieved. Employing a similar method, however, using α-Fe_2O_3 as the initiate pattern, cocoon-like NSPs with double (Fig. 5.2O) or single (Fig. 5.2P) SnO_2 shells can be synthesized [21].

2.2 Meso-macroporous perovskites

A flexible category of materials with a 2D and/or 3D structure architecture and with a mixture of transition metal oxides and non-transition metal oxides is called perovskite mixed-oxides with a general formula ABO_3, which A and B are two different cations, and O is an anion which links to A and B cation. The A cations are bigger than the B cations [24–27]. A is usually an alkaline earth or lanthanide metal ion and B a transition-metal ion [28–31]. Well-known examples are $LaFeO_3$ and $SrTiO_3$. It will be of attention to study the Li cycling features of some of perovskite mixed oxide materials, however due to the reality that the big atomic weight of Lanthanum, the ideal capacities, by the transferreaction of the B cations, will be lower than 500 mAh/g. Li et al. [32] investigated the anodic behaviour of $LaCoO_3$ and $MFeO_3$ (M = Ce, La) and found reversible capacities of 300–400 mAh/g, that lowered gradually up to 100 cycles. The hexagonal unit cell has been described for $Li_xLa_xTiO_{3-\delta}$ ($0 \leq \delta \leq 0.06$, $x = 0.5$) in a latest report [33], the spin due to the tilting of the TiO_6 octahedra. Element cell boundaries were $a = 5.4711(4)$ Å and $c = 13.404(1)$ Å, conforming to one of Glazer's octahedral tilt preparations for ABO_3 materials, that is approved since the tilting angle of the octahedra is small (e.g., $LaMO_3$, M = Al, Cu, Ni). The architecture of material is constituted of approximately ordered TiO_6 (Ti–O 1.943 Å) and La is 12-fold (La–O 2.559–2.911 Å) and Li 4-fold direction by oxygen (Fig. 5.3a). The Lithium ions are located in the central structures, designed by four TiO_6 units, in square-planar formation with Li–O bond (1.81–2.07 Å). Firstly, the replication of the *c* parameter is due to the assembling of the positions along the (001), however the doubling of the parameters *a* and *b* is contributed to the assembling of lanthanum and lithium in the position (110). Recent studies [34] support the previous dual cell $a(\sim 2a_p) \times b(\sim 2a_p) \times c(\sim 2a_p)$, which holds a 3D structure of turning TiO_6 which are tilted along the *b*-axis, instead organized Lanthanum along the *c*-axis, and Lithium in two correspondent off-centred locations at the A site (Fig. 5.3b). The so-called "diagonal-distorted" cell was advised by Várez et al. [35] for the $Li_{3x}La_{(2/3)-x(1/3)-2x}TiO_{3-\delta}$ ($0.06 < x < 0.16$, $0 \leq \delta \leq 0.06$) [36]. The spin was due to ration of Lithium to Lanthanum 1:1 assembling along the *c*-axis. This explanation was declined by Fourquet et

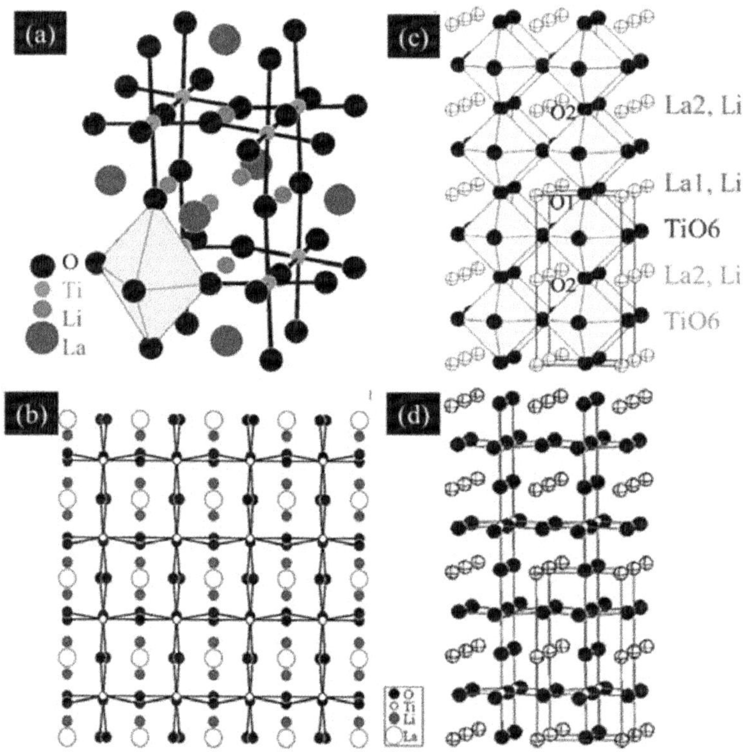

Figure 5.3. (a) Schematic structure of pseudo-cubic perovskite-type $La_{0.5}Li_{0.5}TiO_3$ [37]. (b) Crystal structure of orthorhombic $Li_{3x}La_{(2/3)-x(1/3)-2x}TiO_3$ ($x = 0.05$) [37]. (c, d) Crystal structure of tetragonal $Li_{3x}La_{(2/3)-x(1/3)-2x}TiO_3$ ($x = 0.11$). (Top) ball and stick and (bottom) polyhedral view. Unit cell constant: $a = 3.8741(1)$ Å and $c = 7.7459(5)$ Å: space group *P4/mmm* [36].

al. [36] their group suggested another category of cell on the basis of XRD and HRTEM investigations, that has presently been accepted [36] (Fig. 5.3c,d). The TiO_6 octahedra are connected with one short Ti–O2 (~1.8 Å) bond opposed to one long Ti–O1 (~2 Å) bond and four equal Ti–O3 (~1.94 Å) bonds which interfaced with *c*-axis.

Up to now, the ABO_3 perovskite mixed oxides-type $Li_{3x}La_{(2/3)-x(1/3)-2x}TiO_3$ ($0 < x < 0.16$) shows the maximum Li-ion-conducting solid electrolytes and structurally related materials (Fig. 5.4A) of 10^{-3} S/cm with an activated energy of 0.40 eV at RT [38]. TMOs have been recommended as anode material applicants for LIBs, since it can reversibly react with Li via a movement reaction to deliver two to three times the specific capacity of graphite. In spite of this, the realistic application of TMOs has been frustrated by their problematic working capacity. Hence, the effect of adding

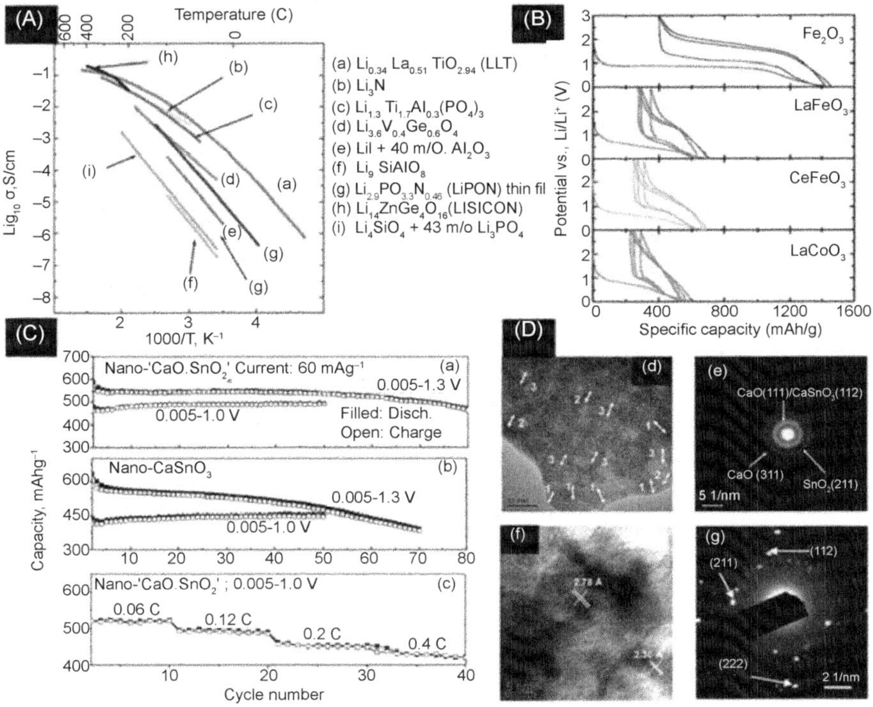

Figure 5.4. (A) Arrhenius plots of the lithium ion conductivities (bulk) of selected metal oxides [38], (B) Colour online potential vs. specific capacity for the first three cycles for coin cells with electrodes made from different powdered perovskite RE–TM–O_3 samples compared to Fe_2O_3, C) Capacity vs. cycle number plots at the current of 60 mA g–1 (0.12C), [39] D) (d) HR-TEM lattice image of X-ray amorphous nano-CaO SnO_2 (e) The SAED pattern. CaO, SnO_2 and $CaSnO_3$, (f) The HR-TEM lattice image of X-ray crystalline nano-$CaSnO_3$, (g) The corresponding SAED pattern of nano-$CaSnO_3$ [39].

of Rare Earth elements (REs) to TMOs as negative-electrode materials was studied. Thin films of pseudo binary misch-metal–Fe–O libraries were prepared by combinatorial sputtering methods. Powdered perovskite-structured RE–TM–O_3 samples, including $LaFeO_3$, $CeFeO_3$ and $LaCoO_3$, were synthesized by solid-state methods (Fig. 5.4B). The framework and electrochemical characterizations of these samples are presented in Fig. 5.4B. It was found that the addition of RE elements reduces the working potential of TMOs. Though, the specific capacity is undesirably lowered. Powdered perovskite mixed-oxide RE–TM–O_3 samples such as $LaFeO_3$,$CeFeO_3$ and $LaCoO_3$ were synthesized using high-energy mechanical heating or milling procedures (300–400 mAh/g). Chowdari et al. [39] fabricated and investigated $CdSnO_3$ (30 nm) described a stable capacity of 475 mAh/g with voltage between 0.005 V and 1.0 V. However, at a greater voltage at 1.3 V, capacity reduced at 1 to 25 cycles from 580 to 245 mAh/g, respectively. Recently, Sharma et al. [39] synthesized X-ray nanocomposite CaO–SnO_2 and X-ray nanophase $CaSnO_3$ by the thermolysis of $CaSn(OH)_6$ in air at 600°C for 6 hours and 24 hours, respectively (Fig. 5.4C). XRD, high resolution-TEM, and selected area (electron) diffraction analysis showed that the X-ray nano-CaO–SnO_2 contains nanoparticle (3–6 nm) of calcium oxide and SnO_2 whereas $CaSnO_3$ contains about 60 nm crystal size (Fig. 5.4D) [39].

2.3 Mesoporous carbon and its composites

With increasing demand for portable electronic devices with high energy capacity, intensive research has been focused on high specific capacity carbonaceous with high reversible capacity. Carbonaceous materials have been intensively studied for anode material in lithium batteries, because of not only the high specific capacity but also more negative redox than most metal oxides, chalcogenides and polymers. Mesoporous (2 nm < pore size < 50 nm) is one of the most promising electrode materials batteries, owing to its unique properties and structures, such as electric conductivity, thermal conductivity, chemical stability, narrow pore-size distributions and high surface areas.

Zhou and his colleagues [40] reported that the ordered mesoporous carbon (CMK-3) in 2003 was synthesized using the SBA-15 as template sucrose as carbon precursor, followed by the impregnation of template with sucrose, carbonization and dissolution of the silica template. Compared to other conventional non-graphitic carbon, CMK-3 contains plenty of uniform mesopores with diameter of 3.9 nm and has a higher surface area of 1030 m^2/g. As shown in Fig. 5.5a, in the first discharge process, CMK-3 exhibits high specific capacity of about 3100 mAh/g, however the reversible capacity only contribute about 1100 mAh/g, which is mainly because of

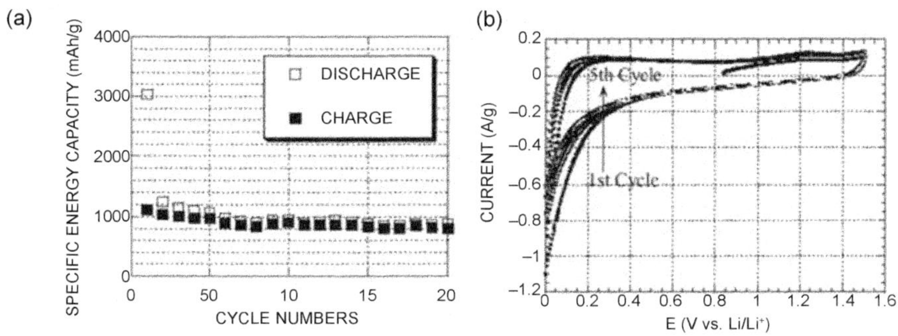

Figure 5.5. (a) Cyclic voltammograms of CMK-3 in 1 M $LiClO_4$ with EC+DMC electrolyte in a potential window from 0.01V to 1.5V (Vs. Li/Li^+) at the scan rate of 0.1 mV/s (b) Discharge and charge cycle performance of CMK-3 at a constant current density of 100 mA/g. Reproduced from Ref. [40] with permission.

the formation of solid electrolyte interface (Fig. 5.5b). After 20 cycles, the reversible capacity can still reach 850 mAh/g which relatively suggests a good cycle performance.

Since the poor conductivity of non-graphitic carbon which may cause the poor cycle stability and rate capability, graphic carbon has been introduced into the mesoporous structure to improve the conductivity. Guo and co-workers [41] used the soft template to synthesize the mesoporous carbon and then introduced the multi-walled carbon nanotubes. The rate capability and cycle performance were improved in comparison to mesoporous carbon.

Lithium alloys have been studied as anode material for lithium ion batteries owing to high capacity and fast lithium mobility [42]. Tin nanoparticles were inserted into mesoporous carbon by Grigoriants and coworkers through a sonochemical process [43]. As shown in Fig. 5.6, a reversible of capacity of 400 mAh/g can be delivered which is contributed by the formation of alloying of Li with tin. In addition, small size of tin nanoparticles could efficiently minimize the volume change during cycling.

Because of carbon material's low capacity and poor rate performance, metal oxide has been widely studied for lithium ion storage due to the mechanism of lithium intercalation. The main problem for the metal oxide is lack of conductivity that may cause loss of the capacity during the charge and discharge process. Cobalt oxide [44], NiO [45], $Li_4Ti_5O_{12}$ [46] and MoO_2 [47] nanoparticles have been reported to impregnate into mesoporous carbon to form the metal oxide nanoparticles/mesoporous carbon composite. Mesoporous carbon plays a crucial role both in formation of nanoparticles and cycle process. During the nucleation at high temperature, the crystal size can be confined inside the mesoporous carbon. During the

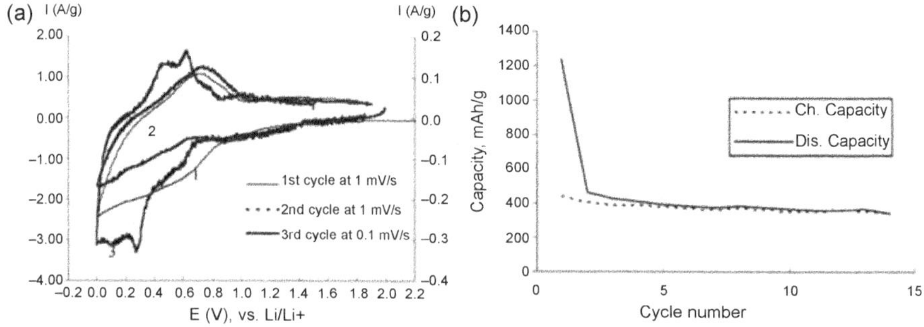

Figure 5.6. Electrochemical measurements of a composite electrode with tin containing MSPC particles as the active mass: (a) cycling voltammograms; (b) galvanostatic cycling, C/20 hour rate, ≈0.05 mA cm^{-2} of geometric area. Reproduced from Ref. [42] with permission.

charge and discharge process, the small size of particles provides a short diffusion pathway for the lithium and mesoporous can prevent the volume expansion of metal oxide during the insertion of lithium. Liu and co-workers synthesized a new type of nano-sized cobalt oxide compounded with Mesoporous Carbon Spheres (MCS) for use as negative material in lithium ion batteries [44]. The composite containing 20 wt.% cobalt oxide shows a high reversible capacity of 703 mAh/g at a current density of 70 mA/g. After 30 cycles, 77% of capacity can be maintained. NiO/CMK-3 nanocomposite was prepared by using spontaneous oxidation of Ni nanoparticles (Fig. 5.7). NiO/CMK-3 nanocomposite delivers a high reversible capacity of 812 mAh/g at a C-rate of 1000 mA/g and the capacity retained 716 mAh/g after 50 cycles. Even at the high C-rate of 3200 mA/g, it shows a high capacity of 413 mAh/g. The mesoporous carbon provides electronic and ionic pathways which suggests a fast kinetics of reaction between NiO and Li^+.

Because of the low cost and abundance of Sodium, sodium ion battery is one of the promising alternatives for lithium ion battery. Although the chemistry of sodium ion batteries is supposed to be similar to lithium ion batteries, commercialized graphite is not suitable for sodium ion battery because of the different intercalation mechanism of sodium ion due its larger size than lithium ion. In recent years, intensive research has been done to explore novel electrode material suitable for sodium ion batteries. Sung-Wool Kim and coworkers investigated the potential anode material for SIBs [48]. Conversion compounds and alloying compounds are also pursued for SIBs. This compound had significant volume change during charge and discharge. Mesoporous carbon has been widely used for confining growth of nanoparticles for lithium ion batteries. The same strategy has been introduced in SIBs.

Xu et al. [49] investigated the mesoporous carbon/Tin anodes both in lithium ion and Sodium ion batteries (Fig. 5.8). Both lithium ion and sodium ion can react with Sn during cycling. However, Na-ion batteries

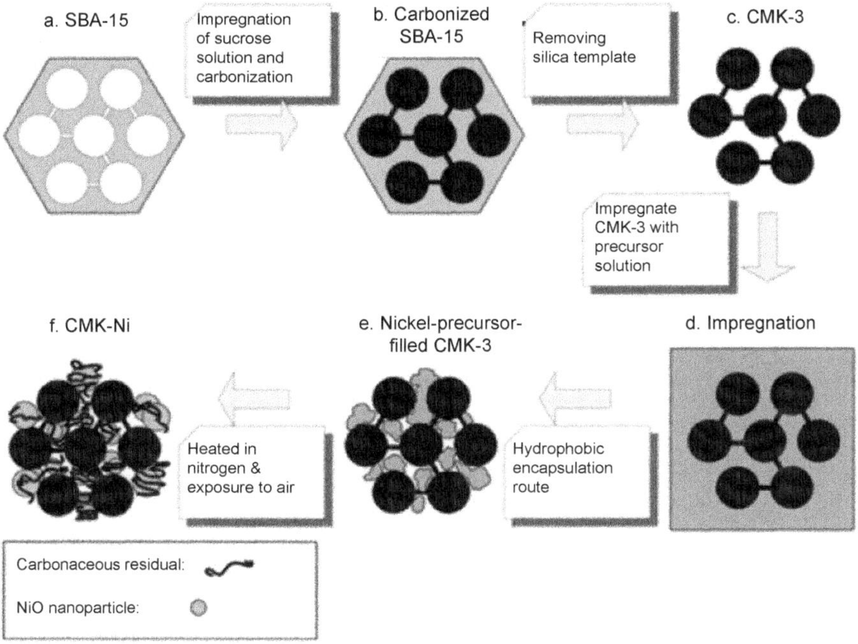

Figure 5.7. Schematic illustration for the synthesis of CMK-Ni. Reproduced from Ref. [45] with permission.

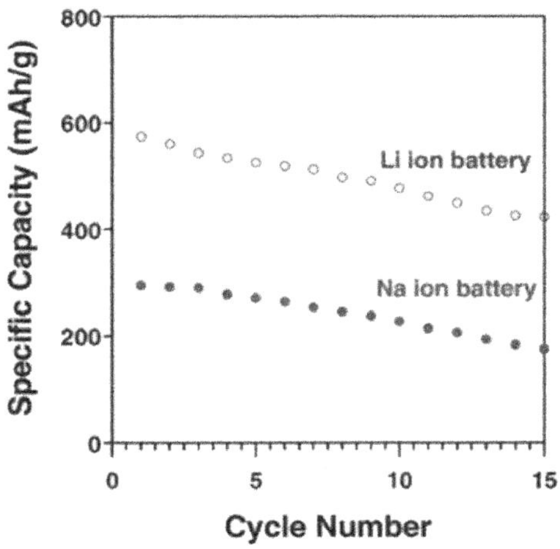

Figure 5.8. Capacity performance of mesoporous C/Sn anodes during charge/discharge cycles at 20 mA/g in Na-ion and Li-ion batteries. Reproduced from Ref. [49] with permission.

show lower capacity and rate performance compared to Li-ion batteries which is mainly because of the slow diffusion of Na ions in Sn and high interfacial resistance of reaction between Na and Sn.

3. Mesoporous Materials for Li-S Batteries

One of the challenging problems of Li-S batteries is rapid capacity fading, which is due to the dissolution of polysulphide intermediates. When the polysulphide ions spread from the cathode to electrolyte and anode, they will eventually react with Li metal, thus causing the loss of active materials. More seriously, the sulphide species can be re-deposed back on the cathode, as a result of which, large agglomerates will be found at the end of discharge. Recently, porous materials, especially micro- (< 2 nm), meso- (2–50 nm) and macro- (> 50 nm) porous materials with their high electronic conductivity, are becoming the promising option for Li-S batteries. Mesoporous materials can not only host high content of sulphur within porous structures, but are favourable for trapping intermediate polysulphides, making electrolytes available, and enhancing the capacity retention in the cathode (Fig. 5.9). Compared with the macroporous materials, the mesoporous one has a high sulphur loading rate up to 70 wt% due to its uniform pore diameter, high volume and interconnected porous structure. However, the microporous materials, which have small pore size and volume, cannot bear fast Li ions diffusion into the cathode, resulting in the low sulphur loading and slow mass transport of molecules. Mesoporous carbon materials also have some other advantages, such as conductivity and are widely studied for Li-S batteries [10, 50–52].

3.1 Ordered mesoporous carbon materials

The trial of using ordered mesoporous carbon to host sulphur was conducted by Nazar et al. [54]. They reported to embed sulphur into the highly ordered mesoporous carbon (CMK-3) by employing a simple melt-diffusion strategy

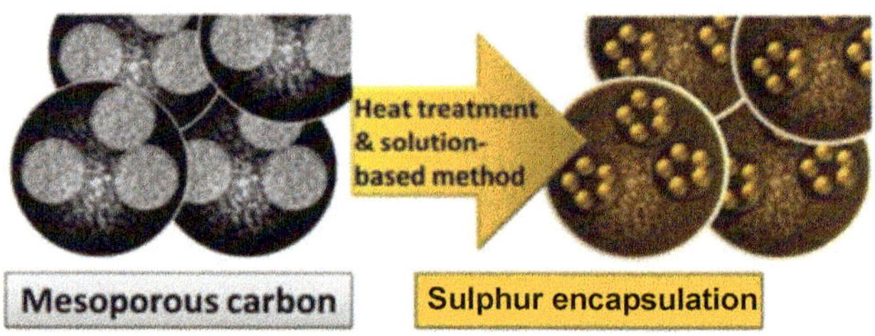

Figure 5.9. Schematic models of the mesoporous carbon host. Reproduced from Ref. [53].

(Fig. 5.10). The results show this kind of C/S composite can load sulphur up to 70 wt% which is nearly the theoretical limit of 79 wt% sulphur/composite ratio as defined by the porosity. The coulombic efficiency of the CMK-3/S for the first charge-discharge is up to 99.94% and the composite can supply enhanced reversible capacity of 1320 mA h g^{-1}, nearly 80% of the theoretical capacity of sulphur. This is because the hierarchical mesoporous carbon can provide pathways for Li ions to transport, and the intimate contact between the active material and liquid electrolyte with the conductive substrate could result in high sulphur utilization. Polymer modified mesoporous carbon materials further demonstrate excellent stability in the Li-S batteries.

Another type of ordered mesoporous carbon material is the one with bimodal pore structure, which has a high surface area and large internal porosities [50]. This uniformly distributed porous carbon was synthesized through a soft-template approach using the KOH activation, exhibiting excellent current density and cycling performance. Generally speaking, the large surface area and structure porosity of the bimodal carbon materials is basic for achieving the high sulphur utilization. The bimodal pores can accommodate the volume change. The microporous structure has a very high surface area and can be functioned as a container for sulphur. In addition, micropores provide high conductivity to the nanocomposites. As for the mesopores, they could promote the transport of Li ions into cathode, and accommodate electrolyte as well as confine polysulphide species in order not to flow out during the electrochemical reaction. However, the electrodes with best electrochemical performance utilize only 11.7% of sulphur [52], which can decrease either specific gravimetric capacity or specific volumetric capacity. It is thus quite significant to develop electrodes with high current density and capacity. Nazar et al. [55] reported hierarchically structured nanocomposites fabricated by a triblock-copolymer-templating approach using an optimized carbon/silica/surfactant ratio. This novel framework served as cathodes in Li-S batteries

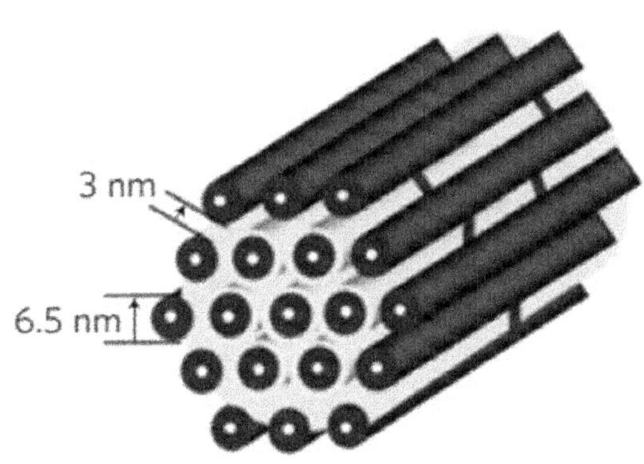

Figure 5.10. A schematic diagram of the CMK-3/S composite. Reproduced from Ref. [54].

and displayed high initial discharge capacities of 1135 mAh g^{-1} and good cycling stability of 550 mAh g^{-1} after 100 cycles at the high current rates of 1675 mA g^{-1}. The perfect electrochemical performance is attributed to the isolated large pores in the structure.

3.2 N-doped ordered mesoporous carbons

Ordered mesoporous carbon materials with nitrogen doping were also conducted recently [56–58]. A novel, nitrogen-doped carbon (MPNC)-sulphur nanocomposite has been studied by Wang et al. [59]. In order to demonstrate the function of nitrogen used in the Li-S batteries, a series of different sulphur-content composites were made (from 80 wt% to 70 wt%). The results showed that MPNC-S70 cathodes have the highest electrochemical performance with the initial specific capacity of 1100 mAh g^{-1} and coulombic efficiency of 99.4% compared to other sulphur-content cathodes (Fig. 5.11). The high performance of such kinds of cathodes can originate from the existence of nitrogen. Figure 5.12 shows the difference of nitrogen/non-nitrogen doped composites, where nitrogen atoms help the formation of bonds between sulphur atoms and oxygen functional groups. This will immobilize sulphur and can enhance the stabilization of sulphur connected with -COOH group in the carbon.

It is reported that the nitrogen-doping cathodes for Li-S batteries could facilitate sulphur reduction and provide high initial capacity and discharge potential. Long et al. [58] reported the Li-S cathode based on nitrogen-doped mesoporous carbon with various nitrogen contents (Fig. 5.13). These nitrogen-enriched nanocomposites were synthesized via a facile colloidal silica assisted sol-gel method. It is concluded that the sulphur content

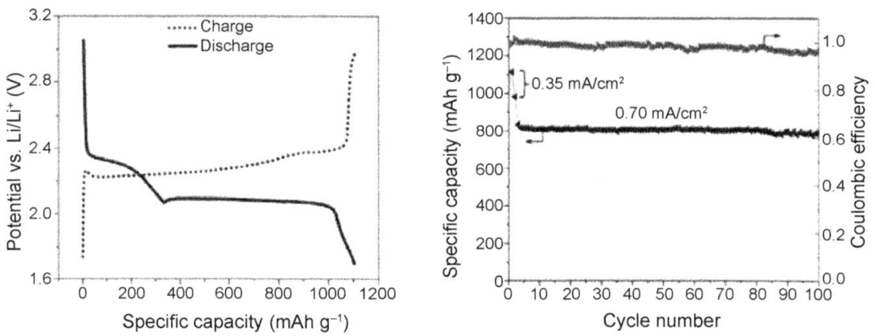

Figure 5.11. (a) The first charge-discharge cycle of MPNC-S70 nanocomposites cathode. (b) Cycling stability and coulombic efficiency of MPNC-S70 for the first two cycles at 0.35 mA cm^{-2} and the remaining cycles at 0.70 mA cm^{-2}. Reproduced from Ref. [59].

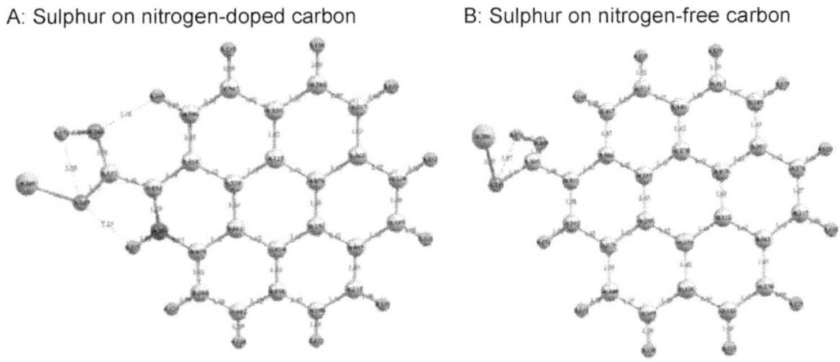

Figure 5.12. Optimized structures of (a) sulphur on nitrogen-doped carbon and (b) sulphur on nitrogen-free carbon. Yellow, red, magenta, silver grey and blue balls represent S, O, N, C, and H atoms, respectively. Reproduced from Ref. [59].

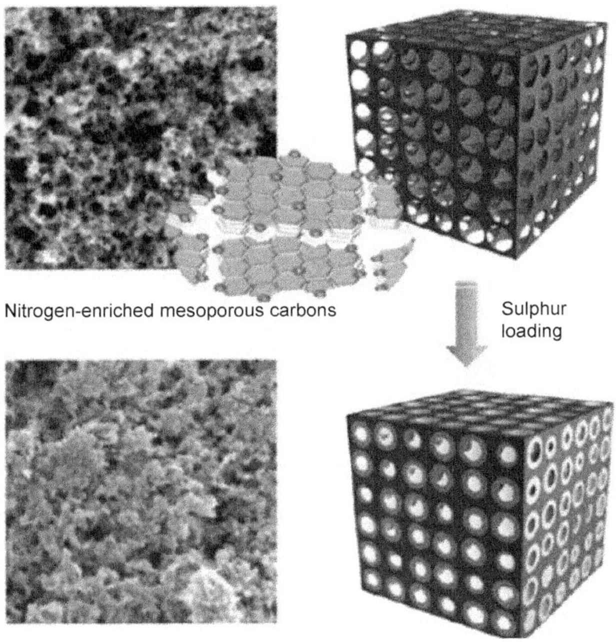

Figure 5.13. A schematic diagram of the nitrogen-doped mesoporous carbon. Reproduced from Ref. [58].

range from 4–8 wt% could greatly enhance the conductivity of carbon and deliver a high reversible discharge capacity of 758 mAh g^{-1} at a 0.2 C rate. The mechanism is postulated that nitrogen could interact with polysulphide species on the carbon.

3.3 Disordered mesoporous carbon materials

Disordered mesoporous carbons can also confine the polysulphides, thus enhancing the cyclic performance of Li-S batteries. Mesoporous materials with different pore sizes and volumes were fabricated by Min-Sik Park et al. [60]. This kind of disordered mesoporous carbons can be easily controlled by various sizes and amount of colloidal silica particles. The colloidal silica templated mesoporous carbons have distinctive advantages in decreasing the impedance when confining the polysulphides in the Li-S batteries [61–63]. Due to the porous structure and conductive characteristic, this kind of mesoporous material can absorb the soluble polysulphides, minimize the loss of sulphur and avoid unexpected side reactions.

3.4 Mesoporous silica and metal oxides as sulphur reservoirs

Not only can mesoporous carbon materials confine active materials, mesoporous silica and metal oxides can also be applied to restrict sulphur species within the cathodes from being dissolved in the electrolyte. Considerable efforts have been devoted into well-developed mesoporous silica which has a high surface area, large pore volume and hydrophilic surface nature [64]. Because the abundance of silica is large and the formation mechanism of mesoporous structure is economically preferable, silica was chosen as the sulphur reservoir. What is more, silica is more chemically stable than other elements, which allows it to be one of the most applicable additives for the batteries. Nazar et al. reported SBA-15, a mesoporous silica composites in which Si-O groups functionalize as internal sulphur reservoir for sulphur cathode [65]. The negatively charged silica surface can accommodate the polysulphide through weak binding, together with the reversible desorption, namely, releasing the active materials when the redox reaction happens. The researchers incorporated SBA-15 into mesoporous carbon (SCM), as a result of which, the compounds exhibited excellent electrochemical performance (Fig. 5.14). The initial discharge capacity of the battery is 960 mA h g^{-1}, larger than the cell without SBA-15 of 920 mAh g^{-1}. This was made possible by utilizing mesoporous silica that is shown to be a highly effective internal polysulphide reservoir for the cathode in Li–S batteries. This not only greatly improves the cycling stability but also eliminates the polysulphide shuttle mechanism to a large degree and allows large-pore carbons to be utilized.

Metal-oxide can also function as the absorbing agent for confining polysulphides into the cathode structure or serve as a supporting active material for emerging extra capacity. There are two kinds of expressions in the application of Li-S batteries. The first one is the metal-oxide additives, which contains aluminum oxides, vanadium oxides, and transition-metal

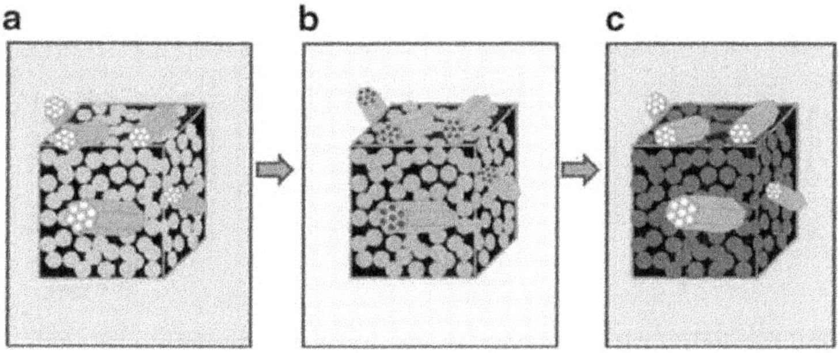

Figure 5.14. Schematic diagram of the 'polysulphide reservoir' provided by the SBA-15 platelets in the SCM/S electrode. The black area represents carbon, yellow pores represent sulphur and grey nanotubes represent SBA-15. Reproduced from Ref. [65].

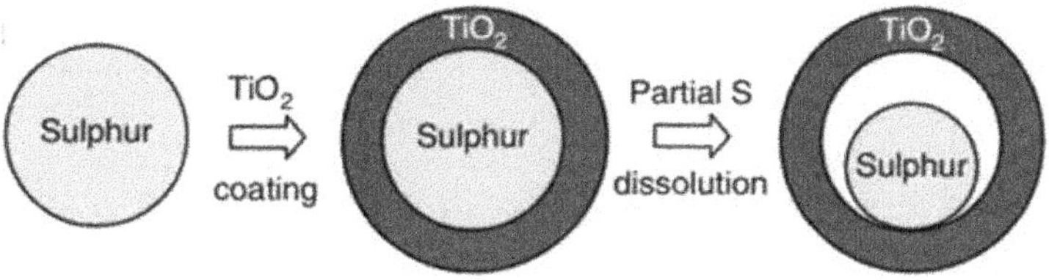

Figure 5.15. Schematic of the synthetic process that involves coating of sulphur nanoparticles with TiO_2 to form sulphur–TiO_2 core–shell nanostructures, followed by partial dissolution of sulphur in toluene to achieve the yolk–shell morphology. Reproduced from Ref. [68].

chalcogenides [66]. However, some requirements need to be met, such as small particle size, porous structure, as well as high surface area, in order not to limit the absorbing ability of the metal-oxide. Lee et al. [67] synthesized nanosized $Mg_{0.6}Ni_{0.4}O$ by using the sol-gel method as an inactive additive for sulphur cathode used in Li/S rechargeable batteries. The electrochemical characterization exhibited good improvement of the initial discharge capacity (1200 mAh g^{-1}), the cycle durability and the rate capability of the sulphur cathode. The second one is metal oxide coating. Yolk-shell morphology composites with sulphur and metal oxide were put forward. Seh et al. [68] designed yolk-shell nanocomposites with sulphur-TiO_2 used in the Li-S batteries (Fig. 5.15). This cathode displayed prolonged cycling stability over 1000 cycles, due to the reason that the extra pore space can not only retain the dissolved polysulphides but also buffer the volumetric expansion during the charge-discharge process. Moreover, this structure can avoid the crack of the TiO_2 spheres, which could prevent the leakage of polysulphides.

4. Mesoporous Materials for Li-Air Batteries

Lithium-Air (Li-Air) batteries are among the most potential next-generation battery technology for use in energy-demanding portable electronics and electric vehicles [69, 70]. As a class of metal-air batteries, Li-Air batteries exhibit enormous energy-storing capacity (theoretical specific energy = 3505 Wh kg^{-1}) through breathing the air oxygen (thus they are also called Li-O_2 batteries) [71, 72]. Since the battery operation involves gas uptake and release, the reaction chemistry almost provide as many serious challenges as the benefits it delivers, especially on the aspects of the "breathing" reactions at the positive electrode (i.e., oxygen reduction reaction, ORR and Oxygen Evolution Reaction, OER). In regard to the electrode reactions, the deposition of insoluble discharge products and the hindered gas diffusion present the major issues that significantly limit the practical performance of Li-Air batteries [73–78]. The solution to these problems requires understanding of the reaction mechanism and sophisticated engineering of the cathode material to tailor the physical and chemical properties. Mesoporous materials are a class of materials with high surface area, large pore volume and interconnected pore channels, which are ideal candidates for gas diffusion and surface redox reactions [73]. Here we will provide a detailed overview of different mesoporous materials for Li-Air batteries; a brief synopsis of the "air-breathing" chemistry at the positive electrode of Li-Air batteries is also provided as a lead into the topic. For the purpose of converging attention to common Li-Air construction, our discussion will be based on the reaction chemistry in a non-aqueous electrolyte system.

Like metal-air batteries, the "air-breathing" reaction at the positive electrode of Li-Air batteries involves complex multi-phase reactions occurring at electrode/electrolyte interfaces [70–72, 79]. At about the interface, the solid electrode in contact with liquid electrolyte should facilitate efficient transport of O_2 (gas) to the electrode to support a steady feed of reactant and consumption of product for an efficient "breathing" of the air. The electrode also has to be electrically conducting, possessing a large surface area and interconnected pore channels with excellent wetting characteristics to facilitate the electrochemical and chemical reactions [71, 77, 80–82]. A combination of these physical properties basically describes the concept of Gas-Diffusion Electrode (GDE) [75–77, 83]. Importantly, the reactions happening on GDE are mostly governed by the physical properties of the electrode. Hence, it is crucial to understand the key reactions and how the physical properties can affect the performance these reactions in Li-Air batteries.

In non-aqueous Li-Air batteries, the cathode reaction undergoes an ORR to form solid Li_2O_2 as the discharged product (Eq. 1) [71, 79, 84]. Even though a further reduction to Li_2O phase (Eq. 2) is possible, it typically forms in minute amounts and this reaction is irreversible.

$$2\,Li^+ + 2\,e^- + O_2 \rightarrow Li_2O_2 \quad (1)$$

$$Li_2O_2 + 2\,Li^+ + 2\,e^- \rightarrow 2\,Li_2O \quad (2)$$

In fact, the formation of solid Li_2O_2 is the major issue in Li-Air batteries [85–87]. Firstly, this material is an insulator and easily deposits on the electrode surface as an amorphous layer or toroid structures, depending on the discharge rate [88]. When the insulating film grows thick enough to block the charge transfer reaction, the electrode reaction occurring at the surface will be terminated [71, 87, 89]. This is demonstrated in Fig. 5.16a; the cell potential tends to drop rapidly when the film thickness reaches a certain limit. Secondly, the growth of the passivating film in pore channels of GDE can clog the pore entrance and inhibit further reaction. The situation is depicted in Fig. 5.16b. Moreover, the incomplete removal of the discharge

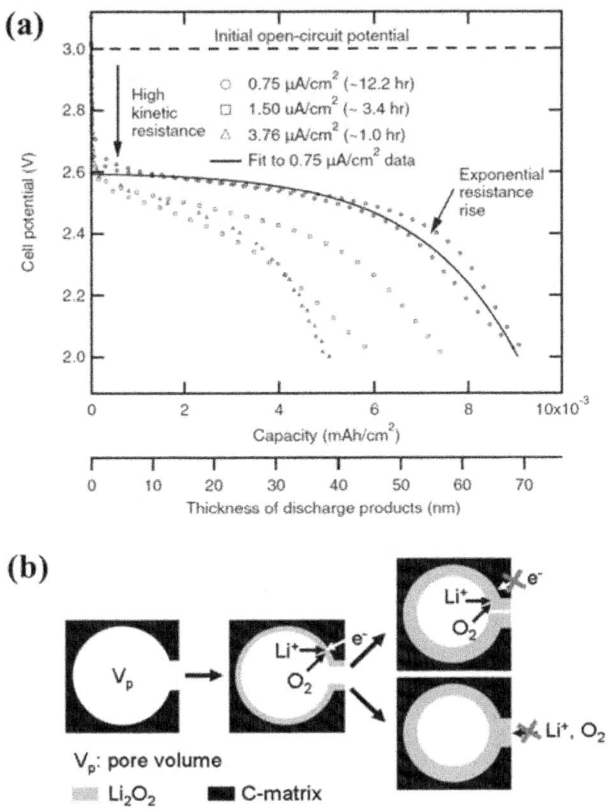

Figure 5.16. (a) A typical discharge profile on porous carbon with a model fit at different discharge current densities and (b) a schematic diagram of pore clogging process during Li-Air battery discharge. The figures are reproduced by permission of The Electrochemical Society from Ref. [71].

product during the charging process due to non-uniform deposition of Li_2O_2 will cause rapid fading of the battery capacity over repeated discharge/charge cycles. It is clear that fine control of Li_2O_2 deposition and uniform film growth during battery discharge is the key to boost up the practical capacity and cycle stability of Li-Air batteries. This is typically assisted by the surface chemistry (to promote the electrocatalysis) and a careful design of the electrode structure (to tune the growth mode of Li_2O_2) [90, 91].

Gas diffusion and transport is another important physical process in Li-Air batteries [76, 77, 83]. When the transport of O_2 gas is restricted within the positive electrode, the discharge product may locally deposit at the electrode surface and block the entrance of reactant gas (as an extreme case of non-uniform deposition). A sophisticate simulation by Albertus et al. has demonstrated the relative impacts of electrode passivation and O_2 transport limitation [81]; the result is shown in Fig. 5.17. In their simulation, the electrode passivation is described by a high electrical resistance of the deposited film, while the transport limitation is described by a low diffusion coefficient. It can be seen that the O_2 transport limit delivers a strong impact (close to two times) to the discharge capacity, although it is less significant than the electrode passivation (about 30 times) [71]. To circumvent the problem of gas transport limitation, the cathode should exhibit high porosity and possess interconnected pore channels to accommodate efficient gas diffusion [92].

In general, Li-Air battery cathodes prefer an optimal porous structure to slow down electrode passivation and facilitate gas transport to attain maximal performance. As a class of structured porous materials,

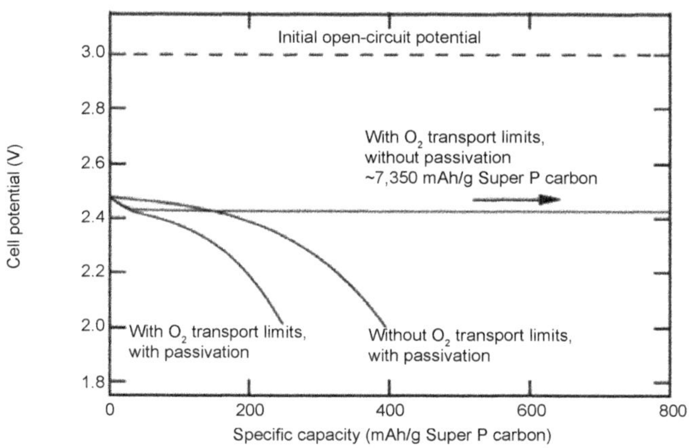

Figure 5.17. A simulation of Li-Air battery discharge profiles demonstrating the relative impacts of electrode passivation and oxygen transport limitation at 0.47 mA/cm^2. The figure is reproduced by permission of The Electrochemical Society from Ref. [81].

mesoporous materials appropriately satisfy the structural requirement for Li-Air battery cathode. Next we will review the recent advances of different types of mesoporous materials for non-aqueous Li-Air batteries and their current status in the field.

4.1 Mesoporous carbons

Carbon materials have significant interest on account of their fascinating physical properties [73, 93–95]. They are of low-cost, light in weight, electrically conducting, structurally strong but flexible enough to accommodate high degrees of porosity and with rich surface chemistry. All these properties are desirable for Li-Air batteries [69, 71, 72]. Being metal-free, carbon materials are usually less catalytically active as compared to metal-based catalysts. Nonetheless, they offer an excellent platform for catalysis and electrochemical reactions given by their good electrical property and large specific surface area from the high porosity. In particular, the meso-porosity in porous carbon is a very important feature that really delivers high surface area as compared to macropores (> 50 nm) while allows efficient electrolyte diffusion into and out of the pores as compared to micropores (< 2 nm) [74, 77, 82, 96–98]. In Li-Air battery application, mesoporous carbons can provide huge electrode/electrolyte interfacial area to host solid Li_2O_2 product and therefore enabling the delivery of enormous discharge capacity.

Strategic modification of mesoporous carbon materials can further advance their performance in Li-Air batteries. There are three general methods: structural engineering, chemical doping and surface functionalization. Each of the strategies focuses on improving an aspect of the challenging issues in the battery reaction. Facilitating diffusion at near the electrode/electrolyte interface is a major challenge and the cathode needs to be highly porous in a specific structure to allow efficient mass transport. Structural engineering of carbons to introduce porosity at multiple levels appears to be a promising way to improve the discharge capacity of the battery. Chemical doping intends to tailor the physicochemical properties of the mesoporous carbon framework. In particular, nitrogen-doping (N-doping) could improve the conductivity of mesoporous carbon, while the nitrogen groups have been reported to exhibit excellent ORR performance in aqueous systems, with an activity comparable to Pt/C catalysts. Similar to chemical doping, surface functionalization by physical or chemical treatment, or grafting catalytic elements can introduce functional units to mitigate particular issues in the mass transport or charge transfer processes. For example, mild oxidation or polyelectrolyte coating can improve the wettability of the carbon surface for entry of electrolyte. A general review on mesoporous carbon materials that fall in the three categories will be provided next.

Structural engineering of carbon at a macro/mesoscopic scale can improve the discharge capability, and perhaps power density of Li-Air batteries by facilitating the mass transport at the electrode/electrolyte interface while securing maximal available surface for product deposition. With no doubt, it is clear that porosity in carbon plays an extremely important role in the discharge mechanism [99]. However, there are questions underlying this principle: (1) what is the optimal extent of porosity, regarding the pore size and the population, and (2) how should the porosity at different levels be structured? In the first point, the extent of porosity should be engineered to adapt with the reaction nature, in which the porosity does not hamper electrolyte diffusion into and out of the pores (or cause pore clogging), while possessing a sufficiently large surface area for product deposition and be able to avoid over-deposition of solid product that possibly prevents the charging reaction [74, 77, 82, 96–99]. Regarding the optimal size of pores for an efficient Li_2O_2-depositing ORR, a number of reports have demonstrated that pores at a meso-scale (20–50 nm) possess the greatest advantage for the discharge process [74, 96, 97]. Large pore volume given by a high pore density has also been shown to be necessary to expand the discharge capacity [100]. On the other hand, electrolyte penetration into the cathode undergoes different modes of diffusion because there would be a stronger resistance along with deeper penetration [77]. Hierarchical porosity and anisotropic gradient porosity hence become important as to the extent of porosity [89, 100–105]. Mesoporous carbons adopting an isotropic hierarchical porosity could provide a rather thorough penetration of electrolyte into the cathode and thus allows better utilization of the available electrode/electrolyte interfaces [99, 100, 104, 105]. With an anisotropic gradient porosity, Tan et al. showed an improved discharge capacity from such a gradient porous structure, which facilitates efficient electrolyte entrance from the large pores at the cathode surface into the inner part of electrode with smaller pores [102]. All of the results to-date have revealed the importance of meso-porosity with a hierarchical porous structure in carbon-based cathode for Li-Air batteries.

The significance of porous structure with multi-scale porosity was very well demonstrated by Xiao et al. that a hierarchical porous reduced graphene oxide (rGO) spheres (Fig. 5.18a) could deliver an exceptionally high first-cycle discharge capacity of 15,000 mAh/g at a current density of ~50 mA/g (Fig. 5.18b) [104]. Figure 5.18c shows that the outstanding discharge capacity comes from the ORR rather than contributed by the oxygen functional groups on rGO (operated in an Ar atmosphere). Their result highlights the enormous benefit brought about by a high surface area hierarchical porous carbon structure. The reasons behind such a large specific energy capacity should be attributed to the multi-scale porosity, high mesopore density and large surface area of rGO, in addition to the synergistic benefits provided by

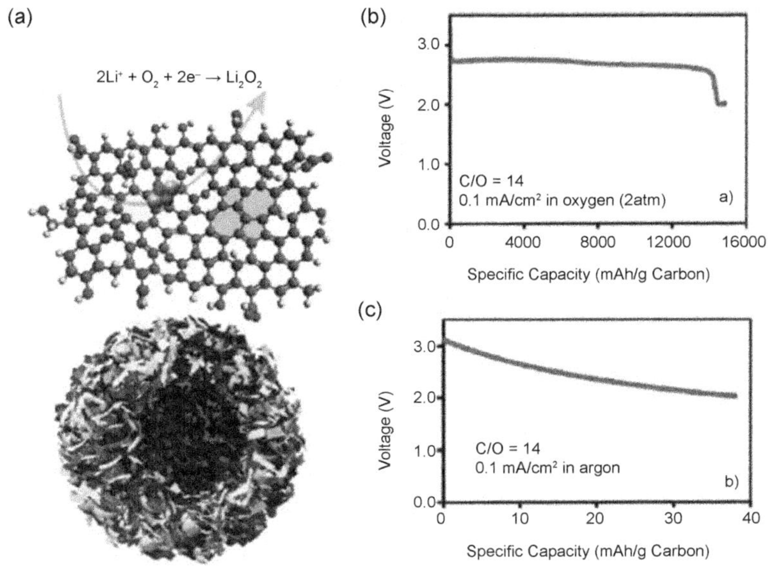

Figure 5.18. (a) Schematic diagram of hierarchically porous rGO and the chemical structure. Discharge profiles of rGO under (b) O_2 and (c) Ar atmosphere, at a current density of 0.1 mA/cm^2. The figures are reproduced from Ref. [104].

the lattice defects and hydrophilic oxygen functional groups. A similar result was also reported on hierarchical N-doped Pierced Graphene Microparticles (N-PGM) [106]. With SiO_2 nanoparticles as hard template for mesoporosity, the N-PGM showed a large first-cycle discharge capacity of 14,300 mAh/g as compared to 8,400 mAh/g of N-doped Graphene Microparticles (N-GM). In another report, Ding et al. showed that rGO by natural self-assembly exhibited a reasonably large surface area with a mesopore size of 10–20 nm and thus could easily achieve a discharge capacity of above 4,000 mAh/g, in comparison to the small discharge capacity given by other kinds of carbons [74]. These reports are excellent examples demonstrating the vast improvement of battery performance by using porosity-tailored carbons.

Heteroatom-doping of mesoporous carbon materials is a promising way to tune the physicochemical properties of carbon. In particular, N-doped carbons have been shown to derive excellent ORR activity comparable to commercial Pt catalysts in aqueous alkaline systems [107]. Despite this there is a huge debate about the true origin of the high activity in N-doped carbons for the ORR, the doped carbons have been shown to be highly effective by numerous reports [107–112]. In reality, N-doping at an appropriate temperature can promote O_2 adsorption and lower the reaction overpotential. Not only does it offer a higher cell voltage, a promoted reaction may enlarge the discharge capacity. In a systematic study by Nie et al., N-doped Mesoporous Carbons (N-MCs) carrying different nitrogen groups (Fig. 5.19a–c) showed varying results in Li-Air batteries [113]. These N-MCs shared similar pore-size distribution (Fig. 5.19d) and pore volume, while the surface area ranged from 566–817 m^2/g along with decreasing thermal treatment temperature (1400°C, 1200°C and 1000°C) [113]. Interestingly, Fig. 5.6e displays that the N-MC carrying only quaternary N-groups provide the best performance among the other three N-MCs containing pyrrolic and pyridinic N-groups, regarding the discharge potential and specific capacity. There are also several reports suggesting the presence of quaternary and/or pyridinic groups in N-doped mesoporous carbons is beneficial to the discharge reaction [101, 106, 109, 114, 115]. Anyway, N-doping appears to be an effective method to enhance the performance of the Li-Air batteries, regardless of its influence to the physical and the chemical property of the carbon. Other heteroatom-doping (e.g., phosphorus-doping) of mesoporous carbon has also been attempted, although none of them exhibited a better activity as compared to N-doped mesoporous carbons [116, 117]. At this point, it is clear that heteroatom-doping, especially N-doping, to mesoporous carbon has a positive impact on the discharge capability when used as cathode material in Li-Air batteries, although how the doping changes the physicochemical

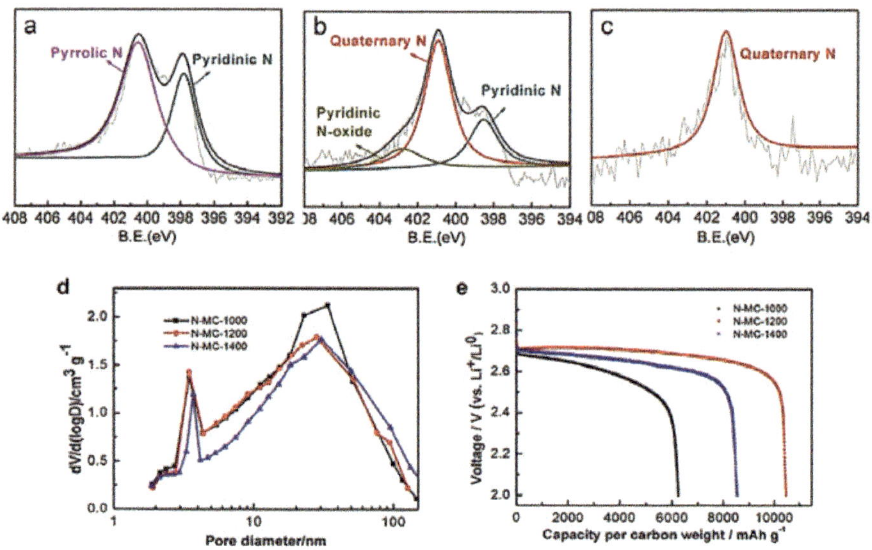

Figure 5.19. XPS N *1s* spectra of N-doped mesoporous carbons (N-MC) treated at (a) 1000°C (N-MC-1000), (b) 1200°C (N-MC-1200) and (c) 1400°C (N-MC-1400). (d) Pore-size distribution of the N-MC materials measured by N2 adsorption isotherm. (e) First-cycle discharge profiles of the N-MC materials at a current density of 300 mAh/g. The figures are reproduced from Ref. [113].

properties (e.g., electrical conductivity, wettability, acidity/basicity and catalytic properties), is still under investigation.

Mesoporous carbon materials with engineered porous structure, modified bulk or surface chemistry provide enormous opportunity to solve the many intrinsic problems in Li-Air battery processes. However, the use of purely carbon-based materials has a major disadvantage that the chemical nature of carbon is less resistant to oxidizing environments (carbon can be oxidized to CO_2 at 0.2 V) [118]. This means that the cycle stability of Li-Air battery would be poor because of the degradation during the charging reaction (i.e., OER), despite the excellent discharge capability attainable by using mesoporous carbon materials. Nevertheless, mesoporous carbon materials have shown great potential as a candidate for one of the cathode reaction. If the drawback of using carbon materials can be alleviated, this class of materials would serve as state-of-the-art cathodes for Li-Air batteries.

4.2 Mesoporous metal oxides

Transition metal oxide is another category of materials that have been widely investigated for Li-Air batteries. These materials contain active metal species and are rigid in structure as compared to carbon materials and are more resistant to oxidizing conditions. Such intrinsic properties have made metal oxide materials particularly suitable for the charging reaction (i.e., OER) in Li-Air batteries. However, metal oxides as electrocatalysts are much denser than carbon materials and are usually of a higher cost. As a solution to this problem, introduction of meso-porosity to these metal oxide materials not only reduces the use of the catalysts but also promotes the physical and electrochemical processes at the positive electrode. In the consequence, mesoporous metal oxide materials stand out as an important class of electrocatalyst for Li-Air batteries.

In the past few years, research has focused on three common types of metal oxides for use in Li-Air batteries: manganese dioxides [119–121], spinel oxides [122–124] and perovskite oxides [125, 126]. These materials are chosen for their low cost, dual catalytic activity for ORR and OER, tunable *d*-orbital structure for favourable oxygen binding and the versatile physical and chemical properties through composition engineering. Next we will review the current progress and prospects of these mesoporous transition metal oxides as cathodes for Li-Air batteries.

Manganese dioxide (MnO_2) is one of the few metal oxides studied during the earlier development of Li-Air batteries [119–121, 127–129]. The reason of the choice was perhaps their high abundance, low toxicity, good stability and high activity for the ORR (despite that the activity was observed in aqueous solutions) [128]. Bruce and co-workers compared a number of common metal oxides (Co_3O_4, Fe_2O_3, CuO and $CoFe_2O_4$) [130] and MnO_2 in different phases (α-MnO_2, β-MnO_2, λ-MnO_2, γ-MnO_2, Mn_2O_3 and Mn_3O_4) and found that α-MnO_2 was among the most active metal oxide in the series, particularly in the form of nanowires (Fig. 5.20) [120]. The α-MnO_2 nanowire could reach an initial discharge capacity of ~3,000 mAh/g (carbon) in a non-aqueous Li-Air battery at a current density of 70 mA/g, even though the capacity dropped down to 50% over 10 cycles [120]. Later, Baeck and co-workers investigated mesoporous MnO_2 catalysts with different morphologies for Li-Air batteries and found that the porous MnO_2 with interpores could exhibit an initial capacity of ~3200 mAh/g (carbon) at a current density of 157 mA/g. There were also other reports on mesoporous MnO_2 composites with small BET surface area (30–130 m^2/g) for non-aqueous Li-Air batteries, although the performances were not even outstanding [121, 127, 131].

Spinel oxides such as Co_3O_4, $MnCo_2O_4$ and other similar derivatives have been investigated for their high stability at versatile compositions and dual capability in performing both ORR and OER [132–137]. However, these metal oxides do not exhibit satisfactory electrical conductivity for use as battery cathode and thus carbon materials are usually incorporated as conductive filler to enable their application. Co_3O_4 is a spinel oxide that has been of particular interest for Li-Air batteries [132, 137–139], because it could deliver decent battery performance among many of the investigated

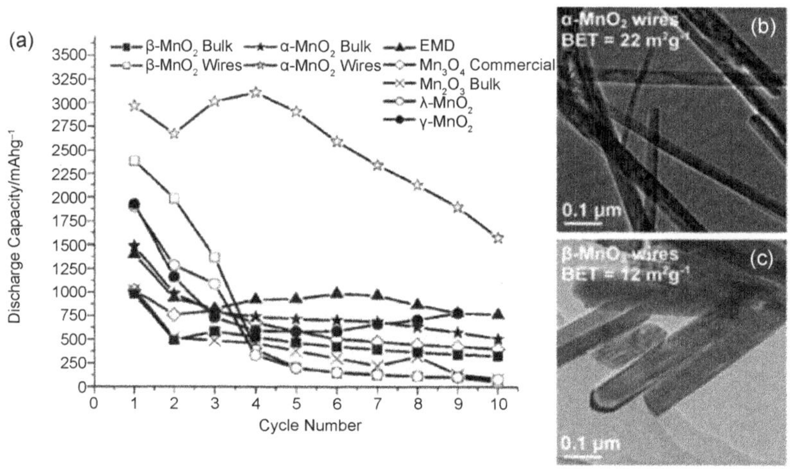

Figure 5.20. (a) Variation of discharge capacity with cycle number for several porous electrodes containing manganese oxides as catalysts: α-MnO_2 in bulk and nanowire form, β-MnO_2 in bulk and nanowire form, γ-MnO_2, λ-MnO_2, Mn_2O_3 and Mn_3O_4. Cycling was carried out at a rate of 70 mA/g in 1 atm of O_2. Capacities are per gram of carbon in the electrode. Cut-off voltages are at 4.15 V and 2.0 V. (b) TEM images of forms of α- and β-MnO_2 nanowire showing their morphologies and surface areas. The figures are reproduced from Ref. [120].

metal oxides, especially when with high porosity [138, 139]. From there, other Co-based spinel oxide derivatives in the form of porous MCo_2O_4 (M = Mn, Fe, Ni, Cu, Zn) have been widely studied [133–135]. To find out an optimal combination of Co-based ternary spinel oxide, Liu and co-workers screened through a range of first-row transition metals and found that porous $FeCo_2O_4$ and $ZnCo_2O_4$ nanorods could demonstrate superior first-cycle discharge capacity than Co_3O_4, $MnCo_2O_4$ and $NiCo_2O_4$ in a similar morphology [134]. However, the porosity in these ternary spinel oxides was not specifically controlled. In a demonstration of how mesoporosity affects the performance of Li-Air battery, Wang and co-workers studied mesoporous $MnCo_2O_4$ microspheres as the cathode catalyst and showed that the mesoporous microspheres (with an average pore-size of 32 nm) in a composite with Super P carbon can demonstrate a first-cycle discharge capacity of 4,681 mAh/g (carbon) at a current density of 200 mA/g (Fig. 5.21a) [135]. The mesoporous $MnCo_2O_4$ microsphere further exhibits good rate capability that a discharge capacity of ~3000 mAh/g (carbon) can be achieved at a rate of 500 mA/g (Fig. 5.21b). There are other examples of using mesoporous mixed spinel oxides for Li-Air batteries with an improved performance [140]. It is clear that introduction of mesoporosity to spinel oxides with optimal compositions can further improve their performance and rate capability.

Perovskite metal oxides recently received significant interest for various applications, including Li-Air batteries [125, 126, 141–143]. This type of metal oxide is known to have excellent structural flexibility to adopt a wide range of substituting rare-earth and transition metal elements. Moreover, perovskite oxides have been considered as bifunctional catalysts for both

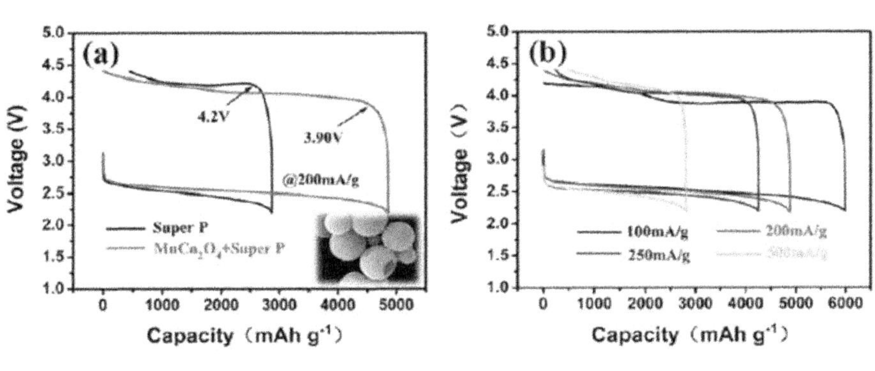

Figure 5.21. (a) First discharge–charge cycle of mesoporous $MnCo_2O_4$ microspheres mixed Super P and pure Super P with the voltage between 2.2 and 4.4 V at a current density of 200 mA/g; the inset is an SEM image of the microsphere. (b) First discharge-charge cycle of mesoporous $MnCo_2O_4$ microspheres mixed Super P at different current densities. The figures are reproduced from Ref. [135].

ORR and OER in alkaline solutions. As shown in Fig. 5.22a, both A-site and B-site in the perovskite structure can allow partial substitution to form $AA'BB'O_3$ [141]. By inserting elements with a varying number of *d*-electrons, the molecular orbital of surface metal ions can be modulated to have a strong orbital overlap with oxygen-related adsorbates (and therefore a high covalency for the metal-oxygen bond) so that the electron transfer in addition to the favourable adsorption can be promoted [141, 142]. In practice, a perovskite oxide catalyst prepared by Sun et al. can easily reach a first-cycle discharge capacity of 3672 mAh/g (carbon) without much porosity (BET surface area = 4.55 m^2/g), although at a current density of 30 mA/g [144]. Learnt from the past experience of other metal oxides, recent reports almost always adopt a porous structure in their perovskite oxides. For example, Wang and co-workers prepared a mesoporous/macroporous $La_{0.5}Sr_{0.5}CoO_{3-x}$ nanotube with a pore-size distribution dominated at 50–100 nm and showed that the first-cycle discharge capacity can reach 5,799 mAh/g (carbon) at a rate of 50 mA/g [125]. Another highlight of using mesoporous perovskite oxide was made by Zhang and co-workers [126]. They have synthesized hierarchical mesoporous $La_{0.5}Sr_{0.5}CoO_{2.91}$ nanowires

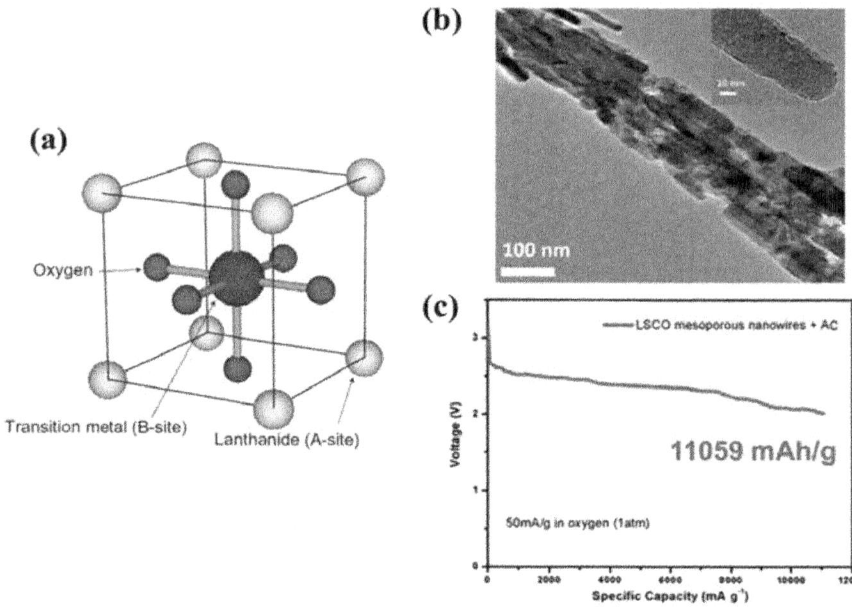

Figure 5.22. (a) The crystal structure of a typical La perovskite oxide (ABO_3). (b) HRTEM images of the hierarchical mesoporous $La_{0.5}Sr_{0.5}CoO_{2.91}$ nanowires and (c) the first-cycle discharge curve of Li-air batteries using hierarchical mesoporous $La_{0.5}Sr_{0.5}CoO_{2.91}$ nanowires (with activated carbon) as the air electrode in oxygen (P_{O2} = 1 atm). The figures are reproduced from Ref. [126, 141].

(Fig. 5.22b) with pore-size ranging from 5 nm to 100 nm (while the main distribution dominated at ~10 nm), and reported an extraordinarily high first-cycle discharge capacity of 11,059 mAh/g (carbon) at a current density of 50 mA/g. This is among the highest reported first-cycle discharge capacity for perovskite oxide cathode to-date. From these examples, there should be no doubt about the importance of meso-porosity and chemistry of the cathode material for Li-Air batteries.

There are also other types of metal oxides with meso-porosity that can deliver excellent Li-Air battery performance [145, 146]. For example, a mesoporous pyrochlore oxide ($Pb_2Ru_2O_{7-x}$) could achieve a first-cycle discharge capacity of 10,400 mAh/g (carbon) at a current density of 70 mA/g [146]. However, the particular oxide has not been well investigated for their special chemistry. For the purpose of converging attention to mesoporous materials, we will not go into detail of other metal oxides.

Certainly, the use of mesoporous transition metal oxide as the battery cathode can improve the discharge capacity and rate capability of Li-Air batteries in addition to their favourable physical/chemical properties, even though the discharge-charge voltage gap usually remains to be ~1.5 V. However, other intrinsic issues such as dissolution in acids, vulnerability to poisons, poor conductivity and structural flexibility to accommodate porous structure remain the problems of using these materials. The search for solutions to these problems is still under intensive research.

4.3 Mesoporous metal (oxide)/carbon hybrids

Mesoporous metal (oxide)/carbon hybrid may be considered as a third class of mesoporous material for Li-Air batteries. Such a hybrid is distinguished from typical metal (oxide)/carbon composite by a covalent linkage between the metal or metal oxide nanoparticles (NPs) and carbon. They are usually prepared by growth, deposition or annealing of NPs on a carbon support. The covalent connection between the NPs and carbon has been shown to derive synergistic benefit to reduce the reaction overpotential in aqueous oxygen reactions [147–149]. Also, this class of materials inherits the advantageous properties from both the metal (oxide) (e.g., high catalytic activity for both OER and ORR) and carbon (e.g., electrical conductivity, framework and high porosity), while mitigates the drawbacks of each material. Hence, the hybrid materials are considered as ideal candidates for Li-Air batteries. At the current stage of development, there are two competing bodies in this class of materials: precious metal NP decorated mesoporous carbon and non-precious metal oxide/mesoporous carbon hybrids.

Non-precious metal oxide/mesoporous carbon hybrids are considered more competitive owing to their low cost, and superior chemical versatility in composition. As discussed earlier, most of metal oxides used in the batteries are grown independent of carbon support to ensure their desired crystal phase, morphology and/or porosity; they are commonly incorporated into the cathode through a blending process with conducting, porous carbons. In the hybrid form, the metal oxide NPs are grown on porous carbons in intimate contact and can derive better stability and sometimes with superior oxygen reaction activity, depending on the form and dimension of the NPs [150]. A work by Zahoor et al., employed a hybrid material of α-MnO_2 and N-doped graphite nanofibre (α-MnO_2/N-GNF) as the air electrode in Li-Air batteries [151]. This hybrid material demonstrates good cycle stability on full-discharge (3000–4500 mAh/g (carbon)) over six cycles, although the discharge-charge voltage gap is ~1.5 V due to the large crystal size (100–200 nm) that probably hinders the electron conduction. A recent study by Yang and co-workers reported a $MnCo_2O_4$/P-doped mesoporous carbon hybrid as superior cathode material for Li-Air batteries (Fig. 5.23a) [117]. They showed that the hybrid material could deliver a high discharge capacity of 13,000 mAh/g (total mass) at a rate of 200 mA/g, even though the discharge-charge voltage gap remained significant (Fig. 5.23b). Moreover, the cycle stability of the cathode was extraordinary; the battery could operate over 200 cycles without degradation when 1000 mAh/g was set as the limited depth of discharge at a rate of 200 mA/g (Fig. 5.23c). When using precious metal oxide hybrid such as a core-shell structured RuO_2-coated CNT catalyst, Jian et al. reported a similar synergistic benefit (a discharge-charge voltage gap of 0.72 V) and good stability (300 mAh/g (catalyst) over 100 cycles at 500 mA/g) in the battery performance [152]. All of these suggest that the covalent linkage between metal oxide NPs and high surface area, porous carbon can derive excellent Li-Air battery performance with good cycle stability.

5. Conclusions

Mesoporous materials are of strong potential for use in Li-ion, Li-S and Li-Air batteries. Their high surface area and the designated porosity are ideal for the bulk/pore/surface-restricted battery chemistry. Apart from this, the physical (e.g., electrical conductivity, crystallographic voids) and chemical (e.g., active sites) properties of these materials should be also optimized to achieve the best performance in terms of discharge capacity, rate capability and cycle stability. Mesoporous carbon materials are electrically conductive and flexible in structural engineering and surface modification. On the other hand, mesoporous metal oxides are rather high activity and exhibit versatile chemistry for the various electrode reactions, although these materials are more resistant, dense and the structural engineering to open up the surface area and porosity can be relatively difficult. In a combination,

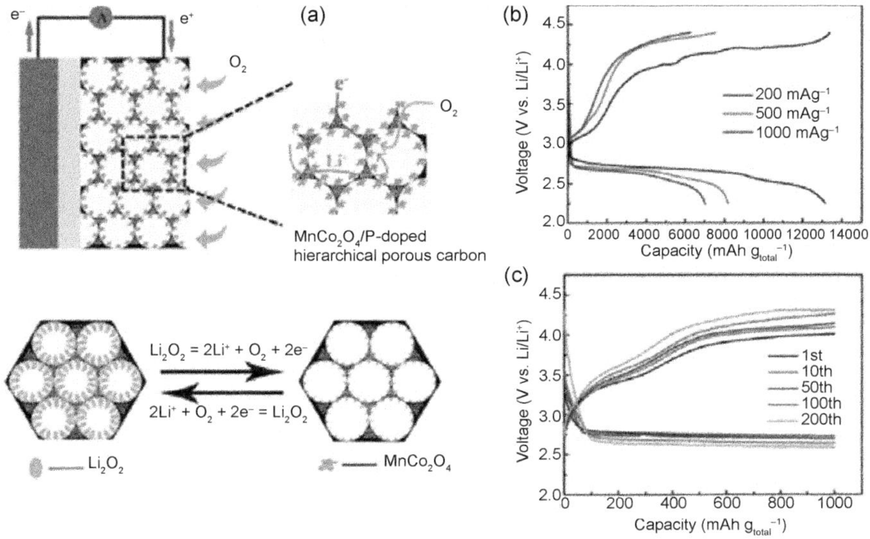

Figure 5.23. (a) A schematic diagram of the discharge-charge mechanism of a Li-Air battery using a $MnCo_2O_4$/P-doped hierarchical porous carbon hybrid cathode. (b) The first three discharge-charge cycles using the hybrid cathode at different current densities and (c) the profiles at different cycles with a limited depth of discharge of 1000 mAh/g (total mass) at a current density of 200 mA/g. The figures are reproduced from Ref. [117].

a covalently linked hybrid of mesoporous carbon and metal oxide may be a way to neutralize the drawbacks of each material while providing synergistic benefit to the battery performance. Finally, it is anticipated that a finely engineered mesoporous material will be an ideal solution to solve the many challenges in advanced Li-based batteries after a long and continuous development in both material and chemistry in the near future for the realization of the technology.

References

[1] Wang, H., Cui, L.-F., Yang, Y. et al. 2010. Mn_3O_4–graphene hybrid as a high-capacity anode material for lithium ion batteries. *J. Am. Chem. Soc.*, 132: 13978–13980.

[2] Zhao, Y., Kuai, L., Liu, Y. et al. 2015. Well-constructed single-layer molybdenum disulphide nanorose cross-linked by three dimensional-reduced graphene oxide network for superior water splitting and lithium storage property. *Sci. Rep.*, 5: 8722.

[3] Liu, Y.-G., Zhao, Y.-Y., Zhang, B.-B. et al. 2015. Assembly of multicomponent nanoframes via the synergistic actions of graphene oxide space confinement effect and oriented cation exchange. *Nanotechnology*, 26: 445601.

[4] Wang, X., Mujtaba, J., Fang, F. et al. 2015. Constructing aligned γ-Fe_2O_3 nanorods with internal void space anchored on reduced graphene oxide nanosheets for excellent lithium storage. *RSC. ADV.*, 5: 91574–91580.

[5] Liu, Y., Cheng, Z., Sun, H. et al. 2014. Mesoporous Co_3O_4 sheets/3D graphene networks nanohybrids for high-performance sodium-ion battery anode. *J. Power. Sources*, 273: 878–884.
[6] Zboril, R., Mashlan, M. and Petridis, D. 2002. Iron(III) oxides from thermal processes—synthesis, structural and magnetic properties, mössbauer spectroscopy characterization, and applications. *Chem. Mater.*, 14: 969–982.
[7] Zhang, J., Huang, T. and Yu, A. 2015. Synthesis and effect of electrode heat-treatment on the superior lithium storage performance of Co_3O_4 nanoparticles. *J. Power. Sources*, 273: 894–903.
[8] Xu, X., Cao, R., Jeong, S. et al. 2012. Spindle-like mesoporous α-Fe_2O_3 anode material prepared from MOF template for high-rate lithium batteries. *Nano. Lett.*, 12: 4988–4991.
[9] Liu, H., Wexler, D. and Wang, G. 2009. One-pot facile synthesis of iron oxide nanowires as high capacity anode materials for lithium ion batteries. *J. Alloy. Compd.*, 487: L24–L27.
[10] Han, Y., Wang, Y., Li, L. et al. 2011. Preparation and electrochemical performance of flower-like hematite for lithium-ion batteries. *Electrochim. Acta*, 56: 3175–3181.
[11] Lei, D., Zhang, M., Qu, B. et al. 2012. α-Fe_2O_3 nanowall arrays: Hydrothermal preparation, growth mechanism and excellent rate performances for lithium ion batteries. *Nanoscale*, 4: 3422–3426.
[12] Reddy, M.V., Yu, T., Sow, C.H. et al. 2007. α-Fe_2O_3 nanoflakes as an anode material for Li-ion batteries. *Adv. Funct. Mater.*, 17: 2792–2799.
[13] Wang, C., Zhou, Y., Ge, M. et al. 2010. Large-scale synthesis of SnO_2 nanosheets with high lithium storage capacity. *J. Am. Chem. Soc.*, 132: 46–47.
[14] Chen, J.S. and Lou, X.W. 2013. SnO_2-based nanomaterials: Synthesis and application in lithium-ion batteries. *Small*, 9: 1877–1893.
[15] Niu, M., Huang, F., Cui, L. et al. 2010. Hydrothermal synthesis, structural characteristics, and enhanced photocatalysis of SnO_2/α-Fe_2O_3 semiconductor nanoheterostructures. *ACS Nano.*, 4: 681–688.
[16] Zhou, W., Cheng, C., Liu, J. et al. 2011. Epitaxial growth of branched α-Fe_2O_3/SnO_2 nano-heterostructures with improved lithium-ion battery performance. *Adv. Funct. Mater.*, 21: 2439–2445.
[17] Chen, J.S., Li, C.M., Zhou, W.W. et al. 2009. One-pot formation of SnO_2 hollow nanospheres and α-Fe_2O_3@SnO_2 nanorattles with large void space and their lithium storage properties. *Nanoscale*, 1: 280–285.
[18] Lou, X.W., Deng, D., Lee, J.Y. et al. 2008. Preparation of SnO2/carbon composite hollow spheres and their lithium storage properties. *Chem. Mater.*, 20: 6562–6566.
[19] Ding, S. and Wen Lou, X. 2011. SnO_2 nanosheet hollow spheres with improved lithium storage capabilities. *Nanoscale*, 3: 3586–3588.
[20] Wang, Z., Luan, D., Boey, F.Y.C. et al. 2011. Fast formation of SnO_2 nanoboxes with enhanced lithium storage capability. *J. Am. Chem. Soc.*, 133: 4738–4741.
[21] Lou, X.W., Yuan, C. and Archer, L.A. 2007. Double-walled SnO_2 nano-cocoons with movable magnetic cores. *Adv. Mater.*, 19: 3328–3332.
[22] Lou, X.W., Li, C.M. and Archer, L.A. 2009. Designed synthesis of coaxial SnO_2@ carbon hollow nanospheres for highly reversible lithium storage. *Adv. Mater.*, 21: 2536–2539.
[23] Ding, S., Chen, J.S., Qi, G. et al. 2011. Formation of SnO2 hollow nanospheres inside mesoporous silica nanoreactors. *J. Am. Chem. Soc.*, 133: 21–23.
[24] Wang, Y., Dai, H., Deng, J. et al. 2013. Three-dimensionally ordered macroporous InVO4: Fabrication and excellent visible-light-driven photocatalytic performance for methylene blue degradation. *Chem. Eng. J.*, 226: 87–94.
[25] Arandiyan, H., Dai, H., Deng, J. et al. 2013. Dual-templating synthesis of three-dimensionally ordered macroporous $La_{0.6}Sr_{0.4}MnO_3$-supported Ag nanoparticles: Controllable alignments and super performance for the catalytic combustion of methane. *Chem. Commun.*, 49: 10748–10750.

[26] Arandiyan, H., Chang, H., Liu, C. et al. 2013. Dextrose-aided hydrothermal preparation with large surface area on 1D single-crystalline perovskite $La_{0.5}Sr_{0.5}CoO_3$ nanowires without template: Highly catalytic activity for methane combustion. *J. Mol. Catal. A-Chem.*, 378: 299–306.
[27] Arandiyan, H., Dai, H., Deng, J. et al. 2014. Three-dimensionally ordered macroporous $La_{0.6}Sr_{0.4}MnO_3$ supported Ag nanoparticles for the combustion of methane. *J. Phys. Chem. C*, 118: 14913–14928.
[28] Arandiyan, H.R. and Parvari, M. 2008. Preparation of La-Mo-V mixed-oxide systems and their application in the direct synthesis of acetic acid. *J. Nat. Gas. Chem.*, 17: 213–224.
[29] Arandiyan, H., Peng, Y., Liu, C. et al. 2014. Effects of noble metals doped on mesoporous LaAlNi mixed oxide catalyst and identification of carbon deposit for reforming CH_4 with CO_2. *J. Chem. Technol. Biot.*, 89: 372–381.
[30] Arandiyan, H.R. and Parvari, M. 2009. Studies on mixed metal oxides solid solutions as heterogeneous catalysts. *Braz. J. Chem. Eng.*, 26: 63–74.
[31] Arandiyan, H., Dai, H., Ji, K. et al. 2015. Enhanced catalytic efficiency of Pt nanoparticles supported on 3D ordered macro-/mesoporous $Ce_{0.6}Zr_{0.3}Y_{0.1}O_2$ for methane combustion. *Small*, 11: 2366–2371.
[32] Li, J., Dahn, H.M., Sanderson, R.J. et al. 2008. Impact of rare earth additions on transition metal oxides as negative electrodes for lithium-ion batteries. *J. Electrochem. Soc.*, 155: A975–A981.
[33] Varaprasad, A.M., Mohan, A.L.S., Chakrabarty, D.K. et al. 1979. Structural and dielectric studies of some perovskite-type titanates. *J. Phys. C: Solid State Phys.*, 12: 465.
[34] Inaguma, Y., Katsumata, T., Itoh, M. et al. 2002. Crystal structure of a lithium ion-conducting perovskite $La_{2/3-x}Li_{3x}TiO_3$ (x = 0.05). *J. Solid State Chem.*, 166: 67–72.
[35] Várez, A., García-Alvarado, F., Morán, E. et al. 1995. Microstructural study of $La_{0.5}Li_{0.5}TiO_3$. *J. Solid State Chem.*, 118: 78–83.
[36] Fourquet, J.L., Duroy, H. and Crosnier-Lopez, M.P. 1996. Structural and microstructural studies of the series $La_{2/3-x}Li_{3x1/3-2x}TiO_3$. *J. Solid State Chem.*, 127: 283–294.
[37] Stramare, S., Thangadurai, V. and Weppner, W. 2003. Lithium lanthanum tritanates: a review. *Chem. Mater.*, 15: 3974–3990.
[38] Inaguma, Y., Liquan, C., Itoh, M. et al. 1993. High ionic conductivity in lithium lanthanum titanate. *Solid State Commun.*, 86: 689–693.
[39] Sharma, Y., Sharma, N., Subba Rao, G.V. et al. 2008. Studies on nano-CaO·SnO_2 and nano-$CaSnO_3$ as anodes for li-ion batteries. *Chem. Mater.*, 20: 6829–6839.
[40] Zhou, H., Zhu, S., Hibino, M. et al. 2003. Lithium storage in ordered mesoporous Carbon (CMK-3) with high reversible specific energy capacity and good cycling performance. *Adv. Mater.*, 15: 2107–2111.
[41] Guo, B., Wang, X., Fulvio, P.F. et al. 2011. Soft-templated mesoporous carbon-carbon nanotube composites for high performance lithium-ion batteries. *Adv. Mater.*, 23: 4661–4666.
[42] Yang, J., Takeda, Y., Imanishi, N. et al. 1999. Study of the cycling performance of finely dispersed lithium alloy composite electrodes under high li-utilization. *J. Power Sources*, 79: 220–224.
[43] Grigoriants, I., Sominski, L., Li, H. et al. 2005. The use of tin-decorated mesoporous carbon as an anode material for rechargeable lithium batteries. *Chem. Commun (Camb).*, 921–923.
[44] Liu, H.-j., Bo, S.-h., Cui, W.-j. et al. 2008. Nano-sized cobalt oxide/mesoporous carbon sphere composites as negative electrode material for lithium-ion batteries. *Electrochim. Acta*, 53: 6497–6503.
[45] Cheng, M.-Y. and Hwang, B.-J. 2010. Mesoporous carbon-encapsulated NiO nanocomposite negative electrode materials for high-rate Li-ion battery. *J. Power Sources*, 195: 4977–4983.

[46] Shen, L., Zhang, X., Uchaker, E. et al. 2012. $Li_4Ti_5O_{12}$ nanoparticles embedded in a mesoporous carbon matrix as a superior anode material for high rate lithium ion batteries. *Adv. Energy Mater.*, 2: 691–698.
[47] Zeng, L., Zheng, C., Deng, C. et al. 2013. MoO_2-ordered mesoporous carbon nanocomposite as an anode material for lithium-ion batteries. *ACS Appl. Mater. Inter.*, 5: 2182–2187.
[48] Kim, S.W., Seo, D.H., Ma, X. et al. 2012. Electrode materials for rechargeable sodium-ion batteries: Potential alternatives to current lithium-ion batteries. *Adv. Energy Mater.*, 2: 710–721.
[49] Xu, Y., Zhu, Y., Liu, Y. et al. 2013. Electrochemical performance of porous carbon/tin composite anodes for sodium-ion and lithium-ion batteries. *Adv. Energy Mater.*, 3: 128–133.
[50] Liang, C., Dudney, N.J. and Howe, J.Y. 2009. Hierarchically structured sulfur/carbon nanocomposite material for high-energy lithium battery. *Chem. Mater.*, 21: 4724–4730.
[51] Schuster, J., He, G., Mandlmeier, B. et al. 2012. Spherical ordered mesoporous carbon nanoparticles with high porosity for lithium–sulphur batteries. *Angew. Chem. Int. Ed. Engl.*, 51: 3591–3595.
[52] Chen, S.-R., Zhai, Y.-P., Xu, G.-L. et al. 2011. Ordered mesoporous carbon/sulphur nanocomposite of high performances as cathode for lithium–sulphur battery. Electrochimi. Acta, 56: 9549–9555.
[53] Manthiram, A., Fu, Y., Chung, S.H. et al. 2014. Rechargeable lithium–sulfur batteries. *Chem. Rev.*, 114: 11751–11787.
[54] Ji, X., Lee, K.T. and Nazar, L.F. 2009. A highly ordered nanostructured carbon–sulphur cathode for lithium–sulphur batteries. *Nat. Mater.*, 8: 500–506.
[55] He, G., Ji, X. and Nazar, L. 2011. High "C" rate Li-S cathodes: Sulfur imbibed bimodal porous carbons. *Energ. Environ. Sci.*, 4: 2878.
[56] Sun, X.G., Wang, X., Mayes, R.T. et al. 2012. Lithium–sulfur batteries based on nitrogen-doped carbon and an ionic-liquid electrolyte. *ChemSusChem.*, 5: 2079–2085.
[57] Mao, Y., Duan, H., Xu, B. et al. 2012. Lithium storage in nitrogen-rich mesoporous carbon materials. *Energ. Environ. Sci.*, 5: 7950.
[58] Sun, F., Wang, J., Chen, H. et al. 2013. High efficiency immobilization of sulfur on nitrogen-enriched mesoporous carbons for Li–S batteries. *ACS Appl. Mater. Interfaces*, 5: 5630–5638.
[59] Song, J., Xu, T., Gordin, M.L. et al. 2014. Nitrogen-doped mesoporous carbon promoted chemical adsorption of sulfur and fabrication of high-areal-capacity sulfur cathode with exceptional cycling stability for lithium-sulfur batteries. *Adv. Funct. Mater.*, 24: 1243–1250.
[60] Park, M.-S., Jeong, B.O., Kim, T.J. et al. 2014. Disordered mesoporous carbon as polysulfide reservoir for improved cyclic performance of lithium–sulfur batteries. *Carbon*, 68: 265–272.
[61] Zakhidov, R.H.B., Anvar, A., Iqbal Zafar. et al. 1998. Carbon structures with three-dimensional periodicity at optical wavelengths. *Science*, 282: 897–901.
[62] Lai, X.P.G.C., Zhang, B., Yan, T.Y. et al. 2009. Synthesis and electrochemical performance of sulfur/highly porous carbon composites. *J. Phys. Chem. C*, 113: 4712–4716.
[63] Shi, Q.-H., Feng, D., Wang, J.-F. et al. 2008. Porous carbon and carbon/metal oxide microfibers with well-controlled pore structure and interface. *J. Am. Chem. Soc.*, 130: 5034–5035.
[64] Zhao, D.-Y., Huo, Q.-S., Melosh, N. et al. 1998. Triblock copolymer syntheses of mesoporous silica with periodic 50 to 300 angstrom pores. *Science*, 279: 548–552.
[65] Ji, X., Evers, S., Black, R. et al. 2011. Stabilizing lithium–sulphur cathodes using polysulphide reservoirs. *Nat. Commun.*, 2: 325.

[66] Han, X., Xu, Y., Chen, X. et al. 2013. Reactivation of dissolved polysulfides in Li-S batteries based on atomic layer deposition of Al2O3 in nanoporous carbon cloth. *Nano Energy*, 2: 1197–1206.
[67] Song, M.-S., Han, S.-C., Kim, H.-S. et al. 2004. Effects of nanosized adsorbing material on electrochemical properties of sulphur cathodes for Li/S secondary batteries. *J. Electrochem. Soc.*, 151: A791.
[68] Wei, S.Z., Li, W., Cha, J.J. et al. 2013. Sulphur–TiO_2 yolk–shell nanoarchitecture with internal void space for long-cycle lithium–sulphur batteries. *Nat. Commun.*, 4: 1331.
[69] Cheng, F. and Chen, J. 2012. Metal–air batteries: From oxygen reduction electrochemistry to cathode catalysts. *Chem. Soc. Rev.*, 41: 2172–2192.
[70] Cairns, E.J. and Albertus, P. 2010. Batteries for electric and hybrid-electric vehicles. *Annu. Rev. Chem. Biol. Eng.*, 1: 299–320.
[71] Christensen, J., Albertus, P., Sanchez-Carrera, R.S. et al. 2012. A critical review of Li/air batteries. *J. Electrochem. Soc.*, 159: R1–R30.
[72] Girishkumar, G., McCloskey, B., Luntz, A.C. et al. 2010. Lithium–air battery: promise and challenges. *J. Phys. Chem. Lett.*, 1: 2193–2203.
[73] Franco, A.A. and Xue, K.-H. 2013. Carbon-based electrodes for lithium air batteries: Scientific and technological challenges from a modelling perspective. *ECS J. Solid State SC.*, 2: M3084–M3100.
[74] Ding, N., Chien, S.W., Hor, T.S.A. et al. 2014. Influence of carbon pore size on the discharge capacity of Li–O_2 batteries. *J. Mater. Chem. A*, 2: 12433–12441.
[75] Schied, T., Ehrenberg, H., Eckert, J. et al. 2014. An O_2 transport study in porous materials within the Li–O_2–system. *J. Power Sources*, 269: 825–833.
[76] Tran, C., Yang, X.-Q. and Qu, D. 2010. Investigation of the gas-diffusion-electrode used as lithium/air cathode in non-aqueous electrolyte and the importance of carbon material porosity. *J. Power Sources*, 195: 2057–2063.
[77] Ye, L., Lv, W., Zhang, K.H.L. et al. 2015. A new insight into the oxygen diffusion in porous cathodes of lithium-air batteries. Energy, 83: 669–673.
[78] Andrei, P., Zheng, J.P., Hendrickson, M. et al. 2010. Some possible approaches for improving the energy density of Li-air batteries. *J. Electrochem. Soc.*, 157: A1287–A1295.
[79] Bruce, P.G., Freunberger, S.A., Hardwick, L.J. et al. 2012. Li-O_2 and Li-S batteries with high energy storage. *Nat. Mater.*, 11: 19–29.
[80] Chen, M., Jiang, X., Yang, H. et al. 2015. Performance improvement of air electrode for Li/air batteries by hydrophobicity adjustment. *J. Mater. Chem. A*, 3: 11874–11879.
[81] Albertus, P., Girishkumar, G., McCloskey, B. et al. 2011. Identifying capacity limitations in the Li/oxygen battery using experiments and modeling. *J. Electrochem. Soc.*, 158: A343–A351.
[82] Xiao, J., Wang, D., Xu, W. et al. 2010. Optimization of air electrode for Li/air batteries. *J. Electrochem. Soc.*, 157: A487–A492.
[83] Springer, T.E. and Raistrick, I.D. 1989. Electrical impedance of a pore wall for the flooded-agglomerate model of porous gas-diffusion electrodes. *J. Electrochem. Soc.*, 136: 1594–1603.
[84] Lee, J.-S., Tai Kim, S., Cao, R. et al. 2011. Metal–air batteries with high energy density: Li–air versus Zn–Air. *Adv. Energy Mater.*, 1: 34–50.
[85] Ogasawara, T., Débart, A., Holzapfel, M. et al. 2006. Rechargeable Li_2O_2 electrode for lithium batteries. *J. Am. Chem. Soc.*, 128: 1390–1393.
[86] Younesi, R., Hahlin, M., Björefors, F. et al. 2013. Li–O_2 battery degradation by lithium peroxide (Li_2O_2): A model study. *Chem. Mater.*, 25: 77–84.
[87] Viswanathan, V., Thygesen, K.S., Hummelshoj, J.S. et al. 2011. Electrical conductivity in Li_2O_2 and its role in determining capacity limitations in non-aqueous Li-O_2 batteries. *J. Chem. Phys.*, 135: 214704–214710.
[88] Aetukuri, N.B., McCloskey, B.D., García, J.M. et al. 2015. Solvating additives drive solution-mediated electrochemistry and enhance toroid growth in non-aqueous Li–O_2 batteries. *Nat. Chem.*, 7: 50–56.

[89] Xue, K.-H., Nguyen, T.-K. and Franco, A.A. 2014. Impact of the cathode microstructure on the discharge performance of lithium air batteries: A multiscale model. *J. Electrochem. Soc.*, 161: E3028–E3035.
[90] Johnson, L., Li, C., Liu, Z. et al. 2014. The role of LiO_2 solubility in O_2 reduction in aprotic solvents and its consequences for Li–O_2 batteries. *Nat. Chem.*, 6: 1091–1099.
[91] Rinaldi, A., Wijaya, O., Hoster, H.E. et al. 2014. History effects in lithium–oxygen batteries: How initial seeding influences the discharge capacity. *ChemSusChem.*, 7: 1283–1288.
[92] Nimon, V.Y., Visco, S.J., De Jonghe, L.C. et al. 2013. Modeling and experimental study of porous carbon cathodes in Li-O_2 cells with non-aqueous electrolyte. *ECS Electrochem. Lett.*, 2: A33–A35.
[93] Li, Q., Cao, R., Cho, J. et al. 2014. Nanostructured carbon-based cathode catalysts for nonaqueous lithium–oxygen batteries. *Phys. Chem. Chem. Phys.*, 16: 13568–13582.
[94] Sun, B., Chen, S., Liu, H. et al. 2015. Mesoporous carbon nanocube architecture for high-performance lithium–oxygen batteries. *Adv. Funct. Mater.*, 25: 4436–4444.
[95] Zhang, L.L., Gu, Y. and Zhao, X.S. 2013. Advanced porous carbon electrodes for electrochemical capacitors. *J. Mater. Chem. A*, 1: 9395–9408.
[96] Ma, S.B., Lee, D.J., Roev, V. et al. 2013. Effect of porosity on electrochemical properties of carbon materials as cathode for lithium-oxygen battery. *J. Power Sources*, 244: 494–498.
[97] Olivares-Marín, M., Palomino, P., Enciso, E. et al. 2014. Simple method to relate experimental pore size distribution and discharge capacity in cathodes for Li/O_2 batteries. *J. Phys. Chem. C*, 118: 20772–20783.
[98] Younesi, S.R., Urbonaite, S., Björefors, F. et al. 2011. Influence of the cathode porosity on the discharge performance of the lithium–oxygen battery. *J. Power Sources*, 196: 9835–9838.
[99] Williford, R.E. and Zhang, J.-G. 2009. Air electrode design for sustained high power operation of Li/air batteries. *J. Power Sources*, 194: 1164–1170.
[100] Kang, J., Li, O.L. and Saito, N. 2014. Hierarchical Meso–macro structure porous carbon black as electrode materials in Li–air battery. *J. Power Sources*, 261: 156–161.
[101] Zhang, Z., Bao, J., He, C. et al. 2014. Hierarchical carbon–nitrogen architectures with both mesopores and macrochannels as excellent cathodes for rechargeable Li–O_2 batteries. *Adv. Funct. Mater.*, 24: 6826–6833.
[102] Tan, P., Shyy, W., An, L. et al. 2014. A gradient porous cathode for non-aqueous lithium-air batteries leading to a high capacity. *Electrochem. Commun.*, 46: 111–114.
[103] Song, Y.-f., Wang, X.-y., Bai, Y.-s. et al. 2013. A gradient porous cathode for non-aqueous lithium-air batteries leading to a high capacity. *T. Nonferr. Metal. Soc.*, 23: 3685–3690.
[104] Xiao, J., Mei, D., Li, X. et al. 2011. Hierarchically porous graphene as a lithium–air battery electrode. *Nano Letters*, 11: 5071–5078.
[105] Liang, H.-W., Zhuang, X., Brüller, S. et al. 2014. Hierarchically porous carbons with optimized nitrogen doping as highly active electrocatalysts for oxygen reduction. *Nat. Commun.*, 5.
[106] Tao, Y., Weimin, Z., Wen-Ting, L. et al. 2015. 2D Materials, 2: 024002.
[107] Gong, K.P., Du, F., Xia, Z.H. et al. 2009. Nitrogen-doped carbon nanotube arrays with high electrocatalytic activity for oxygen reduction. *Science*, 323: 760–764.
[108] Wu, K.-H., Wang, D.-W., Su, D.-S. et al. 2015. A discussion on the activity origin in metal-free nitrogen-doped carbons for oxygen. *ChemSusChem.*, 8: 2772–2788.
[109] Sakaushi, K., Fellinger, T.-P. and Antonietti, M. 2012. Bifunctional metal-free catalysis of mesoporous noble carbons for oxygen reduction and evolution reactions. *ChemSusChem.*, 8: 1156–1160.
[110] Chen, S., Bi, J., Zhao, Y. et al. 2012. Nitrogen-doped carbon nanocages as efficient metal-free electrocatalysts for oxygen reduction reaction. *Adv. Mater.*, 24: 5593–5597.
[111] Qu, L.T., Liu, Y., Baek, J.B. et al. 2010. Nitrogen-doped graphene as efficient metal-free electrocatalyst for oxygen reduction in fuel cells. *Acs Nano.*, 4: 1321–1326.

[112] Yu, D., Zhang, Q. and Dai, L. 2010. Highly efficient metal-free growth of nitrogen-doped single-walled carbon nanotubes on plasma-etched substrates for oxygen reduction. *J. Am. Chem. Soc.*, 132: 15127–15129.
[113] Nie, H., Zhang, Y., Zhou, W. et al. 2014. Nitrogen-containing mesoporous carbon cathode for lithium-oxygen batteries: The influence of nitrogen on oxygen reduction reaction. *Electrochim. Acta*, 150: 205–210.
[114] Wu, G., Mack, N.H., Gao, W. et al. 2012. Nitrogen-doped graphene-rich catalysts derived from heteroatom polymers for oxygen reduction in nonaqueous lithium–O_2 battery cathodes. *ACS Nano.*, 6: 9764–9776.
[115] Nie, H., Zhang, H., Zhang, Y. et al. 2013. Nitrogen enriched mesoporous carbon as a high capacity cathode in lithium–oxygen batteries. *Nanoscale*, 5: 8484–8487.
[116] Yoo, E. and Zhou, H. 2014. Hybrid electrolyte Li-air rechargeable batteries based on nitrogen-and phosphorus-doped graphene nanosheets. *RSC Advances*, 4: 13119–13122.
[117] Cao, X., Wu, J., Jin, C. et al. 2015. $MnCo_2O_4$ anchored on P-doped hierarchical porous carbon as an electrocatalyst for high-performance rechargeable Li–O_2 batteries. *ACS Catal.*, 5: 4890–4896.
[118] Kinoshita, K. 1992. Electrochemical Oxygen Technology, Wiley-Interscience.
[119] Cheng, H. and Scott, K. 2010. Carbon-supported manganese oxide nanocatalysts for rechargeable lithium–air batteries. *J. Power Sources*, 195: 1370–1374.
[120] Débart, A., Paterson, A.J., Bao, J. et al. 2008. α-MnO_2 nanowires: A catalyst for the O_2 electrode in rechargeable lithium batteries. *Angew. Chem. Int. Edit.*, 47: 4521–4524.
[121] Thapa, A.K., Hidaka, Y., Hagiwara, H. et al. 2011. Mesoporous β-MnO_2 air electrode modified with Pd for rechargeability in lithium-air battery. *J. Electrochem. Soc.*, 158: A1483–A1489.
[122] Hamdani, M., Singh, R.N. and Chartier, P. 2010. Co_3O_4 and Co-based spinel oxides bifunctional oxygen electrodes. *Int. J. Electrochem. SC.*, 5: 556–577.
[123] Wang, H., Yang, Y., Liang, Y. et al. 2012. Rechargeable Li–O_2 batteries with a covalently coupled $MnCo_2O_4$–graphene hybrid as an oxygen cathode catalyst. *Energ. Environ. Sci.*, 5: 7931–7935.
[124] Yoon, T. and Park,Y. 2012. Carbon nanotube/Co_3O_4 composite for air electrode of lithium-air battery. *Nanoscale Res. Lett.*, 7 (2012) 28.
[125] Liu, G., Chen, H., Xia, L. et al. 2015. Hierarchical mesoporous/macroporous perovskite $La_{0.5}Sr_{0.5}CoO_3$–x nanotubes: A bifunctional catalyst with enhanced activity and cycle stability for rechargeable lithium oxygen batteries. *ACS Appl. Mater. Inter.*, 7: 22478–22486.
[126] Zhao, Y., Xu, L., Mai, L. et al. 2012. Hierarchical mesoporous perovskite $La_{0.5}Sr_{0.5}CoO_{2.91}$ nanowires with ultrahigh capacity for Li-air batteries. *P. Natl. Acad. Sci. USA.*, 109: 19569–19574.
[127] Thapa, A.K. and Ishihara, T. 2011. Mesoporous α-MnO_2/Pd catalyst air electrode for rechargeable lithium–air battery. *J. Power Sources*, 196: 7016–7020.
[128] Kavakli, C., Meini, S., Harzer, G. et al. 2013. Nanosized carbon-supported manganese oxide phases as lithium–oxygen battery cathode catalysts. *ChemCatChem.*, 5: 3358–3373.
[129] Li, J., Wang, N., Zhao, Y. et al. 2011. MnO_2 nanoflakes coated on multi-walled carbon nanotubes for rechargeable lithium-air batteries. *Electrochem. Commun.*, 13: 698–700.
[130] Débart, A., Bao, J., Armstrong, G. et al. 2007. An O_2 cathode for rechargeable lithium batteries: the effect of a catalyst. *J. Power Sources*, 174: 1177–1182.
[131] Thapa, A.K., Saimen, K. and Ishihara, T. 2010. Pd/MnO_2 air electrode catalyst for rechargeable lithium/air battery. *Electrochem. Solid ST.*, 13: A165–A167.
[132] Cui, Y., Wen, Z. and Liu, Y. 2011. A free-standing-type design for cathodes of rechargeable Li–O_2 batteries. *Energ. Environ. Sci.*, 4: 4727–4734.
[133] Liu, Y., Cao, L.-J., Cao, C.-W. et al. 2014. Facile synthesis of spinel $CuCo_2O_4$ nanocrystals as high-performance cathode catalysts for rechargeable Li–air batteries. *Chem. Commun.*, 50: 14635–14638.

[134] Mohamed, S.G., Tsai, Y.-Q., Chen, C.-J. et al. 2015. Ternary spinel MCo_2O_4 (M = Mn, Fe, Ni, and Zn) porous nanorods as bifunctional cathode materials for lithium–O_2 batteries. *ACS Appl. Mater. Inter.*, 7: 12038–12046.
[135] Ma, S., Sun, L., Cong, L. et al. 2013. Multiporous $MnCo_2O_4$ microspheres as an efficient bifunctional catalyst for nonaqueous Li–O_2 batteries. *J. Phys. Chem. C*, 117: 25890–25897.
[136] Hung, T.-F., Mohamed, S.G., Shen, C.-C. et al. 2013. Mesoporous $ZnCO_2O_4$ nanoflakes with bifunctional electrocatalytic activities toward efficiencies of rechargeable lithium–oxygen batteries in aprotic media. *Nanoscale*, 5: 12115–12119.
[137] Lee, C. and Park, Y. 2015. Characterization and identification of ubiquitin conjugation sites with E_3Ligase recognition specificities. *Nanoscale. Res. Lett.*, 10: 1–8.
[138] Cui, Y., Wen, Z., Sun, S. et al. 2012. Mesoporous Co_3O_4 with different porosities as catalysts for the lithium–oxygen cell. *Solid State Ionics*, 225: 598–603.
[139] Kim, K. and Park, Y. 2012. Catalytic properties of Co_3O_4 nanoparticles for rechargeable Li/air batteries. *Nanoscale Res. Lett.*, 7 (2012) 47.
[140] Li, Y., Zou, L., Li, J. et al. 2014. Synthesis of ordered mesoporous $NiCo_2O_4$ via hard template and its application as bifunctional electrocatalyst for Li-O_2 batteries. *Electrochim. Acta*, 129: 14–20.
[141] Suntivich, J., Gasteiger, H.A., Yabuuchi, N. et al. 2011. Design principles for oxygen-reduction activity on perovskite oxide catalysts for fuel cells and metal–air batteries. *Nat. Chem.*, 3: 546–550.
[142] Suntivich, J., May, K.J., Gasteiger, H.A. et al. 2011. A perovskite oxide optimized for oxygen evolution catalysis from molecular orbital principles. *Science*, 334: 1383–1385.
[143] Zhang, J., Zhao, Y., Zhao, X. et al. 2014. Porous perovskite $LaNiO_3$ nanocubes as cathode catalysts for Li-O_2 batteries with low charge potential. *Sci. Rep.*, 4: 6005.
[144] Sun, N., Liu, H., Yu, Z. et al. 2014. The $La_{0.6}Sr_{0.4}CoO_3$ perovskite catalyst for Li-O_2 battery. *Solid State Ionics*, 268: 125–130.
[145] Horowitz, H.S., Longo, J.M. and Horowitz, H.H. 1983. Oxygen electrocatalysis on some oxide pyrochlores. *J. Electrochem. Soci.*, 130: 1851–1859.
[146] Oh, S.H., Black, R., Pomerantseva, E. et al. 2012. Synthesis of a metallic mesoporous pyrochlore as a catalyst for lithium–O_2 batteries. *Nat. Chem.*, 4: 1004–1010.
[147] Wu, K.-H., Zeng, Q., Zhang, B. et al. 2015. Structural origin of the activity in Mn_3O_4–graphene oxide hybrid electrocatalysts for the oxygen reduction reaction. *ChemSusChem.*, 8: 3331–3339.
[148] Liang, Y., Li, Y., Wang, H. et al. 2011. Co_3O_4 nanocrystals on graphene as a synergistic catalyst for oxygen reduction reaction. *Nat. Mater.*, 10: 780–786.
[149] Liang, Y., Wang, H., Zhou, J. et al. 2012. Covalent hybrid of spinel manganese–cobalt oxide and graphene as advanced oxygen reduction electrocatalysts. *J. Am. Chem. Soci.*, 134: 3517–3523.
[150] Wang, L., Zhao, X., Lu, Y. et al. 2011. $CoMn_2O_4$ spinel nanoparticles grown on graphene as bifunctional catalyst for lithium-air batteries. *J. Electrochem. Soci.*, 158: A1379–A1382.
[151] Zahoor, A., Christy, M., Jang, H. et al. 2015. Increasing the reversibility of Li–O_2 batteries with caterpillar structured α–MnO_2/N–GNF bifunctional electrocatalysts. *Electrochim. Acta*, 157: 299–306.
[152] Jian, Z., Liu, P., Li, F. et al. 2014. Core–shell-structured CNT@ RuO_2 composite as a high-performance cathode catalyst for rechargeable Li–O_2 batteries. *Angew. Chem. Int. Edit.*, 53: 442–446.

CHAPTER 6

Recent Advances in Mesoporous Materials for Lithium Batteries

Yimu Hu,[#] *Hoang Yen*[#] and *Freddy Kleitz**

ABSTRACT

Mesoporous metal oxides, metals, carbons, as well as metal nitrides/sulfides can be advantageously exploited in electrochemical applications because of their attractive features such as short mass and charge pathways, high transport rates of both lithium ions and electrons, intrinsic catalytic properties and hosting capability. The aim of this chapter is to provide an overview of recent developments in the area of mesoporous electrode materials in different kinds of Li batteries, with emphasis on Li-ion batteries. Other Li-based systems (Li-air and Li-S) will be briefly covered. For each battery system, we begin by describing their basic electrochemistry, principle of operation, and the challenges encountered in attempts to achieve better performance. Then, we focus on illustrating the significant contributions of mesoporous materials in advancing battery technologies for energy conversion and storage, with the help of some excellent examples of advanced materials such as electrodes for Li batteries. Finally, suggestions and perspectives on future research directions in this field are also mentioned.

Keywords: Mesoporous materials, Nanostructures, Nanoporous electrode, Templated synthesis, Li-ion battery, Li-air (O_2) battery, Li-S battery, Energy storage

Department of Chemistry, Université Laval, Québec (QC), Canada G1V 0A6.
* Corresponding author: Freddy.Kleitz@chm.ulaval.ca
[#] These authors contributed equally to this work.

1. Introduction and Scope

The global energy demand is soaring. The dependence of energy consumption/production on fossil fuels (petroleum, coal, and nature gas) has led to severe impact on local and global economics and ecology, and has motivated the rapid development of renewable, carbon-neutral energy sources. During the past two decades, a confluence of driving forces has created a sustained and significant world-wide effort to develop alternative energy sources, namely mechanical (hydroelectricity and wind energy), nuclear, solar and chemical energy are being presently investigated. Among these, electrochemical production and storage of energy is an attractive path to convert chemical energy into electrical energy. Systems for this purpose include batteries, fuel cells and supercapacitors. Compared to fuel cells and supercapacitors, batteries have an established market position, and the concept has an increasingly important application market. Common batteries that find wide commercial use are divided into two categories: primary batteries that are designed to be charged once and discarded; and secondary (or rechargeable) batteries that can be charged, discharged into a load and recharged several times. Since the introduction of first commercial Lithium-Ion Battery (LIB) by Sony in the 1990s, rechargeable lithium batteries have revolutionized the portable electronics and small-grid technologies, and later became the technology of choice for emerging applications in large-scale transportation such as Hybrid Electric Vehicles (HEVs), Plug-in Hybrid Electric Vehicles (PHEVs) and all-Electric Vehicles (EVs).

However, compared to fossil fuels such as natural gas (13800 Wh/kg and 5800 Wh/L for specific energy and energy density, respectively), gasoline (12000 Wh/kg and 11000 Wh/L) and coal (6500 Wh/kg and 12000 Wh/L), the inferiority of battery systems including lithium-ion battery (100–265 Wh/kg and 250–676 Wh/L), nickel-metal hydride battery (50 Wh/kg and 140–300 Wh/L), and lead acid battery (33–42 Wh/kg and 60–110 Wh/L) is obvious. Enabling batteries to compete with gasoline is a colossal challenge, that is, to have at least 10-fold increase in energy density over the next decade. The target for the automotive industry, on the other hand, is more realistic and less demanding. Indeed, to satisfy the performance requirements for large-scale transportation, high energy density, long-term stability, low cost and safety are critical. For example, according to the US Department of Energy (DOE), the major goals of the Batteries and Energy Storage Subprogram by 2022 are to develop a battery with a specific energy of 250 Wh/kg and an energy density of 400 Wh/L with a cost of US$125/kwh in order to meet the requirements for a light-duty fleet of HEV or PEV [1].

As energy consumption increases with the growth of global population, electrochemical cells have become tremendously important not only in the mass consumer market, but also in the energy storage market. For instance, energy storage devices have particularly interesting applications in renewable energy generation, such as wind turbines or photovoltaic panels. Supercapacitors and Redox Flow Batteries (RFB) are capable of storing and access energy quickly, and thus control the ramp rates of energy output before presenting it to the grid. They also allow the storage and moving of energy for later use in times of unfavorable weather conditions or on-peak demand, making renewable energy reliable and enabling to adjust their power output on demand, thus allowing the electrical grids to achieve the best performance [1].

Success in the battery market is largely related to the cost, performance and safety of the battery, which depend fundamentally on the properties of the materials that compose the battery system and the reactions that occur on and within these materials. It is beyond the scope of this chapter to discuss all electrochemical systems in detail; instead, this chapter aims at discussing fundamental material science and material chemistry challenges that are applicable to a range of battery systems encountered. While it is not possible to cover all aspects in this broad range of technologies, the focus of this chapter is to discuss the technologies used for electrodes in batteries based on their chemical, mechanical and electrochemical properties, their applications and performance, with particular emphasis on mesoporous materials for lithium-ion battery electrodes. Mesoporous materials draw special attention as electrode candidates for LIBs owing to shorter diffusion length for lithium ions and electronic movement, and high packing density. With this perception, here we discuss some recent interesting results on nanostructured mesoporous electrode materials (both cathode and anode) for applications in high energy LIBs. The electrode material requirements, properties and characterization techniques will also be described with respect to LIBs. Despite the widespread use and rapid development of LIBs, a great need still exists for ameliorating the performance; that is, increasing the cycle life, reducing the cost and improving safety. We will start with a brief overview of the general battery technology, followed by more in-depth discussions of specific electrode materials, their advantages and problems encountered, using crosscutting examples for lithium-ion battery systems. Furthermore, as good candidates with high theoretical energy density that can potentially meet DOE's criteria for large-scale transportation applications, other types of lithium-based batteries, such as lithium-air and -sulfur batteries, will be briefly described.

2. Battery Technologies

2.1 Principles and definitions

A battery is composed of one or several electrochemical cells that are connected in series and/or in parallel to convert chemical energy into electrical energy. Figure 6.1 shows the basic operating principle and key features of batteries. All batteries are closed systems that contain an electronegative electrode (the anode, where oxidation occurs) and an electropositive electrode (the cathode, which associates with reductive reactions) as the charge transfer medium. The anode and cathode are the "active materials" or the "active mass" in the battery, the materials that generate electrical current by means of chemical reaction. Between anode and cathode resides the electrolyte, which provides pure ionic conductivity between the electrodes of the cell. In order to allow ions in electrolyte to flow through the cell and the electrons to go through the external circuit, the electrodes must be good ionic and electronic conductors [2, 3]. Most electrochemically active materials are not good electronic conductors. Practically, electrodes in batteries are composed of particles of the active material, a conductive diluent, such as carbon black and a polymer binder to physically hold the electrode together. Thus, most electrodes are complex porous composites.

When the anode and cathode contact each other, a short-circuit occurs in the compartment; thus a separator is placed between the two electrodes to ensure the physical separation of the electrodes. The separator is a porous inert membrane filled with electrolyte, permeable to ionic flow yet preventing electric contact of the electrodes. When no external current flows through the system, the cell potential or open-circuit voltage (V_{oc}) is determined by the thermodynamic voltage between anode and cathode,

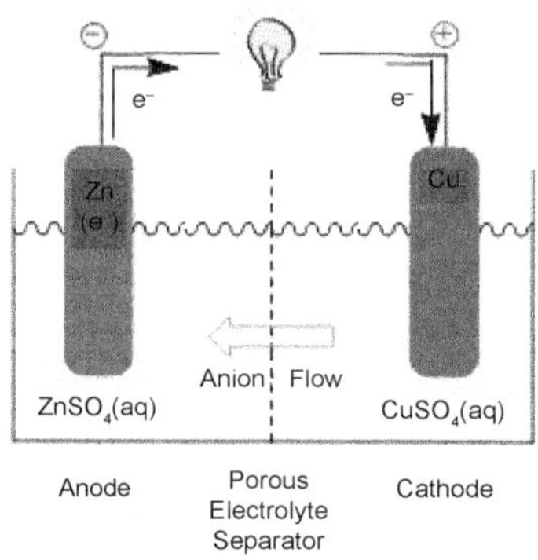

Figure 6.1. Schematic representation of a Daniell cell showing the operation mechanism and key features.

with $\Delta G = -EF$ (where F is the Faraday constant, 96 485.3 C/mol). When the battery is producing current into the external circuit, the voltage of the battery is called closed-circuit voltage, and this operation of delivering electrical energy to an external load is called "discharge". The operation in which the battery is restored to its original charged condition by reversal of the current flow is called "charge". If the electrode discharge reactions are not reversible, the cell is not rechargeable and thus considered as a "primary battery" whereas a rechargeable battery is also called secondary battery.

As introduced above, a primary battery is an electrochemical cell that is designed to be used once to generate electricity until exhausted and discarded, and not recharged with electricity and reused like a secondary cell. Besides daily usage in small and domestic electronics, primary batteries now play an important role in modern society, especially when charging is impractical or impossible. Typical applications for this purpose include military, security and rescue services and car electronics. They also find widely applications for medical uses. For example, the introduction of lithium-iodine ($Li-I_2$) battery in 1975 has greatly extended the pacemaker battery life up to more than 10 years. $Li/SOCl_2$ (lithium thionyl chloride) cells are ideally suited for use in Automatic External Defibrillators (AED). A primary battery system can yield > 50% of its theoretical value in delivered energy. In comparison, rechargeable batteries generally have lower energy storage capacity; the practical energy content of a rechargeable battery is approximately 25% of its theoretical value. Several different combinations of electrode materials and electrolytes are used, including lead-acid (Pb-acid), nickel-cadmium (Ni-Cd), nickel-metal hydride (Ni-MH), lithium ion (Li-ion), and lithium ion polymer (Li-ion polymer). Different batteries rely on different chemistries, depending on what they are used for. Advances in these chemistries have allowed larger and more powerful batteries to be used in vehicles and other applications. Focusing on vehicular applications, Table 6.1 lists four major types of batteries that are most commonly used, along with their advantages and limitations [2, 3].

Certain terms are used to describe and compare the performance of the system to generate electrical energy: "specific energy" (expressed per unit of weight, Wh/kg), "energy density" (expressed per unit of volume, Wh/L) and "specific capacity" (Ah/kg), all of which are directly related to the chemistry of the system. Taking cathode as an example, the energy density (E) is calculated by Equation (1):

$$E = \frac{\Delta V \times C}{1 + m_c} \tag{1}$$

where ΔV is the operating voltage of the full cell, C is the capacity of the cathode material, and m_c is the matching weight of the anode material,

Table 6.1. Main applications, advantages and limitations of different rechargeable battery systems that are most commonly used in vehicular technologies.

Technology	Main Application	Advantages	Limitations
Pb-acid battery	automobile starting, lighting, ignition	inexpensive; reliable chemistry	short cycle life (500–800 cycles); low specific energy (33–42 Wh/kg); operable only at low temperature (–35–40°C)
Ni-MH battery	electric batteries for most currently-available HEVs	environmentally friendly; simple storage and transportation	more expensive than Pd-acid batteries; high self-discharge; limited service life
Li-ion battery	widely used in PHEVs and EVs; some HEVs	high specific energy (100–265 Wh/kg, in practice), high energy efficiency (89–90%); long life; relatively low self-discharge	expensive; safety issues due to poor heat management
Li-polymer battery	HEVs	light weight; improved safety compared to Li-ion batteries	expensive; lower energy density and decreased life cycles compared to Li-ion

whereas the specific capacity (C) of the cathode can be calculated by Equation (2):

$$C = \frac{N \times F}{A_r} \tag{2}$$

where N is the number of valency of material, F is the Faraday constant, and A_r is the atomic weight of the element that consists of the cathode.

Since the purpose of a battery is to store energy and release it at the appropriate time in a controlled manner, its ability to deliver all available energy to an equipment without leaving energy behind is critical. The charge and discharge current of a battery is measured in C-rate, the rate at which a battery is charged or discharged relatively to its maximum capacity. Most portable batteries, with the exception of the lead acid, are rated at 1C, which means that the discharge current will discharge the entire battery in 1 hour. For example, for a battery with a capacity of 100 Amp/hr, 1C equates to a discharge current of 100 Amps. A 5C rate for this battery would be 500 Amps, and a 0.5C rate would be 50 Amps. In practice, the discharge current is described by direct measurement of the instantaneous current-voltage characteristics on a discharge curve, known as charge/discharge profile, or galvanostatic cyclic performance, as shown in Fig. 6.2. Since the capacity is expressed in Ampere per hour, calculating the current necessary to charge or discharge a battery is straightforward. This curve can also be

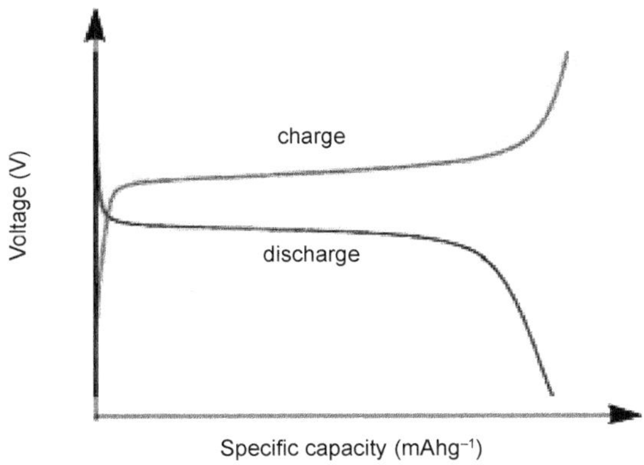

Figure 6.2. Typical charge/discharge curve of a battery.

used to determine the cell capacity, the effect of charge-discharge rate and temperature on the state of the battery.

2.2 Lithium-ion Batteries (LIBs)

Figure 6.3 compiles the theoretical values of specific capacity for a series of common electrochemically active elements. Being the most electropositive (–3.04 V vs. SHE), lightest metal (equivalent weight 6.94 g/mol) and the least dense solid element (specific gravity 0.53 g/cm^3) in the periodic table, these unique characteristics render lithium a promising candidate material that is capable of delivering high energy density per electron.

Among various types of batteries (including carbon-zinc, alkaline and nickel-cadmium batteries), the lithium battery industry is undergoing rapid expansion and innovation, now dominating the largest segment of power sources for portable electronics (accounting for more than 63% of worldwide sales values in computer, cell phone, smart-grid technologies) and electric vehicles.

The advantage in using Li metal was first demonstrated in the 1970s with the assembly of primary Li cells. Meanwhile, the discovery of intercalation compounds played a critical role in the development of high-energy rechargeable Li batteries. Figure 6.4a represents a schematic operating principle of rechargeable Li-metal battery, using Li metal as the anode and TiS_2 (an intercalation compound) as the cathode [4]. During the discharge process, the Li electrode is discharged by oxidation to form Li^+ cation, which is going into the solution and providing outer electron flow. The reaction is reversible by redeposition of the lithium. However, the redeposition of the Li is rough and mossy, and dendrite growth on the

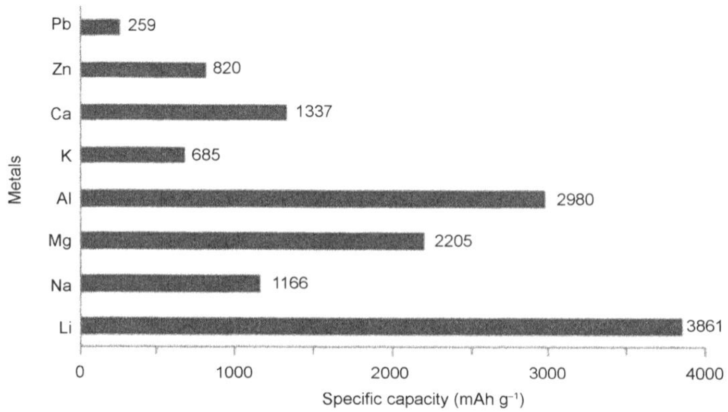

Figure 6.3. The specific capacity of a number of electrochemically active elements.

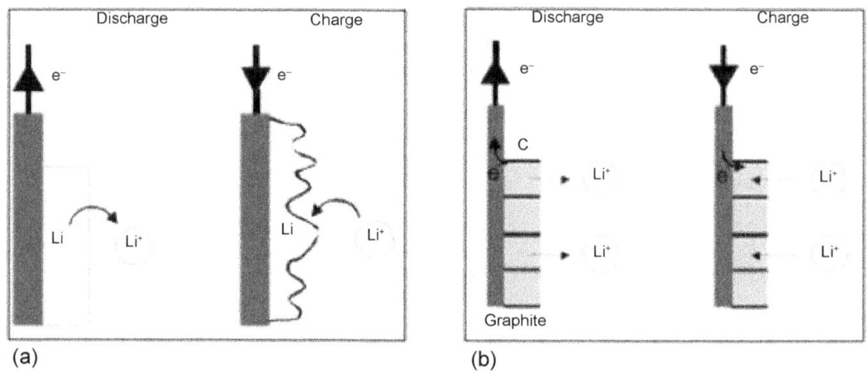

Figure 6.4. Schematic representations showing discharge and charge mechanisms of battery electrodes using various electrode materials, as discussed in the text. Reproduced with permission [2]. Copyright 2004, American Chemical Society.

metal surface after repeated Li plating can lead to internal short circuit, causing severe safety issues. Therefore, lithium metal as an anode material in rechargeable batteries was ultimately rejected due to safety concerns. One alternative approach to tackle the safety issues caused by the use of Li metal electrodes is to substitute the Li metal anode by a second insertion material, as shown in Fig. 6.4b [5, 6].

In 1991, Sony introduced the first commercial lithium-ion technology, which solved the dendritic Li growth in Li metal cells and subsequent safety concerns. Nowadays, in the case of LIBs, the anode is the source of lithium ions and is composed of lithium insertion/conversion compound such as graphitic carbon, which has a planar sheet of carbon atoms arranged in a hexagonal array as basic building unit. When charging, lithium intercalates into the carbon layer to form Li_xC alloy. The cathode is made up of another Li^+ host material possessing a much more positive redox potential such as conventional Lithium Cobalt Oxide ($LiCoO_2$), Lithium Manganese Oxide ("LMO", $LiMn_2O_4$), Lithium Iron Phosphate ("LFP", $LiFePO_4$) or Lithium Nickel Manganese Cobalt Oxide ("NMC", $LiNi_xMn_yCo_zO_2$) (*vide infra*). In LIB systems, the performance of the battery is mainly governed by the capacity of the cathode. Calculations suggest that increase of anode capacity by a factor of 10 only yields 47% of increase of overall energy density whereas 57% of improvement can be achieved by doubling the capacity of cathode [7]. The chemical reactions involved in the conventional LIB cell can be briefly described as:

On cathode (marked as +): $$LiM_{1-x}O_2 + xLi^+ + xe^- \underset{\text{discharge}}{\overset{\text{charge}}{\rightleftharpoons}} LiMO_2$$

On anode (marked as -): $$Li_xC \underset{\text{discharge}}{\overset{\text{charge}}{\rightleftharpoons}} C + xLi^+ + xe^-$$

Overall: $$Li_xC + Li_{1-x}MO_2 \underset{\text{discharge}}{\overset{\text{charge}}{\rightleftharpoons}} LiMO_2 + C$$

The performance of rechargeable LIBs strongly depends on the active materials employed for Li storage. That is, materials for electrodes should reversibly accommodate a large amount of lithium to achieve high specific capacity, whereas high current density requires fast ionic/electronic transfer in the cell.

During the past three decades, great efforts and progress have been made to improve the performance of LIBs. Two approaches have focused on the design and synthesis of nanostructured electrode materials: (1) identifying new high theoretical energy density and high capacity cathode materials, and (2) exploring new synthesis method to improve the stability and reliability of anode materials with high capacity. Given the large array of techniques available for synthesizing porous electrodes, it is important to consider the influence of the structural features of porous electrodes on their electrochemical properties in order to choose the most suitable ones for LIB applications. As mentioned above, improvement in the performance and cost of the cathode material is a very efficient way to improve the overall performance and cost of Li-ion batteries. For the anode in LIBs, on the other hand, Si is a promising candidate because of its large specific capacity and low discharge potential as compared to conventional graphite. However, its practical use has been limited due to large volume expansion and contraction associated with intercalation/deintercalation of lithium during cycling, which leads to fast capacity fading

[8]. Significant research efforts have focused on developing nanostructured Si to overcome this barrier, as well as alternative active materials with better structural stability, such as TiO_2 and mesoporous carbon/silicon composite [9–11].

Another ongoing effort is to improve the reaction kinetics. One of the key challenges of the current LIB technology for vehicular applications is its poor rate performance, which cannot meet the peak power demands to power the EV/HEV for starting, accelerating and uphill driving. Therefore, a high rate charge/discharge, i.e., satisfying performance in kinetics is a must. In LIBs, the kinetics can be categorized into three steps: diffusion in bulk, surface reactions and diffusion in the electrolyte. Hence, the power capability of a lithium battery depends critically on the rate at which the lithium ions migrate through the electrolyte and enhancing the diffusion of lithium ions is one of the most important factors in maximizing the utilization of active materials at high energy density. Large surface area extending of the electrode can be achieved throughout the porosity of the electrode. The incorporation of nanoporous electrode structure reduces the path length over which the Li^+ and electrons have to move to the reaction site. The pore size, porosity, uniformity and tortuosity of pore space have critical impacts on the electrochemical storage performance of the active materials. In this perspective, strategies to increase the rate performance have focused on synthesizing mesoporous electrode materials. In general, mesoporous electrode materials provide short transport lengths (10–20 nm) for lithium ions owing to their nano-sized grains, thus facilitating easy access for electrolytes due to their nanopores (5–10 nm) during insertion/extraction of lithium ions within the bulk particles [12, 13]. Such materials have mesoporous interconnection inside the particles, which provide large contact area between the active materials and the electrolyte, facilitating Li ion diffusion. Ordered mesoporous materials, as an extension from ordinary mesoporous materials, were applied to further improve Li ion diffusion through the ordered mesoporous framework in the active materials [14–17]. For the aforementioned reasons, such mesoporous materials have generated significant attention due to both fundamental scientific interest as well as commercial applications. Recent applications of mesoporous solids as components for lithium-ion battery electrodes will be described next.

3. Mesoporous Materials for Li ion Batteries (LIBs)

Here we will focus on mesoporous materials as electrodes for LIBs. It would require several volumes to provide a comprehensive summary of the immense body of work that are now being carried out worldwide in the field of lithium battery materials; excellent reviews on these topics can be found elsewhere [12, 14–17]. We will first briefly discuss relevant methods of synthesizing mesoporous electrodes with various pore sizes and pore architectures. We will then provide specific examples from the literature that address the performance parameters benefiting from mesostructures.

3.1 Synthesis of mesoporous electrodes for LIBs

The term "mesoporous materials" refers either to ordered or disordered networks with broad or narrow distribution of pores in the range between 2 and 50 nm [18, 19]. The synthetic methods for preparing mesoporous solids can be classified as templated and non-templated. In this context only mesoporous battery materials exhibiting well-defined pore sizes or architectures, which are best achieved via templating methods, will be addressed. However, numerous template-free methods of fabricating mesoporous electrodes are also available, such as solvothermal/hydrothermal syntheses, electrodeposition, aerogel synthesis, ultrasonication and so forth [12, 16, 17, 20]. Although these methods tend to produce materials with disordered mesostructures and broad pore size distributions, some of them are simpler than templating pathways, and they can yield materials with sufficiently porous features that enhance the performance of an electrode [21–23]. In the following discussion about synthetic methods for mesoporous solids, the influence of synthetic parameters on the pore size, pore morphology, and wall thickness in the products will be emphasized, as these meso-dimensions influence the electrochemical performance of the materials.

3.1.1 Soft templating methods for preparation of mesoporous electrodes

Various mesoporous materials, including oxides, carbon and metals, suitable for electrodes in lithium batteries have been prepared via soft templating approaches, using different surfactants as organic templates, or also called Structure-Directing Agents (SDAs), used in the presence of given inorganic precursors (Fig. 6.5a) [14, 16, 17, 24]. The necessary chemistry has been well developed for silica synthesis, where the organized mesophases are often obtained via a cooperative self-assembly process taking place between the template species and the silica network precursors [19, 25]. Removal of the surfactants by calcination or extraction produces the mesoporous materials. Syntheses are generally carried out in an aqueous solution under mild or hydrothermal conditions, and the resulting materials are usually obtained in powdered form. Alternatively, a generalized non-aqueous solvent method to prepare thin films, designated as Evaporation-Induced Self-Assembly (EISA) is also widely employed (Fig. 6.5b) [26, 27].

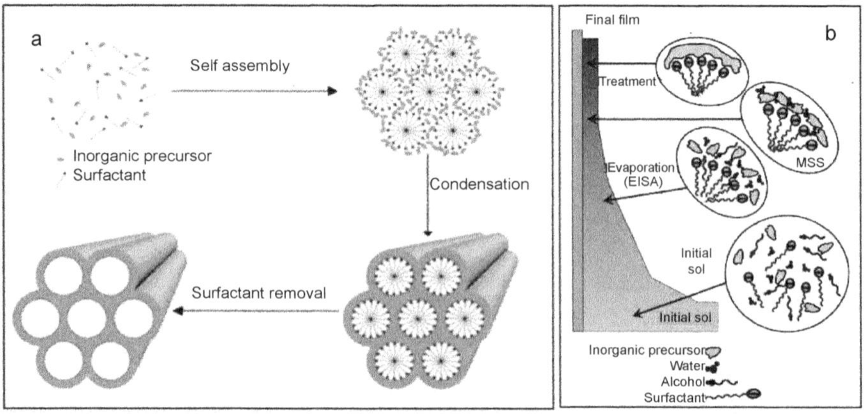

Figure 6.5. (a) Schematic representation of the synthesis of mesoporous solids by soft templating. Reproduced with permission [28]. Copyright 2005, American Chemical Society. (b) Formation of mesostructured thin films via Evaporation-Induced Self-Assembly (EISA). Reproduced with permission [26]. Copyright 2004, Wiley-VCH Verlag GmbH & Co. KGaA.

These mesoporous thin films offer some advantages for the fabrication of electrodes without intergrain effects or without using organic polymer binders (e.g., Nafion 177, poly(vinyl alcohol), polystyrene) to strengthen the mechanical stability of the electrode surface as in the case of powdered mesoporous materials [15]. The pore architecture of materials derived from soft templating methods can be controlled by the choice of surfactants, solvents and synthesis conditions. By now, many different mesostructured phases have been reported including lamellar, hexagonal, cubic, double gyroid or other symmetries [18, 19, 24, 25]. The pore size and wall thickness are adjustable through several reliable methods, such as by choosing longer alkyl chain surfactants, using large block copolymers as templates, adding swelling agents or increasing the temperature of the hydrothermal treatment. Controlling the hydrophobic volume of the surfactant aggregates is a prominent way to adjust the pore sizes of soft-templated ordered mesoporous materials. When surfactants with low molecular weight are used as templates, such as Sodium Dodecyl Sulfate (SDS) or hexadecylamine, mesoporous materials with small pore sizes (normally below 4 nm) are usually obtained [29–31]. Alternatively, block copolymers (e.g., PEO-PPO-PEO) are used to synthesize mesoporous materials with large pore size ranging from 5 to about 15 nm [32–36]. Diblock copolymers (for example, KLE122–132 PI-b-PEO, PS-b-PEO and PB-b-PEO) even lead to materials with larger pore size (up to ~40 nm) compared with triblock copolymers with similar molecular weights or hydrophilic chain lengths [37–40]. The versatility of the soft templating approach is well illustrated by the various potential mesoporous electrode materials synthesized so far, for example, mesoporous cathode (e.g., $Li_3Fe_2(PO_4)_3$ [41], V_2O_5 [42, 43]) and mesoporous anode (e.g., carbon [17, 44], TiO_2 [45], SnO_2 [30], $Sn_2P_2O_7$ [46], $Li_4Ti_5O_{12}$ [47, 48]) structures.

3.1.2 Hard templating (Nanocasting) method for preparation of mesoporous electrodes

Soft templating is a versatile method to prepare mesoporous materials with a variety of compositions and mesoporous structures. However, it is often difficult to obtain ordered non siliceous mesoporous materials, particularly transition metal oxides with highly crystalline walls when the crystallization temperatures of the inorganic phases are higher than the temperatures at which the organic surfactant templates are removed [49]. On the other hand, in the case of metal oxide compositions, the difficulty in controlling the hydrolysis and polymerization of the metal precursors still often leads to the formation of materials with very poor structural order [50]. To overcome these limitations of the soft templating route, the hard templating strategy has been developed [51]. In this method, a preformed porous solid material (usually mesoporous silica, carbon or assembly of nanoparticles) is used as a rigid matrix, i.e., the pores of the template are filled with one or more precursor species which will react inside the pores to form the desired material (Fig. 6.6).

The hard template is subsequently removed by thermal treatment for carbon or by alkaline or HF solution for silica, to yield the product as a sort of inverted replica of the original hard template. Thus, a hard template with bicontinuous pore/wall structures produces 3D-interconnected mesoporous replicas, whereas templates with 2D hexagonal arrangement produce nanowire arrays, which may be interconnected only if the template contains some complementary pores between the cylindrical mesopores [52–55]. Mesoporous silica materials KIT-6 (pore structure with *Ia3d* cubic symmetry)

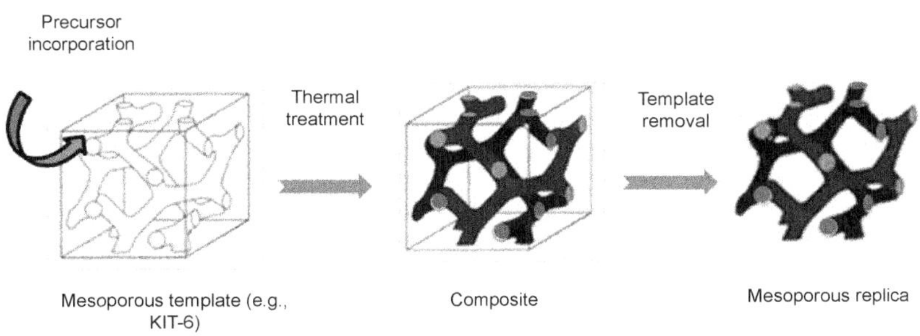

Figure 6.6. Schematic representation of hard templating method for the synthesis of mesostructured materials.

and SBA-15 (2D hexagonal mesopore structure) are the most commonly used hard templates for nanocasting of mesoporous electrodes [56–61], as their porosity characteristics are readily controlled through monitoring the temperature of the hydrothermal treatment during their synthesis, and their mesopores are large enough to accommodate precursor fluids for the desired electrode. Depending on the synthesis conditions, the mesopore size and wall thickness of these silica materials can easily be tailored from 4.5–12 nm and 2–5 nm, respectively [34, 35]. The appropriate experimental conditions to obtain high-quality replica structures have been discussed elsewhere [14, 24, 62, 63]. Briefly, the pore size and wall thickness of nanocast materials can be tuned via controlling the pore size and wall thickness of the hard template used and by changing the crystallization temperature of the mesoporous metal oxides. A number of ordered mesoporous electrode materials have been reported using mostly KIT-6 and SBA-15 silicas as hard templates, including cathode materials, such as β-MnO_2 (tetragonal rutile structure) [57, 58, 64], $LiFePO_4$ (olivine structure) [65], $LiCoO_2$ spinel [60] and $Li_{1.12}Mn_{1.88}O_4$ spinel [61]; as well as anode materials, such as Cr_2O_3 [59, 66], Si [10], Co_3O_4 [56, 67], TiO_2 [11, 68, 69], NiO [53, 70], SnO_2 [71–73], WO_{3-x} [74], Fe_2O_3 [75], MoO_2 [76] and $Cu_{1-x}Fe_{2+x}O_4$ [77].

Crystallographic phases of the active electrode materials play an important role in the electrochemical performance of the electrodes. A crystallized wall structure can be expected to provide better thermal and mechanical stability, as well as superior electrical properties compared to amorphous analog. The crystallographic nature obtained in the syntheses mentioned above depends, in part, on thermal history, material dimensions as well as precursors. With soft templates, the calcination temperatures are chosen by considering the thermal decomposition of surfactant and the rate of nucleation and crystal growth of the active material [24]. Hard templating procedures enable synthesis of mesoporous materials with highly crystalline walls because the hard templates are stable enough to tolerate high temperatures allowing many metal oxides to crystallize [14, 62]. In syntheses of multicomponent electrode materials, control of impurity and phase is critical during synthesis. In such cases, the presence of chelating agents (e.g., citric acid, glycine) may be needed to achieve precise control over product stoichiometry [63].

3.2 Electrochemical properties of mesoporous electrodes for LIBs

As mentioned above, several parameters, including diffusion of electrolyte, charge transfer and specific capacity at high charge/discharge rates as well as cycling stability, may benefit from mesoporous electrode architectures. Here the influence of structural features of mesoporous electrodes on electrochemical performance for LIBs will be discussed.

3.2.1 Mesoporous materials for improvement of capacity, charge transfer and electrolyte diffusion

Specific capacity, rate capability and structural stability during lithiation/delithiation processes of an electrode material are influenced by its pore size, pore architecture as well as the nature of the material. Considering the internal battery resistance, the pores will impact mostly the internal electrolyte resistance [12, 13]. In general, the presence of pores in an electrode facilitates electrolyte penetration and inherently produces large surface area, leading to rapid lithium ion transport and increased electrolyte/electrode contact area, and therefore, fast electrode reactions. Moreover, thin walls provide short path lengths for lithium ion diffusion in the solid phase with the time for lithiation/delithiation reduced by a factor of 10^6 as the particle dimensions decrease from the micrometer range to the nanometer range. The data indicate that mesoporosity is important for improving the high rate performance and can be optimized to ensure efficient transport of Li^+ without waste of volumetric energy density. Nevertheless, it is difficult to isolate the effects of each porosity parameters on electrochemical performance, as the challenge in this kind of study lies in the fact that one mesopore dimension cannot always be varied without concomitant modification of other properties. However, some general trends have been observed as illustrated below.

A few good examples of the influence of pore size on specific capacities have been demonstrated. In a study of mesoporous β-MnO_2 synthesized by hard templating with KIT-6 template, the pore size and wall thickness both played an important role in determining the rate, however the latter was more sensitive than the former [58]. The rate capability decreased with increasing pore wall thickness from 5 to 8.5 nm. For materials with a similar wall thickness of 7.5 nm and bimodal mesopore distributions, the rate capabilities depended on the ratio of pore volumes from large mesopores (11 nm pores) to smaller mesopores (3.3–3.5 nm). The exact dependence appeared to be complex, but the pore size had a significant effect on current densities in the range 100–1500 mAh/g (Fig. 6.7). The results showed that the thin walls could accommodate the strain of the phase boundary between MnO_2 and Li_xMnO_2 despite the large volume expansion on intercalation (55.7–72.3 $Å^3$). In addition, an important criterion for an efficient electrolyte distribution evenly throughout a porous electrode is the need for pores of uniform size and interconnectivity. For example, the first discharge capacity for bicontinuous mesoporous structure $LiCoO_2$ nanocast, obtained from mesoporous silica KIT-6, is some 20 mAh/g higher than that for the equivalent interconnected nanowire nanocast from mesoporous silica SBA-15, and the fade of discharge capacity on cycling is less marked for the former analogs [60]. Advantages of ordered mesoporous electrodes have been reported in comparison to bulk, nanoparticles and disordered

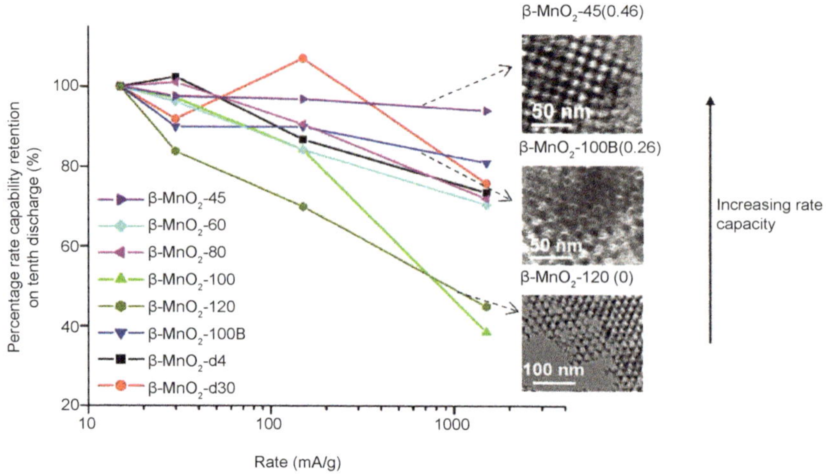

Figure 6.7. Tenth discharge capacities for the different mesoporous β-MnO_2 electrodes at different rates as a percentage of the discharge capacity at the lowest rate (15 mA/g). The active materials mass loading is 2.5 mg/cm^2. Ratios of large (11 nm) to small (3.3–3.5 nm) pores volumes are given in parentheses. Textural properties of the other samples can be found in the original reference. Reproduced with permission [58]. Copyright 2010, American Chemical Society.

mesoporous analogs. An example is in the case of mesoporous anatase TiO_2 anode [11], for which the packing efficiency was greater compared to anatase nanoparticles, and less conductive carbon was needed to maintain electron transport pathways through the electrode at high rates (greater than 1500 mA/g) because of the micrometer sized mesoporous particles. Interestingly, the ordered mesoporous anatase showed even higher rate performance of 125 mAh/g at 12 A/g (35.7 C) compared to disordered mesoporous anatase with a capacity of 91 mAh/g at 10 A/g.

The volumetric capacity of an electrode is critical in battery applications where space is limited, for instance, in portable electronics and electric vehicles. In this context, a particularly striking example is the mesoporous WO_{3-x} anode nanocasted from mesoporous KIT-6 silica [74]. This electrode exhibited a reversible capacity of 748 mAh/g, corresponding to 6.5 Li/W and a remarkable volumetric capacity of ~1500 mAh/cm^3, which is comparable to Li metal (~2000 mAh/cm^3). In another study, mesoporous $Li_4Ti_5O_{12}$ thin film electrode synthesized by the soft-templating method maintained high capacity of ca. 150 mAh/g with excellent cycling stability at the high rate of 64 C, which is superior to ordinary bulk nanocrystalline $Li_4Ti_5O_{12}$ [48]. Another recent impressive example is ordered mesoporous $TiNb_2O_7$ (TNO) with large pore size (~40 nm) synthesized by block copolymer-assisted self-assembly [78]. The resultant $TiNb_2O_7$ anode exhibited a high capacity of 289 mAh/g and excellent performance at high rates of 162 mAh/g at 20 C and 116 mAh/g at 50 C (which equals to 19.35 A/g), which outperformed other previous Ti- and Nb-based anode materials reported so far (e.g., $Li_4Ti_5O_{12}$, TiO_2, Nb_2O_5, $TiNb_2O_7$) as displayed in Fig. 6.8.

The concomitant benefit of a mesoporous structure is a large specific surface area, leading to a large electrolyte/electrode contact area. As a consequence, current density per unit surface area decreases, i.e., reducing electrode polarization and improving charge transfer. For example, ordered mesoporous NiO nanocast materials synthesized from KIT-6 were tested as an anode material [70], which showed much lower charge transfer resistance compared to the bulk counterpart. The kinetic studies revealed that the activation energies for lithium ion intercalation were 20.8 and 45.0 kJ/mol for mesoporous and bulk NiO, respectively. On the other hand, high surface area may enhance the formation of a Solid-Electrolyte Interface (SEI) at the anode side. Mesoporous Cr_2O_3 anodes replicated from KIT-6 silica are an example where significant consumption of available

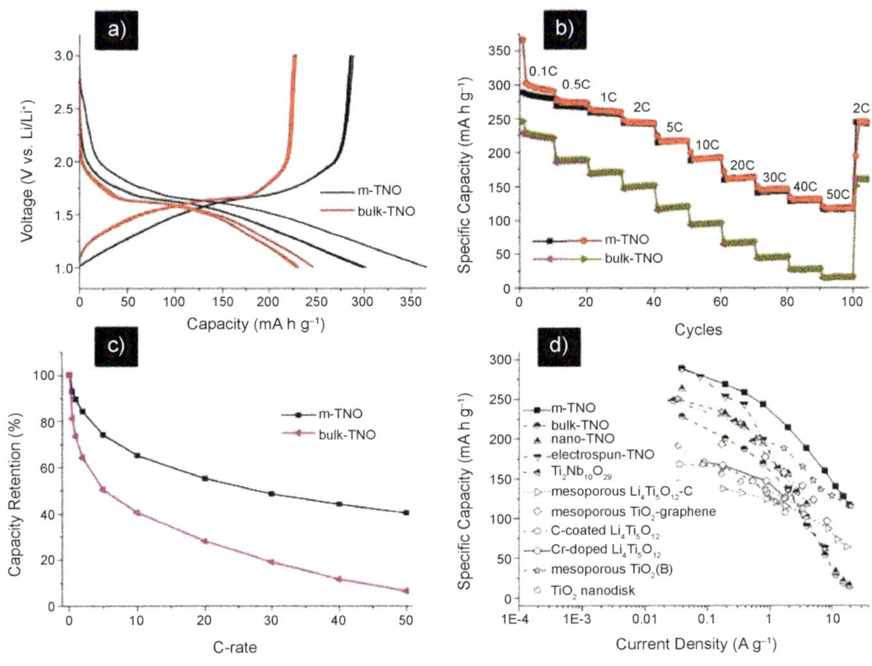

Figure 6.8. (a) Galvanostatic charge–discharge curves (initial three cycles) of m-TNO and bulk-TNO electrodes at 0.1 C rate conditions (= 38.7 mA/g). (b) Cycle and rate performance plots and (c) capacity retention plots of m-TNO and bulk-TNO electrodes obtained at various current densities, 0.1 to 50 C-rate. (d) Capacity retention of various TNO electrodes and other titanium- and niobium-based electrodes as a function of rate. Reproduced with permission [78]. Copyright 2014, American Chemical Society.

lithium ions during the early cycles occurred as a consequence of mesoporosity [59, 66], resulting in large irreversible capacity of the first cycle for the mesoporous Cr_2O_3 electrode. This disadvantage has been observed in the case of ordered mesoporous carbon termed as CMK-3, with high surface area values of 1030 m^2/g and pore sizes of about 4 nm, fabricated using ordered silica SBA-15 as a hard template [79]. The material, when tested as an anode material for LIBs, displayed a very high reversible specific capacity (850–1100 mAh/g). However, there was a large initial irreversible capacity loss of about 2000 mAh/g (Fig. 6.9), which was primarily ascribed to extensive SEI formation associated with the high surface area of the material.

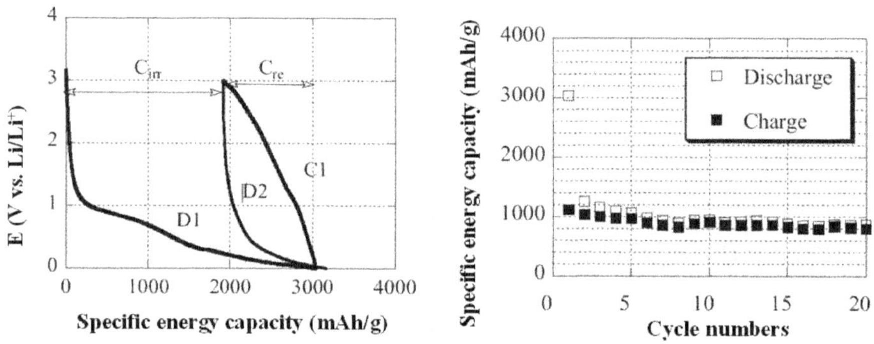

Figure 6.9. (Left) Galvanostatic charging and discharging processes of CMK-3 at a constant current of 100 mA/g. D1, C1 and D2 are the first discharge, first charge and second discharge, respectively. (Right) Discharge and charge cycle performance of CMK-3 at a constant current of 100 mA/g. Reproduced with permission [79]. Copyright (2003) Wiley-VCH Verlag GmbH & Co. KGaA.

3.2.2 Mesoporous materials as buffer matrix during cycling and as supports to improve stability and conductivity of the active phase

Carbon negative graphite in LIBs displays a practical capacity of 350 mAh/g compared to the theoretical value of 372 mAh/g for the end compound of LiC_6. To improve the anode performance, in addition to tailoring porosity and nanostructured morphology, ongoing research efforts are focused on several alloy compounds with higher theoretical specific capacities of lithium as carbon alternatives, including Li-Si, Li-Sn, Li-Al, Li-Pb, Li-Cd and other alloys [8, 9, 80, 81]. While advantageous in terms of gravimetric capacity (e.g., as high as 4200 mAh/g for Si), most of Li alloys suffer large volumetric expansion/contraction on charge/discharge (as much as 300%), which causes pulverization of particles, and therefore, loss of mechanical integrity and electrical contacts between particles. Introducing pores into an alloy or combination of such alloy active materials within a porous matrix (generally non-electrochemically active metals or carbon) can alleviate these stresses to some extent by providing some room for expansion and improving electric conductivity of the electrode. The micrometer-sized mesoporous electrodes are less prone to disintegration caused by volume swings on cycling than discrete nanoparticles. In addition, mesoporous electrodes with nano-sized crystals in the pore walls can better tolerate strains from lithium insertion compared to micron-sized particles, and therefore can suppress irreversible phase transformations that take place in microcrystalline materials. For example, ordered mesoporous α-Fe_2O_3 thin film electrodes, produced by soft-templating using large block dicopolymer KLE, exhibited excellent cycling stability. The enhancement of cyclability was attributed to the ability of the mesoporous framework to better accommodate volume changes by suppressing of the irreversible phase transition from α-Fe_2O_3 to cubic $Li_xFe_2O_3$ which occurs upon lithiation in bulk hematite [82].

Similarly, a cathode made from nonporous micrometer-sized β-MnO_2 particles did not show much, if any, lithium intercalation capacity. In contrast, the extent of intercalation increased markedly in mesoporous β-MnO_2 of similar external particle sizes with a reversible capacity of 284 mAh/g (corresponding to the composition $Li_{0.92}MnO_2$) (Fig. 6.10) [57]. It has been demonstrated that the MnO_2 crystal structure of the walls was preserved in the form of β-Li_x-MnO_2 on cycling, rather than converted to $LiMn_2O_4$ spinel, like what otherwise happens to bulk β-MnO_2 or nanoparticulate β-MnO_2. This enhanced performance once again demonstrated the unique property of mesoporous materials capable of accommodating the volume changes during cycling process.

In terms of volume buffering effects, the porosity resulting from internal pore systems is more effective than porosity between nanoparticles, because in the former, the pores are uniform in size and shape, as well as being interconnected throughout the material. For instance, mesoporous Si@C anode templated from SBA-15 silica exhibited a high charge capacity of 3163 mAh/g for the first cycle and was stable up to 80 cycles with capacity of 2738 mAh/g (87% capacity retention), whereas carbon-coated Si nanoparticles showed rapid capacity fade after 50 cycles from 4000 mAh/g to 1000 mAh/g (25% capacity retention) due to the occurrence of particle pulverization on cycling [10]. Ordered mesoporous Li-Mn-O spinel cathode exhibited capacity retention 50% higher than the equivalent bulk phase at a rate of 3000 mA/g (30C), and higher gravimetric and volumetric energy density than those fabricated from the corresponding bulk material [61].

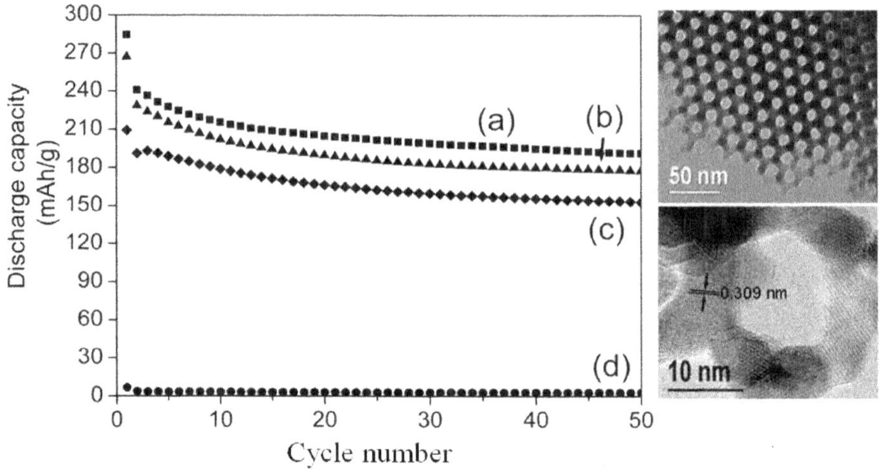

Figure 6.10. (Left) Capacity retention for mesoporous β-MnO_2 cycled at (a) 15, (b) 30 and (c) 300 mA/g; (d) bulk β-MnO_2 cycled at 15 mA/g. (Right) TEM and high resolution TEM (HRTEM) images of β-MnO_2. Reproduced with permission [57]. Copyright (2007) Wiley-VCH Verlag GmbH & Co. KGaA.

In addition to engineering nanostructures, another approach is to encapsulate the active components within the pore channels or into the walls of a mesoporous support that provides additional stability and/or conductivity. Various nanocomposites of ordered mesoporous carbons loaded with Sn or SnO_2 and used as anode materials for LIBs have been reported [83–85]. Mesoporous carbon CMK-3 with 17 wt% of SnO_2 loaded inside the pores exhibited higher reversible capacity and cycling stability compared to pristine CMK-3 electrode, while SnO_2 suffered poor cycling stability arising from the large volume change (about 358%) upon charge/discharge processes [83]. Here, non-electrochemically active phases could enhance stability of active components significantly.

In mesoporous SnO_2 synthesized from SBA-15 silica template by melt-infiltration of $SnCl_2 \cdot 2H_2O$, where the amounts of residual SiO_2 were controlled in the range of 0.9–17.4 wt% by using different concentrations of NaOH or HF for the silica etching, the material containing 6.0 wt% of silica remained structurally stable up to 700°C. This optimum amount of residual silica greatly improved cyclability, and the material with 3.9 wt% of silica exhibited a specific capacity of 600 mAh/g after 30 cycles at a rate of 50 mA/g (Fig. 6.11) [72].

The low intrinsic electronic conductivities of several potential alternative electrode materials limit their capacities at high rates. Mesoporous structured electrodes offer possibilities to functionalize a conductive phase into their frameworks. One useful approach involves carbon coating on the

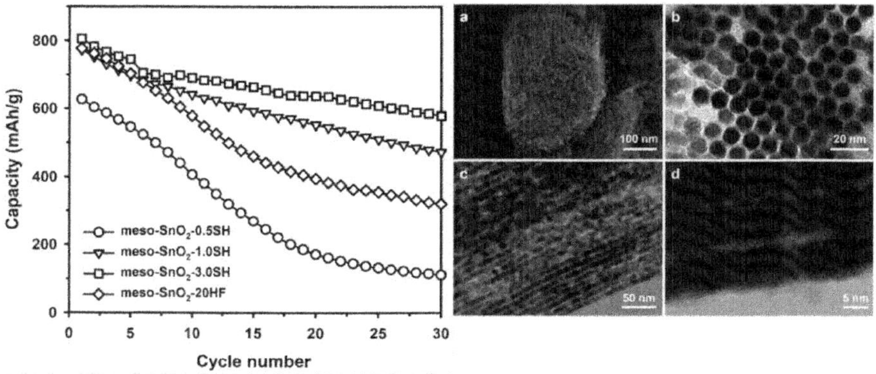

Figure 6.11. (Left) Cycle-life performances of meso–SnO_2 materials depending on the amount of residual silica species: meso-SnO_2-0.5SH (17.4 wt%), meso-SnO_2-1.0SH (6.0 wt%), meso-SnO_2-3.0SH (3.9 wt%), meso-SnO_2-20HF (0.9 wt%). (Right) High-resolution SEM image, TEM images in the [100] direction (b), TEM image in the [110] direction (c), and high-resolution TEM image for the meso–SnO_2–1.0SH material (d). Reproduced with permission [72]. Copyright 2009, Royal Society of Chemistry.

inner surface of a porous electrode material by pyrolysis of polymeric templates in inert atmosphere. By this synthetic pathway mesoporous $Li_4Ti_5O_{12}$/carbon and TiO_2/carbon composites were achieved via carbonization of block copolymer polyisoprene-block-poly(ethylene oxide) (PI- b -PEO) template [47, 86]. The conductive carbon-coated mesostructure $Li_4Ti_5O_{12}$/C retained a reversible capacity of 115 mAh/g at 10C rate (corresponding to only 6 minutes for full charge and discharge), with 90% capacity retention after 500 cycles. Electrochemical Impedance Spectroscopy (EIS) revealed that charge transfer resistance is much smaller in carbon-coated mesoporous $Li_4Ti_5O_{12}$/C compared to bulk $Li_4Ti_5O_{12}$ [47]. Similarly, for the mesoporous TiO_2/carbon nanocomposite, the *in situ* pyrolyzed conductive carbon layer helped to enhance the capacity and stable cycle performance, which were inefficient without the carbon coating [86]. Alternatively, deposition of metallic conductive phases, such as a thin film of Cu or Sn (with less than 1 wt%) on the surface of mesoporous anatase TiO_2 electrodes was also observed to mitigate polarization and improve the electrode capacity compared to uncoated TiO_2 electrodes [87].

4. Mesoporous Materials in Li-air (O_2) and Li-S Batteries

Although LIBs will continue to be instrumental in powering portable electronics for a number of years, in the longer term new types of batteries will be required to meet the demands of key markets such as transport and energy grid. Two lithium-based batteries: Li-O_2 and Li-S are receiving intense interest [88–92] and will be briefly discussed here. Li-O_2 and Li-S batteries have theoretical energy densities several times greater than that of commercial LIBs (ca. 2600 Wh/kg for Li-S and 3500 Wh/kg for Li-O_2), which could satisfy a travelling distance of 500 km on a single charge for electric vehicles. As a rule of thumb, there is always a substantial reduction in practical energy stored compared to theoretical values as presented in Fig. 6.12. The practical values for relevant technologies are well established, however it is a very rough estimate for Li-O_2. Furthermore, the abundance and non-toxicity of S and O_2 provide these battery systems with improved energy economy and sustainability.

Li–O_2 and Li–S share the same lithium metal anode, but have different active cathode components (O_2 and S, respectively) (Fig. 6.13). Therefore, in both batteries, at the anode side, the Li-metal anode is oxidized on discharge, releasing Li^+ into the electrolyte, and the process is reversed on charge. At the positive electrode, in the case of Li-O_2, O_2 the porous cathode enters from the air, dissolves in the electrolyte in the pores, and is reduced. If a non-aqueous electrolyte is used, Li_2O_2 is formed as the final discharge product in the pores of the cathode. Aqueous electrolytes involve formation of LiOH at the cathode on discharge. These discharge

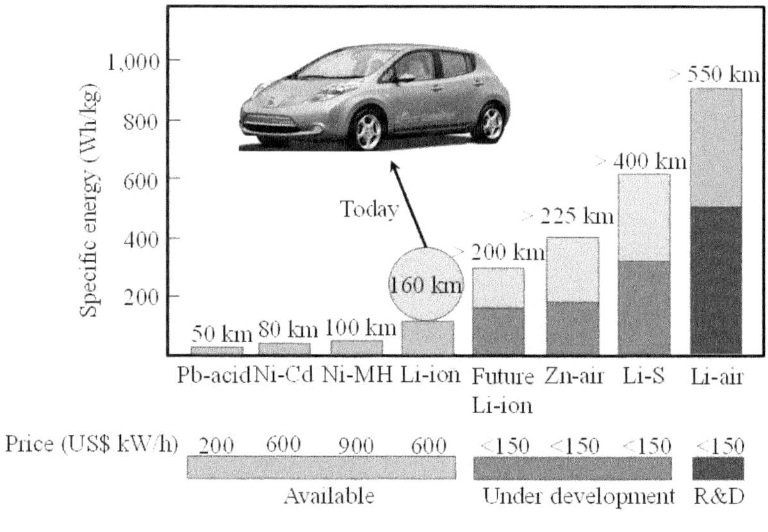

Figure 6.12. Practical specific energies for some rechargeable batteries, along with estimated driving distances and pack prices. For future technologies, a range of anticipated specific energies are given as shown by the lighter shaded region on the bars in the chart for rechargeable batteries under development and R&D. The values for driving ranges are based on the minimum specific energy for each technology and scaled on the specific energy of the Li-ion cells (140 Wh/kg) and driving range (160 km) of the Nissan Leaf. The prices for technologies under development represent targets set by the US Advanced Battery Consortium. Adapted from reference [88]. Copyright 2012, American Chemical Society.

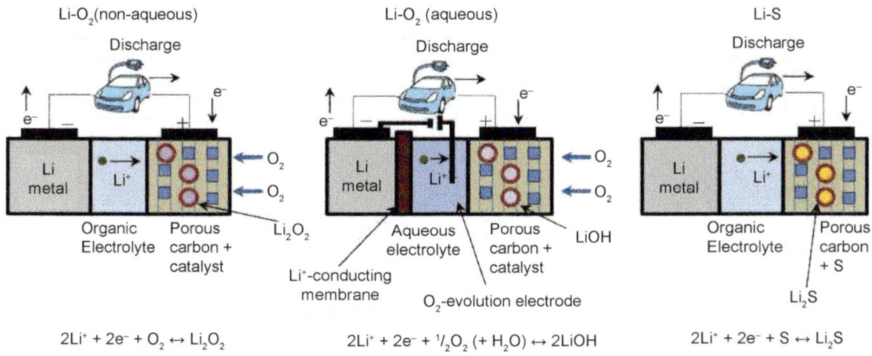

Figure 6.13. Schematic representations of non-aqueous and aqueous Li–O_2 and Li–S cells. Adapted from reference [88]. Copyright 2012, American Chemical Society.

products are oxidized to form Li^+ and O_2 on charging. In the case of Li-S, at the cathode, S is reduced combined with Li^+ to form various polysulfide intermediates (Li_2S_x, x = 2–8), and to ultimately produce Li_2S at the end of discharge. The decomposition of these products back to Li^+ and S occurs upon the charging process.

Both Li-O_2 and Li-S are facing formidable challenges related to the anode, cathode, electrolyte and many other aspects as discussed in details in some recent reviews [81, 88–93]. In the limited space available, it is hard to incorporate all aspects and review a vast amount of work on these two battery technologies. Instead, emphasis will be placed on key problems that can be solved by engineering porosity of cathode materials, since both Li-O_2 and Li-S rely on porous cathodes.

For Li-O_2, the electrochemical reactions involve the diffusion of O_2 and the accumulation of discharge products in the cathode. In addition to conductivity and catalytic activity, rational design of the cathode structure is important for improvement of charge transfer and mass transport as well as minimizing clogging of the pores by solid discharge products [88–90, 92]. The porous electrode must ensure facile transport of O_2 as much as possible through the gas phase before reaching the electrolyte/electrode interface where it is reduced. If pores are too small, pore clogging is likely to occur, whereas pores that are too large may lead to an ineffective use of volume. Moreover, morphology and particle size of discharge products, such as Li_2O_2, have substantial impacts on the battery performance. Usually, it is difficult to decompose such bulky discharge products into Li^+ and O_2 upon charging and the deposition of these insulating species on the electrode surface could result in the passivation of the electrode surface. Therefore, controlling the morphology and size of the discharge products becomes important, which could be achieved by tailoring the pore architecture of the electrode. It is expected that mesoporosity will be optimal for this, and enable accommodating active sites evenly on the pore surface. Oxygen electrochemistry in Li-O_2 batteries is a very complex process associated with Oxygen Reduction Reaction (ORR) and Oxygen Evolution Reaction (OER) [94]. Thus, a variety of materials have been investigated as cathodes, including metals, metal oxides, functional carbon materials, metal-carbon hybrids, metal nitrides [90, 92]. For example, mesoporous Co_3O_4 materials of different porosities prepared by hard templating approach were tested as electrochemical catalysts mixed with carbon on cathode in Li-O_2 batteries [95]. It was found that the mesoporous structure significantly improved round-trip efficiency and specific capacity over the bulk catalyst, particularly in the case of a continuous cubic mesopore structure. A large specific capacity of 2250 mAh/g carbon and a high round-trip efficiency of 81.4% were obtained for the cubic mesoporous Co_3O_4 nanocasted from KIT-6 silica which was aged at 40°C. In another example, 3D ordered mesoporous carbon decorated with FeO_x or Pd/FeO_x was shown to allow for control over the size and deposition of Li_2O_2 within the large pores of the mesoporous carbon, which leads to a high capacity (6000 mAh/g carbon) and improves cyclability of more than 68 cycles [96]. Similarly, in a study related to the effect of mesostructure, large mesoporous carbon MSU-F-C with 3D interconnected pores (~30 nm) was found to lower the charging potentials at a high current density and increase the rate capabilities, in comparison to the results obtained with a CMK-3 carbon having small pores (~3 nm) [97].

For Li-S, the first major problem is poor ionic and electronic conductivities of discharge products, resulting in limited rate capacity and poor rechargeability of the battery. Second, various soluble polysulfide Li_2S_x ($3 \leq x \leq 6$) intermediates give rise to significant capacity fading during cycling and a so-called shuttle effect due to migration of these dissolved polysulfides onto the Li anode [88, 91, 93]. Thus, several strategies have been considered to design porous cathodes that enable efficient electron transport to the S as well as trapping the soluble polysulfides. Mesoporous materials offer sufficient loading rate (up to 70 wt% in cathode composite) and better confinement of polysulfides with stronger interaction between S and pore walls compared to macroporous materials. Mesoporous carbon materials have been widely studied in combination with S as cathodes, owing to their low cost, high conductivity and light weight [98–104].

Cathodes based on ordered mesoporous carbon (CMK-3)-sulfur composite have been demonstrated to provide high reversible capacities up to 1320 mAh/g. It was also proven that placing a hydrophilic polymer coating (e.g., polyethylene glycol) on the surface of the carbon further retards the diffusion of polysulfides out of the electrode (Fig. 6.14) [98]. Using mesoporous silica or metal-organic additives as additional polysulfide

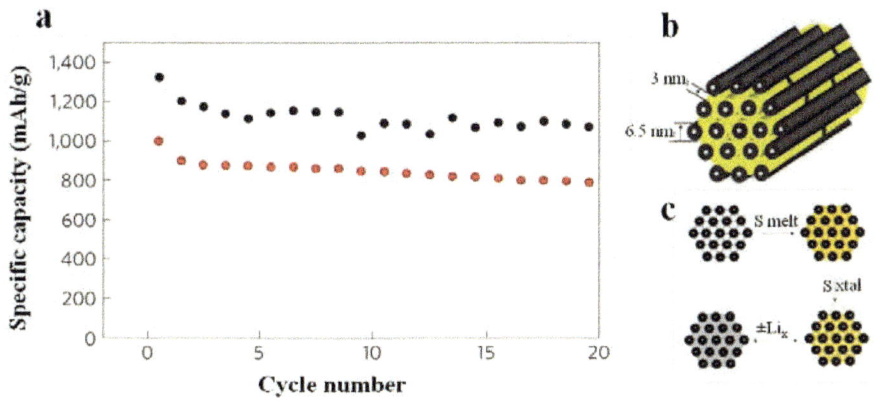

Figure 6.14. (a) Cycling stability comparison of CMK-3/S-PEG (upper points, in black) versus CMK-3/S (lower points, in red) at 168 mA/g at room temperature. (b) A schematic diagram of the sulfur (yellow) confined in the interconnected pore structure of mesoporous carbon, CMK-3, formed from carbon tubes that are propped apart by carbon nanofibers. (c) Schematic diagram of composite synthesis by impregnation of molten sulfur, followed by its densification on crystallization. The lower diagram represents subsequent discharging–charging with Li, illustrating the strategy of pore-filling to tune for volume expansion/contraction. Reproduced with permission [98]. Copyright 2012, American Chemical Society.

reservoirs embedded in the carbon-sulfur composite can further enhance the cycling stability of the cathode [100]. Moreover, improvement in performance of Li-S cells can be achieved by functionalization of the mesoporous carbon with nitrogen [99, 102]. Although diverse in porous structure and chemical composition, the general idea of engineering porosity and functionality is to achieve the best sulfur retention through the intimate contact between the insulating sulfur and the conductive carbon framework, while ensuring facile transport of the electrolyte.

5. Conclusions and Future Outlooks

As energy consumption grows rapidly with the increasing global population, batteries as an energy source or electricity storing device, will become increasingly important. No matter which battery system is being considered, the performance (i.e., cell voltage, energy density, cycle-life or lifetime) and safety are directly related to the intrinsic properties of the materials that form the electrodes.

The importance of engineered porosity in developing high performance lithium batteries with higher capacity and longer cycle life was highlighted. The great progress in synthesis of mesoporous materials with a high control over pore architecture has offered opportunities to achieve novel energy storage materials for lithium batteries, particularly for high rate applications in transportation. Regarding the pore size, it appears that medium to large size mesopores are most beneficial in terms of rate capabilities and volumetric energy. In the case of viscous electrolytes, mesoporous/ macroporous hierarchical materials have also become more advantageous, as discussed elsewhere [12, 105–109]. Concerning the pore morphology, bicontinuous structures with well-defined pore sizes are desirable as they provide continuous transport paths for charge carriers through the walls and the pores. However, determining specific pore architecture and pore size for optimal performance is a complex question that requires additional insight studies and well controlled experimental conditions. The greatest challenge facing the commercialization of mesoporous electrodes remains the high cost and complexity of synthesis procedures. Novel concepts for the facile synthesis of mesoporous solids will certainly be of significant value, not only in this field but also for many other applications.

Li-O_2 and Li-S batteries present high theoretical specific energy and energy density and are more sustainable than LIBs. However, reaching beyond LIBs calls for intensive research to address the challenging problems of these battery technologies. Although significant advances have been made, the present systems are still not sufficiently performing for most practical applications. In particular, it is important to better understand the fundamental electrochemistry taking place in the Li-O_2 and Li-S cells to facilitate the design of new electrode materials.

Overall, finding the best performing system can only be achieved by selecting appropriate—whether existing or new—materials. Future advances in this domain will require the integration of new materials, physics, electrochemistry, engineering and interdisciplinary cooperation at a level of sophistication unprecedented in the battery technology sector.

Acknowledgements

The authors acknowledge financial support from the National Science and Engineering Research Council (Canada).

References

[1] Liu, J., Zhang, J.-G., Yang, Z. et al. 2013. Materials science and materials chemistry for large scale electrochemical energy storage: From transportation to electrical grid. *Adv. Funct. Mater.*, 23: 929–46.

[2] Winter, M. and Brodd, R.J. 2004. What are batteries, fuel cells, and supercapacitors? *Chem. Rev.*, 104: 4245–69.

[3] Park, J.-K. 2012. Principles and applications of lithium secondary batteries. John Wiley & Sons.

[4] Whittingham, M.S. 1976. Electrical energy-storage and intercalation chemistry. *Science*, 192: 1126–27.
[5] Murphy, D.W., Disalvo, F.J., Carides, J.N. et al. 1978. Topochemical reactions of rutile related structures with lithium. *Mate. Res. Bull.*, 13: 1395–402.
[6] Lazzari, M. and Scrosati, B. 1980. Cyclable lithium organic electrolyte cell based on 2 intercalation electrodes. *J. Electrochem. Soc.*, 127: 773–74.
[7] Tarascon, J.M. 2010. Key challenges in future li-battery research. *Phil. Trans. R. Soc. A*, 368: 3227–41.
[8] Besenhard, J.O., Yang, J. and Winter, M. 1997. Will advanced lithium-alloy anodes have a chance in lithium-ion batteries? *J. Power. Sources*, 68: 87–90.
[9] Reddy, M.V., Rao, G.V.S. and Chowdari, B.V.R. 2013. Metal oxides and oxysalts as anode materials for Li ion batteries. *Chem. Rev.*, 113: 5364–457.
[10] Kim, H. and Cho, J. 2008. Superior lithium electroactive mesoporous Si@carbon core-shell nanowires for lithium battery anode material. *Nano Lett.*, 8: 3688–91.
[11] Ren, Y., Hardwick, L.J. and Bruce, P.G. 2010. Lithium intercalation into mesoporous anatase with an ordered 3D pore structure. *Angew. Chem. Int. Ed.*, 49: 2570–74.
[12] Vu, A., Qian, Y.Q. and Stein, A. 2012. Porous electrode materials for lithium-ion batteries - how to prepare them and what makes them special. *Adv. Energy Mat.*, 2: 1056–85.
[13] Goodenough, J.B. 2007. General concepts. *In*: Lithium Ion Batteries. Wiley-VCH Verlag GmbH. pp. 1–25.
[14] Ren, Y., Ma, Z. and Bruce, P.G. 2012. Ordered mesoporous metal oxides: Synthesis and applications. *Chem. Soc. Rev.*, 41: 4909–27.
[15] Walcarius, A. 2013. Mesoporous materials and electrochemistry. *Chem. Soc. Rev.*, 42: 4098–140.
[16] Zhu, C., Du, D., Eychmüller, A. et al. 2015. Engineering ordered and nonordered porous noble metal nanostructures: Synthesis, assembly, and their applications in electrochemistry. *Chem. Rev.*, 115: 8896–943.
[17] Roberts, A.D., Li, X. and Zhang, H.F. 2014. Porous carbon spheres and monoliths: Morphology control, pore size tuning and their applications as li-ion battery anode materials. *Chem. Soc. Rev.*, 43: 4341–56.
[18] Kleitz, F. 2008. Ordered Mesoporous Materials. *In*: Ertl, G., Knözinger, H., Schüth, F., Weitkamp, J. Handbook of Heterogeneous Catalysis, Volume 1. Wiley-VCH.
[19] Soler-illia, G.J.D., Sanchez, C., Lebeau, B. et al. 2002. Chemical strategies to design textured materials: From microporous and mesoporous oxides to nanonetworks and hierarchical structures. *Chem. Rev.*, 102: 4093–138.
[20] Liu, R., Duay, J. and Lee, S.B. 2011. Heterogeneous nanostructured electrode materials for electrochemical energy storage. *Chem. Commun.*, 47: 1384–404.
[21] Kennedy, T., Mullane, E., Geaney, H. et al. 2014. High-performance germanium nanowire-based lithium-ion battery anodes extending over 1000 cycles through *in situ* formation of a continuous porous network. *Nano Lett.*, 14: 716–23.
[22] Sun, H., Xin, G., Hu, T. et al. 2014. High-rate lithiation-induced reactivation of mesoporous hollow spheres for long-lived lithium-ion batteries. *Nat. Commun.*, 5: 4526.
[23] Hu, Y.S., Guo, Y.G., Sigle, W. et al. 2006. Electrochemical lithiation synthesis of nanoporous materials with superior catalytic and capacitive activity. *Nat. Mater.*, 5: 713–17.
[24] Gu, D. and Schuth, F. 2014. Synthesis of non-siliceous mesoporous oxides. *Chem. Soc. Rev.*, 43: 313–44.
[25] Wan, Y. and Zhao, D. 2007. On the controllable soft-templating approach to mesoporous silicates. *Chem. Rev.*, 107: 2821–60.
[26] Grosso, D., Cagnol, F., Soler-Illia, G.J.d.A.A. et al. 2004. Fundamentals of mesostructuring through evaporation-induced self-assembly. *Adv. Funct. Mater.*, 14: 309–22.

[27] Lu, Y., Ganguli, R., Drewien, C.A. et al. 1997. Continuous formation of supported cubic and hexagonal mesoporous films by sol-gel dip-coating. *Nature*, 389: 364–68.
[28] Hatton, B., Landskron, K., Whitnall, W. et al. 2005. Past, present, and future of periodic mesoporous organosilicas—the PMOs. *Acc. Chem. Res.*, 38: 305–12.
[29] Beck, J.S., Vartuli, J.C., Roth, W.J. et al. 1992. A new family of mesoporous molecular-sieves prepared with liquid-crystal templates. *J. Am. Chem. Soc.*, 114: 10834–43.
[30] Ulagappan, N. and Rao, C.N.R. 1996. Mesoporous phases based on SnO_2 and TiO_2. *Chem. Commun.*, 1685–86.
[31] Banerjee, S., Santhanam, A., Dhathathreyan, A. et al. 2003. Synthesis of ordered hexagonal mesostructured nickel oxide. *Langmuir*, 19: 5522–25.
[32] Kipkemboi, P., Fogden, A., Alfredsson, V. et al. 2001. Triblock copolymers as templates in mesoporous silica formation: Structural dependence on polymer chain length and synthesis temperature. *Langmuir*, 17: 5398–402.
[33] Soler-Illia, G., Crepaldi, E.L., Grosso, D. et al. 2003. Block copolymer-templated mesoporous oxides. *Curr. Opin. Colloid Interface Sci.*, 8: 109–26.
[34] Kleitz, F., Choi, S.H. and Ryoo, R. 2003. Cubic Ia3d large mesoporous silica: Synthesis and replication to platinum nanowires, carbon nanorods and carbon nanotubes. *Chem. Commun.*, 2136–37.
[35] Kleitz, F., Berube, F., Guillet-Nicolas, R. et al. 2010. Probing adsorption, pore condensation, and hysteresis behavior of pure fluids in three-dimensional cubic mesoporous KIT-6 silica. *J. Phy. Chem. C*, 114: 9344–55.
[36] Kleitz, F., Czuryszkiewicz, T., Solovyov, L.A. et al. 2006. X-ray structural modeling and gas adsorption analysis of cagelike SBA-16 silica mesophases prepared in aF127/butanol/H_2O system. *Chem. Mater.*, 18: 5070–79.
[37] Zhao, D., Huo, Q., Feng, J. et al. 2014. Nonionic triblock and star diblock copolymer and oligomeric surfactant syntheses of highly ordered, hydrothermally stable, mesoporous silica structures. *J. Am. Chem. Soc.*, 136: 10546–46.
[38] Zhang, J., Deng, Y., Gu, D. et al. 2011. Ligand-assisted assembly approach to synthesize large-pore ordered mesoporous titania with thermally stable and crystalline framework. *Adv. Energy Mat.*, 1: 241–48.
[39] Kuemmel, M., Grosso, D., Boissière, C. et al. 2005. Thermally stable nanocrystalline γ-alumina layers with highly ordered 3D mesoporosity. *Angew. Chem. Int. Ed.*, 44: 4589–92.
[40] Brezesinski, T., Groenewolt, M., Antonietti, M. et al. 2006. Crystal-to-crystal phase transition in self-assembled mesoporous iron oxide films. *Angew. Chem. Int. Ed.*, 45: 781–84.
[41] Zhu, S.M., Zhou, H.S., Miyoshi, T. et al. 2004. Self-assembly of the mesoporous electrode material $Li_3Fe_2(PO4)_3$ using a cationic surfactant as the template. *Adv. Mater.*, 16: 2012–17.
[42] Luca, V. and Hook, J.M. 1997. Study of the structure and mechanism of formation through self-assembly of mesostructured vanadium oxide. *Chem. Mater.*, 9: 2731–44.
[43] Liu, P., Lee, S.H., Tracy, C.E. et al. 2002. Preparation and lithium insertion properties of mesoporous vanadium oxide. *Adv. Mater.*, 14: 27–30.
[44] Meng, Y., Gu, D., Zhang, F.Q. et al. 2005. Ordered mesoporous polymers and homologous carbon frameworks: Amphiphilic surfactant templating and direct transformation. *Angew. Chem. Int. Ed.*, 44: 7053–59.
[45] Fattakhova-Rohlfing, D., Wark, M., Brezesinski, T. et al. 2007. Highly organized mesoporous TiO_2 films with controlled crystallinity: A Li-insertion study. *Adv. Funct. Mater.*, 17: 123–32.
[46] Kim, E., Son, D., Kim, T.-G. et al. 2004. A mesoporous/crystalline composite material containing tin phosphate for use as the anode in lithium-ion batteries. *Angew. Chem. Int. Ed.*, 43: 5987–90.

[47] Kang, E., Jung, Y.S., Kim, G.H. et al. 2011. Highly improved rate capability for a lithium-ion battery nano-$Li_4Ti_5O_{12}$ negative electrode via carbon-coated mesoporous uniform pores with a simple self-assembly method. *Adv. Funct. Mater.*, 21: 4349–57.

[48] Haetge, J., Hartmann, P., Brezesinski, K. et al. 2011. Ordered large-pore mesoporous $Li_4Ti_5O_{12}$ spinel thin film electrodes with nanocrystalline framework for high rate rechargeable lithium batteries: Relationships among charge storage, electrical conductivity, and nanoscale structure. *Chem. Mater.*, 23: 4384–93.

[49] Boettcher, S.W., Fan, J., Tsung, C.K. et al. 2007. Harnessing the sol-gel process for the assembly of non-silicate mesostructured oxide materials. *Acc. Chem. Res.*, 40: 784–92.

[50] Yang, P.D., Zhao, D.Y., Margolese, D.I. et al. 1998. Generalized syntheses of large-pore mesoporous metal oxides with semicrystalline frameworks. *Nature*, 396: 152–55.

[51] Lu, A.H., Zhao, D. and Wan, Y. 2010. Nanocasting: A versatile strategy for creating nanostructured porous materials. *Royal Society of Chemistry*.

[52] Shi, Y.F., Meng, Y., Chen, D.H. et al. 2006. Highly ordered mesoporous silicon carbide ceramics with large surface areas and high stability. *Adv. Funct. Mater.*, 16: 561–67.

[53] Jiao, F., Hill, A.H., Harrison, A. et al. 2008. Synthesis of ordered mesoporous NiO with crystalline walls and a bimodal pore size distribution. *J. Am. Chem. Soc.*, 130: 5262–66.

[54] Yen, H., Seo, Y., Kaliaguine, S. et al. 2012. Tailored mesostructured copper/ceria catalysts with enhanced performance for preferential oxidation of CO at low temperature. *Angew. Chem. Int. Ed.*, 51: 12032–35.

[55] Yen, H., Seo, Y., Guillet-Nicolas, R. et al. 2011. One-step-impregnation hard templating synthesis of high-surface-area nanostructured mixed metal oxides ($NiFe_2O_4$, $CuFe_2O_4$ and Cu/CeO_2). *Chem. Commun.*, 47: 10473–75.

[56] Rumplecker, A., Kleitz, F., Salabas, E.-L. et al. 2007. Hard templating pathways for the synthesis of nanostructured porous Co_3O_4. *Chem. Mater.*, 19: 485–96.

[57] Jiao, F. and Bruce, P.G. 2007. Mesoporous crystalline β-MnO_2—a reversible positive electrode for rechargeable lithium batteries. *Adv. Mater.*, 19: 657–60.

[58] Ren, Y., Armstrong, A.R., Jiao, F. et al. 2010. Influence of size on the rate of mesoporous electrodes for lithium batteries. *J. Am. Chem. Soc.*, 132: 996–1004.

[59] Liu, H., Du, X.W., Xing, X.R. et al. 2012. Highly ordered mesoporous Cr_2O_3 materials with enhanced performance for gas sensors and lithium ion batteries. *Chem. Commun.*, 48: 865–67.

[60] Jiao, F., Shaju, K.M. and Bruce, P.G. 2005. Synthesis of nanowire and mesoporous low-temperature $LiCoO_2$ by a post-templating reaction. *Angew. Chem. Int. Ed.*, 44: 6550–53.

[61] Jiao, F., Bao, J., Hill, A.H. et al. 2008. Synthesis of ordered mesoporous Li-Mn-O spinel as a positive electrode for rechargeable lithium batteries. *Angew. Chem. Int. Ed.*, 47: 9711–16.

[62] Wagner, T., Haffer, S., Weinberger, C. et al. 2013. Mesoporous materials as gas sensors. *Chem. Soc. Rev.*, 42: 4036–53.

[63] Nair, M.M., Yen, H. and Kleitz, F. 2014. Nanocast mesoporous mixed metal oxides for catalytic applications. *Comptes Rendus Chimie*, 17: 641–55.

[64] Luo, J.-Y., Zhang, J.-J. and Xia, Y.-Y. 2006. Highly electrochemical reaction of lithium in the ordered mesoporosus beta-MnO_2. *Chem. Mater.*, 18: 5618–23.

[65] Lim, S., Yoon, C.S. and Cho, J. 2008. Synthesis of nanowire and hollow $LiFePO_4$ cathodes for high-performance lithium batteries. *Chem. Mater.*, 20: 4560–64.

[66] Dupont, L., Laruelle, S., Grugeon, S. et al. 2008. Mesoporous Cr_2O_3 as negative electrode in lithium batteries: Tem study of the texture effect on the polymeric layer formation. *J. Power. Sources*, 175: 502–09.

[67] Wang, G., Liu, H., Horvat, J. et al. 2010. Highly ordered mesoporous cobalt oxide nanostructures: Synthesis, characterisation, magnetic properties, and applications for electrochemical energy devices. *Chem. Eur. J.*, 16: 11020–27.

[68] Yue, W., Randorn, C., Attidekou, P.S. et al. 2009. Syntheses, Li insertion, and photoactivity of mesoporous crystalline TiO_2. *Adv. Funct. Mater.*, 19: 2826–33.
[69] Yue, W., Xu, X., Irvine, J.T.S. et al. 2009. Mesoporous monocrystalline TiO_2 and its solid-state electrochemical properties. *Chem. Mater.*, 21: 2540–46.
[70] Liu, H., Wang, G., Liu, J. et al. 2011. Highly ordered mesoporous NiO anode material for lithium ion batteries with an excellent electrochemical performance. *J. Mater. Chem.*, 21: 3046–52.
[71] Kim, H. and Cho, J. 2008. Hard templating synthesis of mesoporous and nanowire SnO_2 lithium battery anode materials. *J. Mater. Chem.*, 18: 771–75.
[72] Shon, J.K., Kim, H., Kong, S.S. et al. 2009. Nano-propping effect of residual silicas on reversible lithium storage over highly ordered mesoporous SnO_2 materials. *J. Mater. Chem.*, 19: 6727–32.
[73] Fan, J., Wang, T., Yu, C. et al. 2004. Ordered, nanostructured tin-based oxides/carbon composite as the negative-electrode material for lithium-ion batteries. *Adv. Mater.*, 16: 1432–36.
[74] Yoon, S., Jo, C., Noh, S.Y. et al. 2011. Development of a high-performance anode for lithium ion batteries using novel ordered mesoporous tungsten oxide materials with high electrical conductivity. *Phys. Chem. Chem. Phys.*, 13: 11060–66.
[75] Jiao, F., Bao, J. and Bruce, P.G. 2007. Factors influencing the rate of Fe_2O_3 conversion reaction. *Electrochemical and Solid State Letters*, 10: A264–A66.
[76] Shi, Y.F., Guo, B.K., Corr, S.A. et al. 2009. Ordered mesoporous metallic MoO_2 materials with highly reversible lithium storage capacity. *Nano Lett.*, 9: 4215–20.
[77] Jiao, F., Yen, H., Hutchings, G.S. et al. 2014. Synthesis, structural characterization, and electrochemical performance of nanocast mesoporous Cu-/Fe-based oxides. *J. Mater. Chem. A*, 2: 3065–71.
[78] Jo, C., Kim, Y., Hwang, J. et al. 2014. Block copolymer directed ordered mesostructured $TiNb_2O_7$ multimetallic oxide constructed of nanocrystals as high power Li-ion battery anodes. *Chem. Mater.*, 26: 3508–14.
[79] Zhou, H.S., Zhu, S.M., Hibino, M. et al. 2003. Lithium storage in ordered mesoporous carbon (CMK-3) with high reversible specific energy capacity and good cycling performance. *Adv. Mater.*, 15: 2107–10.
[80] Bruce, P.G. 2008. Energy storage beyond the horizon: Rechargeable lithium batteries. *Solid State Ion.*, 179: 752–60.
[81] Choi, N.-S., Chen, Z., Freunberger, S.A. et al. 2012. Challenges facing lithium batteries and electrical double-layer capacitors. *Angew. Chem. Int. Ed.*, 51: 9994–10024.
[82] Brezesinski, K., Haetge, J., Wang, J. et al. 2011. Ordered mesoporous α-Fe_2O_3 (hematite) thin-film electrodes for application in high rate rechargeable lithium batteries. *Small*, 7: 407–14.
[83] Qiao, H., Li, J., Fu, J. et al. 2011. Sonochemical synthesis of ordered SnO_2/CMK-3 nanocomposites and their lithium storage properties. *ACS Appl. Mater. Interfaces*, 3: 3704–08.
[84] Guo, R., Lin, R., Yue, W.B. et al. 2015. Enhanced electrochemical performance of mesoporous carbon with increased pore size and decreased pore wall thickness. *Electrochim. Acta*, 174: 1050–56.
[85] Xu, G.-L., Chen, S.-R., Li, J.-T. et al. 2011. A composite material of SnO_2/ordered mesoporous carbon for the application in lithium-ion battery. *J. Electroanal. Chem.*, 656: 185–91.
[86] Lee, J., Jung, Y.S., Warren, S.C. et al. 2011. Direct access to mesoporous crystalline TiO_2/carbon composites with large and uniform pores for use as anode materials in lithium ion batteries. *Macromol. Chem. Phys*, 212: 383–90.
[87] Mancini, M., Kubiak, P., Wohlfahrt-Mehrens, M. et al. 2010. Mesoporous anatase TiO_2 electrodes modified by metal deposition: Electrochemical characterization and high rate performances. *J. Electrochem. Soc.*, 157: A164–A70.

[88] Bruce, P.G., Freunberger, S.A., Hardwick, L.J. et al. 2012. Li-O_2 and Li-S batteries with high energy storage. *Nat. Mater.*, 11: 19–29.
[89] Yin, Y.-X., Xin, S., Guo, Y.-G. et al. 2013. Lithium–sulfur batteries: Electrochemistry, materials, and prospects. *Angew. Chem. Int. Ed.*, 52: 13186–200.
[90] Wang, Z.-L., Xu, D., Xu, J.-J. et al. 2014. Oxygen electrocatalysts in metal-air batteries: From aqueous to nonaqueous electrolytes. *Chem. Soc. Rev.*, 43: 7746–86.
[91] Manthiram, A., Chung, S.-H. and Zu, C. 2015. Lithium–sulfur batteries: Progress and prospects. *Adv. Mater.*, 27: 1980–2006.
[92] Ma, Z., Yuan, X.X., Li, L. et al. 2015. A review of cathode materials and structures for rechargeable lithium-air batteries. *Energ. Environ Sci.*, 8: 2144–98.
[93] Manthiram, A., Fu, Y.Z. and Su, Y.S. 2013. Challenges and prospects of lithium-sulfur batteries. *Acc. Chem. Res.*, 46: 1125–34.
[94] Hong, W.T., Risch, M., Stoerzinger, K.A. et al. 2015. Toward the rational design of non-precious transition metal oxides for oxygen electrocatalysis. *Energ. Environ Sci.*, 8: 1404–27.
[95] Cui, Y.M., Wen, Z.Y., Sun, S.J. et al. 2012. Mesoporous Co_3O_4 with different porosities as catalysts for the lithium-oxygen cell. *Solid State Ion.*, 225: 598–603.
[96] Xie, J., Yao, X., Cheng, Q. et al. 2015. Three dimensionally ordered mesoporous carbon as a stable, high-performance Li–O_2 battery cathode. *Angew. Chem. Int. Ed.*, 54: 4299–303.
[97] Park, J., Jeong, J., Lee, S. et al. 2015. Effect of mesoporous structured cathode materials on charging potentials and rate capability of lithium–oxygen batteries. *ChemSusChem.*, 8: 3146–52.
[98] Ji, X., Lee, K.T. and Nazar, L.F. 2009. A highly ordered nanostructured carbon-sulphur cathode for lithium-sulphur batteries. *Nat. Mater.*, 8: 500–06.
[99] Qu, Y.H., Zhang, Z.A., Zhang, X.H. et al. 2015. Highly ordered nitrogen-rich mesoporous carbon derived from biomass waste for high-performance lithium-sulfur batteries. *Carbon*, 84: 399–408.
[100] Ji, X., Evers, S., Black, R. et al. 2011. Stabilizing lithium-sulphur cathodes using polysulphide reservoirs. *Nat. Commun.*, 2: 325.
[101] Schuster, J., He, G., Mandlmeier, B. et al. 2012. Spherical ordered mesoporous carbon nanoparticles with high porosity for lithium–sulfur batteries. *Angew. Chem. Int. Ed.*, 51: 3591–95.
[102] Zhao, X., Liu, Y., Manuel, J. et al. 2015. Nitrogen-doped mesoporous carbon: A top-down strategy to promote sulfur immobilization for lithium–sulfur batteries. *ChemSusChem.*, 8: 3234–41.
[103] He, G., Ji, X. and Nazar, L. 2011. High "C" rate Li-S cathodes: Sulfur imbibed bimodal porous carbons. *Energ. Environ Sci.*, 4: 2878–83.
[104] Choudhury, S., Agrawal, M., Formanek, P. et al. 2015. Nanoporous cathodes for high-energy Li-S batteries from gyroid block copolymer templates. *ACS Nano.*, 9: 6147–57.
[105] Doherty, C.M., Caruso, R.A., Smarsly, B.M. et al. 2009. Colloidal crystal templating to produce hierarchically porous $LiFePO_4$ electrode materials for high power lithium ion batteries. *Chem. Mater.*, 21: 2895–903.
[106] Ren, Y., Ma, Z., Morris, R.E. et al. 2013. A solid with a hierarchical tetramodal micro-meso-macro pore size distribution. *Nat. Commun.*, 4: 2015.
[107] Ma, Y., Tai, C.-W., Younesi, R. et al. 2015. Iron doping in spinel $NiMn_2O_4$: Stabilization of the mesoporous cubic phase and kinetics activation toward highly reversible Li^+ storage. *Chem. Mater.*, 27: 7698–709.
[108] He, G. and Manthiram, A. 2014. Nanostructured Li_2MnSiO_4/C cathodes with hierarchical macro-/mesoporosity for lithium-ion batteries. *Adv. Funct. Mater.*, 24: 5277–83.
[109] Guo, Z., Zhou, D., Dong, X. et al. 2013. Ordered hierarchical mesoporous/macroporous carbon: A high-performance catalyst for rechargeable Li-O_2 batteries. *Adv. Mater.*, 25: 5668–72.

CHAPTER 7

Controlling the Photoanode Mesostructure for Dye-sensitized and Perovskite-sensitized Solar Cells

Wu-Qiang Wu,[1] *Dehong Chen*[1] and *Rachel A. Caruso*[2,*]

ABSTRACT

This chapter summarizes the progress made in the use of multifunctional mesoporous semiconducting metal oxide materials as photoanodes for high performance dye-sensitized solar cells and perovskite-sensitized solar cells. The impact of the mesoscopic properties (surface area, pore sizes, pore volumes and porosity) and morphologies (mesoporous nanoparticles, hierarchically mesoporous spheres, ordered mesoporous film and one dimensional porous structures) on device performance are highlighted to elucidate the advantages of enhanced light harvesting and charge collection for such mesostructured nanomaterials. The chapter concludes with an outlook on the future development of mesoporous materials towards highly efficient optoelectronic applications.

[1] Particulate Fluids Processing Centre, School of Chemistry, The University of Melbourne, Melbourne, Victoria, Australia.

[2] Particulate Fluids Processing Centre, School of Chemistry, The University of Melbourne, Melbourne, Victoria, Australia & CSIRO Manufacturing, Clayton South, Victoria, Australia.

* Corresponding author: rcaruso@unimelb.edu.au

Keywords: Nanorod, Nanowire, Nanotube, Nanoparticle/nanoparticulate, Sphere/spherical, Hierarchical/hierarchically, Shell, Inject/injected/injection (relating to charge), Extract/extracted/extraction (relating to charge), Collect/collected (relating to charge), Recombination (relating to charge), Harvest/harvested/harvesting (relating to light), Scattering, Reflect

1. Introduction

Our predicted energy needs as well as environmental pollution related to current methods for energy production have called for the rapid development of renewable energy sources. Solar energy is well recognized as one of the most promising choices to meet the predicted energy demand in the near future. This has inspired advances in solar cell research in an attempt to improve the direct conversion of solar radiation energy into electricity [1]. Among the various types of solar cells, Dye-sensitized Solar Cells (DSCs) and Perovskite-sensitized Solar Cells (PSCs) are of great interest as alternatives to conventional silicon-based solar cells because of their low cost manufacturing processes and relatively high efficiency [2–4].

To develop solar cells with high power conversion efficiency, intensive investigations have focused on the optimization of both the materials and the configuration of the devices [5]. In solar cells using the sensitization concept, including DSCs and PSCs, the photoanode materials serve as scaffolds for anchoring sensitizers and provide transport paths for photo-generated electrons. An ideal photoanode should possess a large surface area to maximize sensitizer loading, fast electron mobility for effective electron transport and charge extraction, porosity for efficient electrolyte penetration, as well as a light scattering ability to confine the incident photons within the cell [6]. Therefore, mesoporous semiconductor metal oxide-based photoanodes with high surface area, variable pore sizes, pore volume, porosity and an interconnected porous network have exhibited advantages, such as increasing the light harvesting efficiency, improving electrolyte permeation and ensuring fast electron transport. This makes them suitable materials for photoelectrodes in solar cells [7, 8].

This chapter starts with a brief outline of the way in which DSCs and PSCs function before recent variations in the mesostructured metal oxide photoanodes are reviewed. Photoanodes prepared from nanoparticles, ordered mesoporous networks, mesoporous beads or one Dimensional (1D) porous structures are described. Correlations between the photoanode morphology and properties, such as ability to capture light, produce and transfer electrons and therefore overall cell performance, are made. A short perspective and outlook for mesostructured photoanode-based solar cells is given at the end of the chapter.

2. Configuration and Working Mechanism of DSCs and PSCs

A DSC device has a sandwich configuration, consisting of a photoanode with a dye molecule-sensitized mesoporous film of a semiconductor oxide (TiO_2, ZnO, SnO_2, etc.) deposited on transparent Fluorine-doped Tin Oxide (FTO) glass and a counter electrode of FTO coated with platinum nanoparticles. The redox electrolyte, typically containing iodide and triiodide ions (I^-/I_3^-), is injected between these two electrodes, and the device sealed [3, 9]. The concept of the DSC device was inspired by the photosynthesis of green plants. As shown in Fig. 7.1a, when irradiated by light, the dye molecules anchored on the surface of the semiconductor capture photons of sufficient energy, thereby creating excitons with the electrons being injected into the conduction band of the semiconductor. The electrons transfer through the semiconductor layer and migrate through the external circuit. The oxidized dye molecules are reduced by the redox couple in the liquid electrolyte. The I_3^- is also reduced to I^- by the electrons at the counter electrode. Thus a complete closed circuit is constructed to continuously convert incident solar light to electricity. However, the charge injection into the semiconductor and transport processes are accompanied by charge recombination at surface defects or by interaction with oxidized dye molecules or species in the electrolyte. This limits the performance of the DSC.

PSCs originated from the DSC concept using organometallic halide perovskite compounds, $CH_3NH_3PbX_3$ (X = Cl, Br or I), as light harvesters. The PSCs have shown rapid progress as an alternative low-cost solution-processed solar energy technology [12, 13]. A common mesoscopic PSC consists of several parts and works as follows (Fig. 7.1b): (i) The FTO glass covered with a very thin compact titania layer, allows light transmittance and prevents direct electric contact between the FTO and perovskite material, thus reducing charge recombination; (ii) on top of the compact layer is a thin mesoporous titania layer that transfers injected electrons; (iii) a perovskite film adsorbed on the surface and filling the pores of the mesoporous oxide layer, harvests light and achieves efficient charge separation; (iv) a layer of Hole Transporting Material (HTM) on top of the perovskite film is required for hole migration; and (v) a thin metal film (Au or Ag) is used as a counter electrode to collect the generated charge.

3. Mesostructured Photoanodes for DSCs

The pioneering report of DSCs obtaining a Power Conversion Efficiency (PCE) as high as 7% was published in 1991 [14]. O'Regan and Grätzel employed nanosized, wide band gap titania crystals for transporting photo-induced electrons and a ruthenium complex (N719) as the sensitizer to absorb the solar light [14]. In the following decades, this work stimulated

(a)

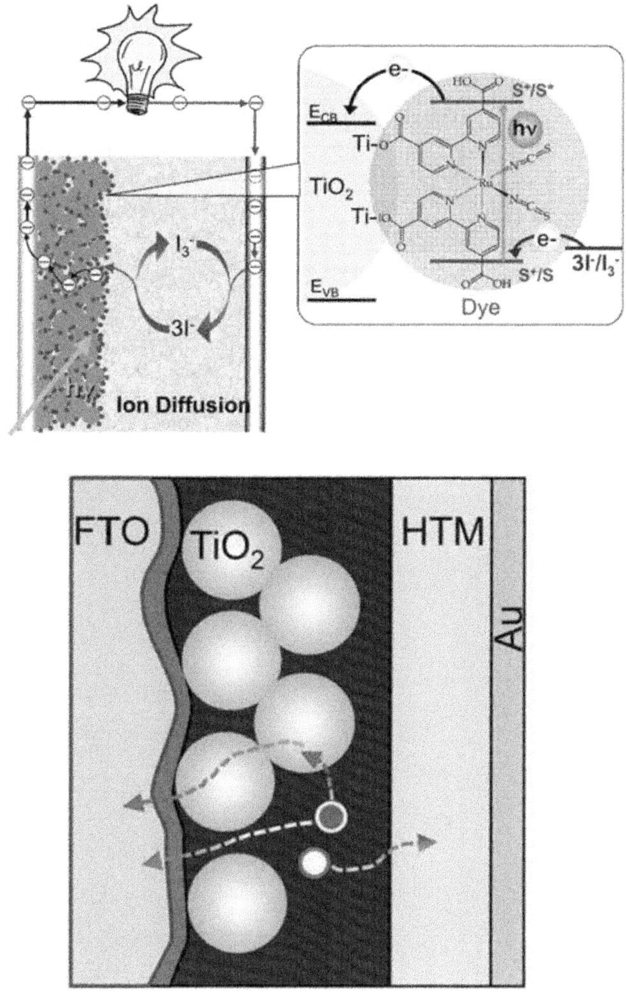

Figure 7.1. Sketches illustrating the structures and functioning of: (a) a typical DSC [10], in which the dye molecules anchored on TiO_2 nanocrystals act as light harvesters, the iodide/triiodide redox couple regenerates the dye, and I_3^- accepts electrons from the counter electrode and then is reduced to I^-. (b) a mesoscopic PSC device, in which the electron (grey dot) will be collected (upper arrows) and transferred through the TiO_2 to the FTO front electrode and the hole (white dot) will be collected (lower arrow) by the HTM and Au back contact [11]. Images reproduced with permission [10, 11]. Copyright 2005, American Chemical Society, and Copyright 2015, Wiley-VCH Verlag GmbH & Co.

numerous follow-up studies that varied and/or optimized each component of the device. For instance, various materials and nanostructures have undergone investigation, as have novel dyes that can absorb light over broad wavelength regions with high extinction coefficients, and liquid or solid state electrolytes with good stability [15, 16]. As a key component of the DSCs, the photoanode material plays an important role in determining the dye loading and charge injection, transport and collection efficiencies. Hence the structural properties of the photoanode film influence the photocurrent density (J_{sc}), photovoltage (V_{oc}) and the PCE of the solar cells. To fabricate highly efficient solar cells, the photoanode requires: a large internal surface area for a high dye loading capacity that will ensure a high light harvesting efficiency; superior light scattering functions that can prolong the light pathway within the device, thus improving light utilization; fast electron transport for effective charge collection and suppressed charge recombination; and well interconnected porosity for electrolyte infiltration and diffusion [17]. The design and optimization of diverse nanostructured photoanodes, particularly mesoporous metal oxides, have been widely studied to improve the DSC efficiency. Next nanoparticles, mesoporous spherical particles and 1D porous structures are described and discussed in reference to their application in DSCs.

3.1 Nanoparticulate films

The most commonly used mesostructured photoanode is composed of ~20 nm anatase titania nanoparticles, giving it a good surface area to adsorb dye for efficient light harvesting. A porous film consisting of nanoparticles (15 nm in size) can provide a surface area as high as 780 cm^2 per centimetre square of film. These nanoparticle films were deposited on the FTO glass via a screen-printing or a doctor-blading method. A record PCE surpassing 13% has been achieved by using a photoanode consisting of 20 nm anatase titania nanocrystals sensitized with a zinc porphyrin dye [15]. However, further improvement in the DSC performance based on this kind of mesoscopic, nanoparticle-based films has been restricted by disadvantages associated with the nanoparticles. First is the high transparency in the visible light range of the films made of nanometre-sized titania particles. Over 65% incident sunlight can transmit through such a film that lacks light scattering ability, thereby limiting the portion of incident light that can be harvested by the dye molecules [18, 19]. Second, random and disordered electrical pathways, as well as short electron diffusion lengths (10–35 μm) induced by excessive trapping and detrapping events, occur within defects, surface states and grain boundaries of the nanoparticles. This slows the electron transport and enhances the probability of charge recombination and thus inhibits charge collection [20, 21]. Hence, increasing the light scattering and reducing the recombination reactions for photo-induced electrons are key points to improving DSCs containing nanoparticulate photoanodes.

3.2 Mesoporous spherical particles

As the conventional particulate films are disordered networks, with defects and electron trapping sites between particles, approaches to improve the light harvesting and charge collection characteristics of photoanodes for DSCs have been investigated. Various mesoporous spherical particles have been synthesized and fabricated into mesoporous films for DSCs with significant progress.

As mentioned above, the photoanode composed of nanometre-sized titania particles (~20 nm in diameter) suffers from poor light scattering, which limits the light harvesting efficiency and thus the PCE. To resolve this issue, a scattering layer made from ~400 nm titania (rutile) or zirconia particles with high reflective index can be coated over the transparent nanocrystalline titania film to enhance the light harvesting [19]. However, such large particles possess low surface area and apart from scattering light, do not contribute to the electrode function. To overcome this disadvantage, bi-functional nano-embossed hollow spherical titania particles (1–3 μm in diameter) were synthesized that exhibited superior light scattering and higher dye loading capability (~5 times higher) than the traditional 400 nm scattering particles. An improved J_{sc} and a PCE of 10.34% were obtained when these particles were used as the light scattering layer on top of a nanocrystal based photoanode [22].

Sub-micrometre sized spheres composed of nanoparticles have been assembled into photoanodes and studied for enhancement of both the light harvesting and charge collection efficiencies without sacrificing the total specific surface area of the films [7]. A hexadecylamine (HDA)-induced sol-gel self-assembly and ammonia-assisted solvothermal process were developed to obtain monodisperse mesoporous anatase beads with high specific surface area, variable pore size, high crystallinity and well-interconnected nanocrystal networks [7]. By carefully adjusting the HDA:H_2O:Ti molar ratio and ammonia concentration, control over the monodispersity, crystal size, bead diameter (300–800 nm), pore size (14–46 nm) and crystallinity of the mesostructured hybrid beads was shown [23]. The abundant mesopores in the beads (830 ± 40 nm in diameter, Figs. 7.2a and 7.2b) allows access to a high surface area (~108.0 $m^2\ g^{-1}$) for dye attachment. This in combination with their strong light scattering ability, means that DSCs fabricated using these beads as the photoanode show an improvement in J_{sc}, 12.79 mA cm^{-2} when compared to 10.60 mA cm^{-2} for a device constructed using commercial Evonik (Degussa) P25 nanoparticles for the photoanode [7].

The DSC devices were optimized, with a 12 μm thick anatase bead (~830 nm in diameter) film composed of 23 nm-pore sized spheres (Figs. 7.2c and 7.2d) and sensitized with a C101 Ru(II)-based dye to obtain a PCE of 10.6% (Fig. 7.3a). This was a ~25% increment in PCE as compared

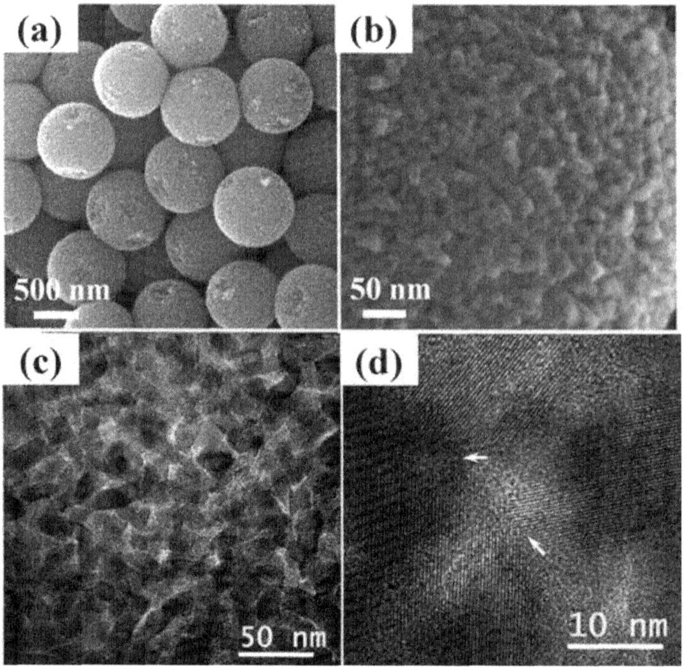

Figure 7.2. (a–b) SEM images and (c) TEM image of the monodisperse mesoporous anatase titania beads obtained using a sol-gel and subsequent solvothermal process. (d) HRTEM image showing the intergrowth between the anatase crystals within the bead, as indicated by the white arrows. Images reproduced with permission [7, 23]. Copyright 2009, Wiley-VCH Verlag GmbH & Co. and Copyright 2010, American Chemical Society.

to the cells made with P25 nanocrystal electrodes of similar film thickness [8]. The light harvesting enhancement achieved using the anatase bead films was demonstrated by the corresponding Incident Photon-to-current Conversion Efficiency (IPCE) curves (Fig. 7.3b). From 400 to 800 nm, the anatase bead electrodes showed a much higher IPCE (~92% at 570 nm) than the P25 films (~70% at 570 nm), indicating the improved light harvesting characteristics that originate from the combination of an increased dye uptake on the beads and the stronger diffuse reflectance (light scattering) properties. These photoanodes containing mesoporous beads also exhibited faster electron transport and a longer electron lifetime than the commercial P25 nanoparticle-based photoanode counterparts. As indicated in Figs. 7.3c and 7.3d, the electron lifetime of the DSCs containing anatase bead films was almost two times longer than that of P25-based DSCs; additionally the electron diffusion length was longer. This excellent electron transport can be attributed to the good contact between adjacent nanoparticles arising from the crystal intergrowth inside the beads, which would decrease the grain

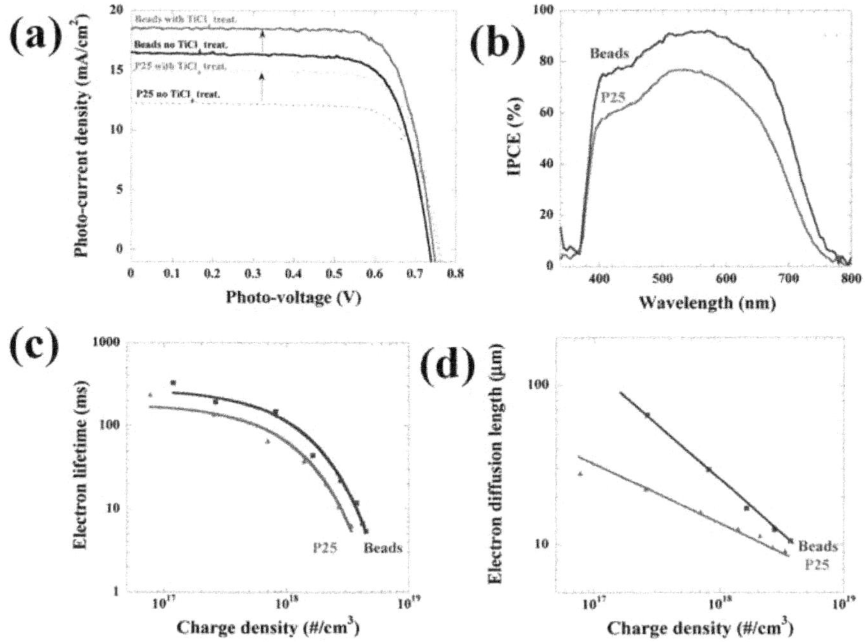

Figure 7.3. (a) *J-V* curves, (b) IPCE, (c) electron lifetime and (d) electron diffusion length of the DSCs based on 12 µm thick films composed of anatase titania beads or P25 and sensitized with C101 dye [8]. Images reproduced with permission [8]. Copyright 2010, American Chemical Society.

boundary crossing and thus extend electron lifetime and reduce charge recombination within the cells [8].

A similar concept was adopted by Lee et al. who employed mesoporous titania spheres with a diameter of ~250 nm consisting of ~10 nm nanoparticles (Fig. 7.4a), which provided a high BET specific surface area, allowing for a high dye loading, and an excellent light scattering effect [24]. A multi-scale porous structure with the large pores generated from the interstitial voids among the spheres (denoted as A, ~55 nm for deformed spheres) and small internal mesopores formed by the stacking of primary nanoparticles (denoted as B, ~9.1 nm for deformed spheres) can be visualized in Fig. 7.4b. These pores can facilitate electrolyte permeation and diffusion. As a result, DSCs containing this hierarchical film achieved an impressive PCE of 8.44% [24]. Apart from titania, ZnO spherical aggregates (Figs. 7.4c and 7.4d) were also reported to efficiently improve the performance of ZnO based DSCs by providing both high surface area and superior light scattering to boost light harvesting. The particle size-dependent light scattering effect was carefully investigated on ZnO spherical aggregates with diverse diameters. The polydisperse ZnO spheres with a large average

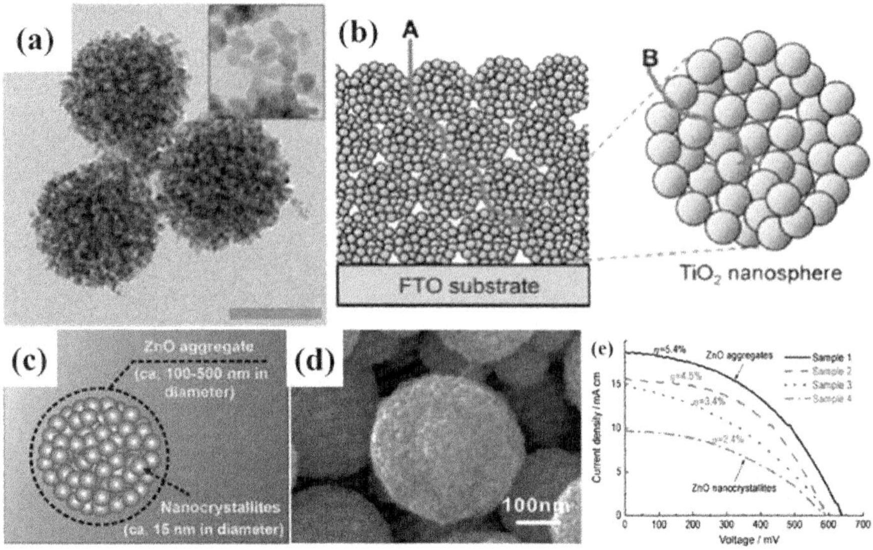

Figure 7.4. (a) TEM image of well-defined titania nanospheres composed of titania nanoparticles and (b) sketch showing the electrolyte diffusion through the external large pores (A) and internal (B) small pores in the mesoporous titania sphere film [24]. (c) Schematic diagram and (d) SEM image of bi-functional mesoporous ZnO spheres consisting of densely packed nanoparticles. (e) A comparison of the *J-V* characteristics of ZnO spheres prepared at different temperatures (Sample 1, 2, 3, 4 are ZnO prepared at 160, 170, 180 and 190°C, respectively) [25]. Images reproduced with permission [24, 25]. Copyright 2008 and 2009, Wiley-VCH Verlag GmbH & Co.

size or broad size distribution gave excellent light scattering and thus better light harvesting. As a result, a PCE as high as 5.4% (Fig. 7.4e) was obtained for the optimized ZnO aggregates [25].

Porous spheres with hollow structure also showed promising capability in enhancing light trapping by multiple reflection and scattering within the electrodes. This concept has been proven by photoanodes containing hollow structures, yolk-shell structures, shell-in-shell structures or multi-shelled hollow spheres [26–29]. Wang et al. reported a new type of shell-in-shell titania hollow sphere (Fig. 7.5a) featuring superior light scattering capabilities, which made it suitable for use as the scattering layer in DSCs, achieving a PCE of 9.10% [26]. Fang et al. reported hierarchical yolk-shell anatase titania beads with nearly 90% of the exposed facets being {001} and a mesoporous inner sphere (Fig. 7.5b) with a high specific surface area of 245.1 m^2g^{-1}. When used to prepare the photoanode in DSCs, a PCE of 6.01% was attained, which was higher than that of the standard Degussa P25 counterpart (4.46%), owing to higher dye loading capability, longer lifetime of the injected electrons, as well as superior light scattering in the

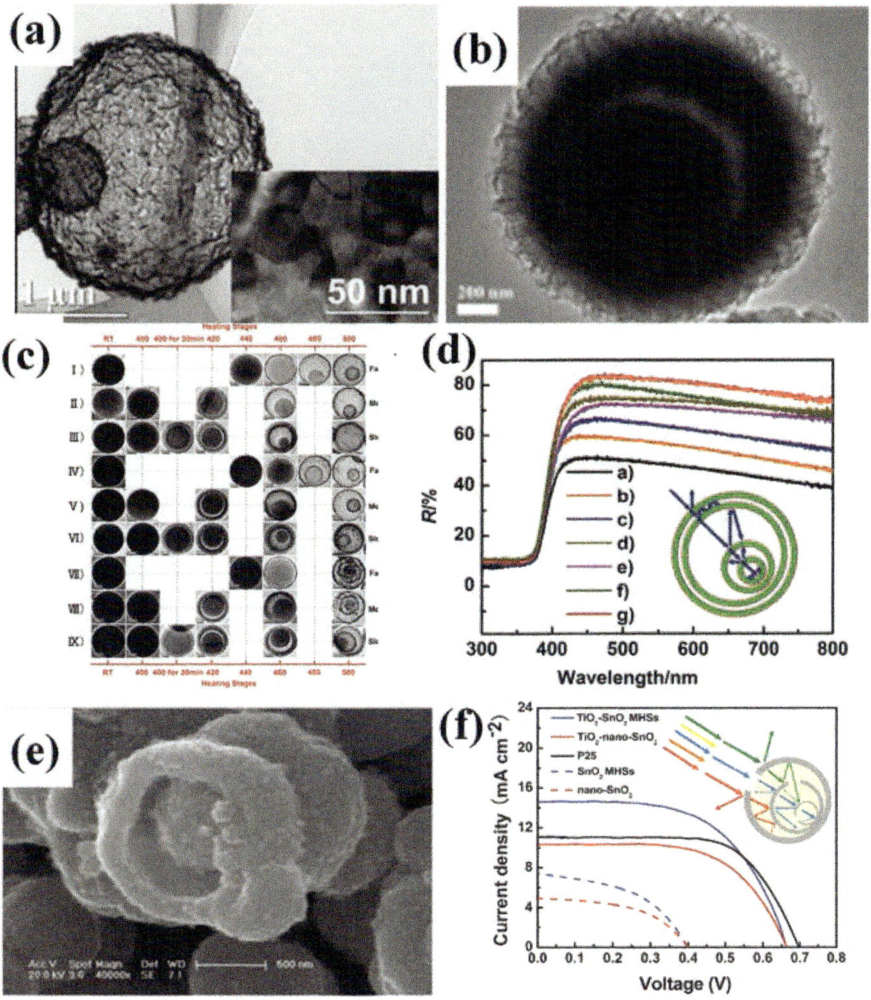

Figure 7.5. TEM images of (a) shell-in-shell titania hollow spheres (inset at higher magnification) [26] and (b) yolk-shell titania spheres [29]; (c) Evolution process of a family of multi-shelled ZnO hollow microspheres; (d) diffuse reflectance spectra of the different ZnO hollow spheres with different shell layers: a) single-shelled, b) double-shelled, c) triple-shelled, d) quadruple-shelled, e) double-shelled (with close double shells in the exterior), f) triple-shelled (with close double shells in the exterior and a smaller hollow core) and g) quadruple-shelled (with close double shells in the exterior and double-shelled hollow core). Inset indicates light reflection within a multishelled sphere [27]; (e) SEM image of a multilayered SnO_2 hollow microsphere (SnO_2 MHSs); (f) *J-V* curves of DSCs based on P25, nanocrystalline SnO_2 (nano-SnO_2), TiO_2-coated nanocrystalline SnO_2 (TiO_2-nano-SnO_2), SnO_2 MHSs and TiO_2-SnO_2 MHSs photoanodes. The inset in (f) illustrates the multiple reflections and scattering of light in the multilayered hollow spheres [28]. Images reproduced with permission [26–29]. Copyright 2011 and 2012, The Royal Society of Chemistry [26, 27], Copyright 2012 and 2009, Wiley-VCH Verlag GmbH & Co. [28, 29].

visible light region in the yolk-shell bead films [29]. Recently, a general strategy to produce multi-shelled hollow metal oxide microspheres by using carbonaceous microspheres as sacrificial templates and a simple programmable calcination process has been reported by Wang et al. [27]. As an example, Fig. 7.5c illustrates the evolution of a ZnO hollow sphere with different numbers of defined shells and inter-shell spacings. Augmenting the number of shells is beneficial to increase the light capturing efficiency. PCEs up to 5.6% were achieved for DSCs containing quadruple-shelled (two close double shells in the exterior and double-shelled hollow interior cores) hollow ZnO microspheres due to the optimized surface area achieved by manipulating the inter-shell spaces and maximized light harvesting through scattering and multiple reflection inside the hierarchically structured spheres (Fig. 7.5d) [27]. Similar to titania and ZnO, multilayered SnO_2 hollow microspheres (Fig. 7.5e) were also reported to reflect and scatter the incident solar light in DSCs device. In addition, a thin layer of titania (~50 nm) was introduced as a blocking layer on the outer shell to suppress the interfacial charge recombination on the surface of these multilayered SnO_2 hollow spheres. As a result, an improved J_{sc} (14.6 mA cm^{-2}), V_{oc} (664 mV) and PCE (5.64%) were achieved [28].

Since a large number of grain boundaries still exist within nanoparticle-based spherical architectures that are adverse for charge collection within the cells, spiky beads assembled to include epitaxial 1D building blocks have been fabricated to provide enhanced electron transport characteristics in the devices [30, 31]. A metastable ammonium titanate-mediated synthesis method was reported to fabricate spiky mesoporous anatase titania beads, which are composed of 15 nm nanocrystals in the core and much larger elongated single crystal spikes on the surface (Fig. 7.6a). Consequently, an impressive PCE of 10.30% was obtained for the DSCs containing these spiky mesoporous titania beads as the photoanode (Fig. 7.6b) [30]. Furthermore, these spiky mesoporous titania beads with high surface area, strong light scattering capability and high porosity for improved mass transport properties, were beneficial in the construction of DSCs using bulky cobalt-based redox couples and/or viscous electrolytes [32]. DSCs based on an optimized film composition consisting of a transparent mesoporous titania nanocrystal underlayer, a spiky mesoporous titania beads (~800 nm in diameter) top layer, and a cobalt redox electrolyte yielded PCEs up to 11.4% under AM 1.5G one sun illumination [32]. Kuang and co-workers reported tri-functional 3D hierarchical anatase titania microspheres (HTS) consisting of single crystal nanorods and nanoparticles (Figs. 7.6c and 7.6d) prepared by using a one-pot acid-thermal method [31]. In DSC applications these 3D HTS exhibited high dye loading and light scattering ability, and higher charge-collection efficiency (faster electron transport and longer electron lifetime, seen in Figs. 7.6e and 7.6f) compared to the commercial P25 nanoparticles, leading to an impressive

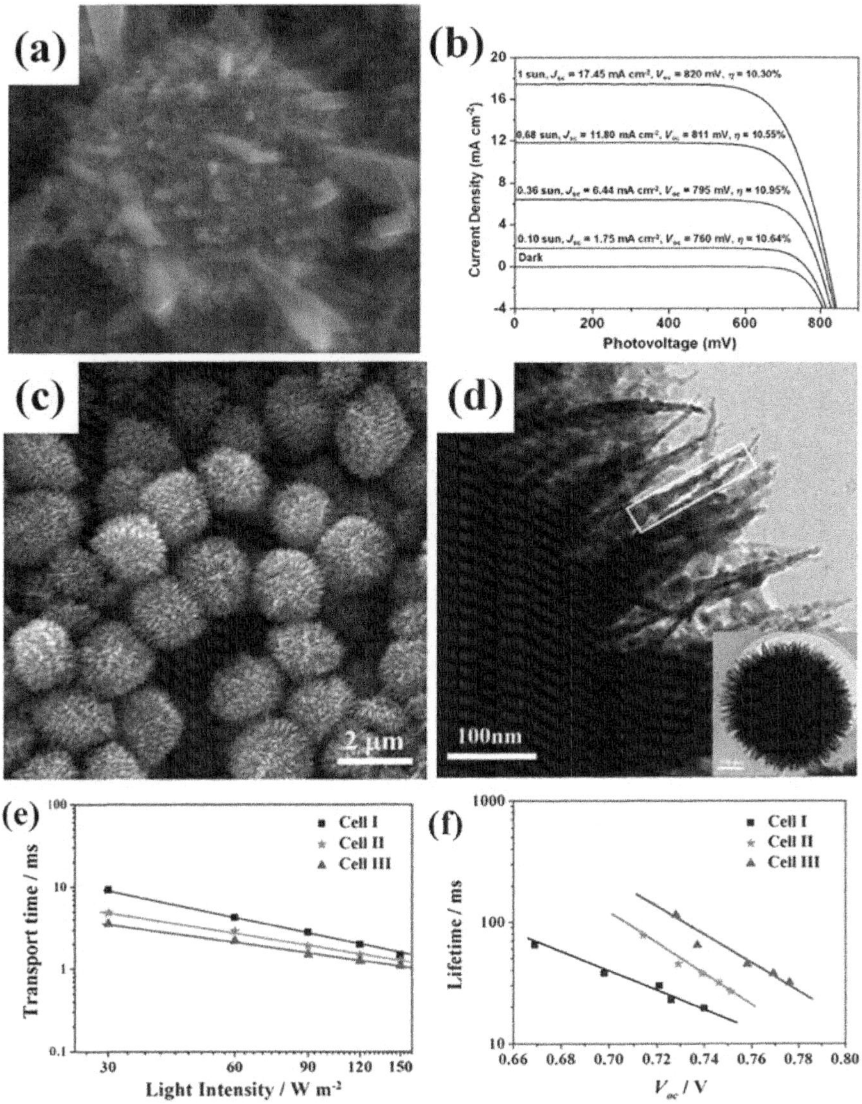

Figure 7.6. (a) SEM image of the spiky titania bead and (b) *J*-*V* curves of DSCs based on C106 dye-sensitized spiky titania mesoporous beads under varying light intensities [30]. (c) SEM image and (d) TEM image of the hierarchical anatase titania microsphere (HTS). Inset is a lower magnification image showing the complete sphere (e) Electron transport time and (f) electron lifetime for DSCs based on cell I (P25), II (P25 +HTS), and III (HTS) [31]. Images reproduced with permission [30, 31]. Copyright 2012, Wiley-VCH Verlag GmbH & Co., and Copyright 2011, The Royal Society of Chemistry.

PCE of 10.34% [31]. Very recently, optimized hyperbranched HTS with interconnected and interwoven networks and a high porosity, have been obtained via subsequent modification. A novel tri-layered titania photoanode using these hyperbranched HTS as the top scattering layer have been applied in DSCs, achieving an impressive PCE of 11.0% [33].

Zhao and co-workers have recently reported the 3D open mesoporous titania microspheres (~800 nm in diameter, Fig. 7.7a) with well-controlled, radially oriented hexagonal mesochannels and single crystal-like anatase walls (Figs. 7.7b–7.7d) synthesized via a simple evaporation-driven oriented assembly protocol. Such radially oriented mesoporous titania microspheres feature promising advantages in terms of well-controlled pore size, porosity, crystallinity, as well as mesostructure ordering. The spheres have a large accessible surface area of 112 $m^2\ g^{-1}$, a large pore volume of 0.164 $cm^3\ g^{-1}$ and single crystal-like anatase walls with dominant (101) exposed facets for fast electron transport. Hence these materials are ideal for constructing mesoscopic photoanode films for DSCs. Solar cells constructed using such titania spheres and N719 dye yield a PCE as high as 12.1% (Fig. 7.7e). The electron diffusion coefficient indicated that the charge collection within the radially oriented mesoporous titania microspheres is much greater than that of either the bulk spheres or P25 films. This can be explained by the diffusion of the electrolyte through the mesochannels, as well as fast electron transport and suppressed charge recombination within the highly-crystalline and well-oriented titania [34].

As discussed above, the micrometre or sub-micrometre sized spheres with structures assembled from the primary 0D nanoparticles or 1D nanometre sized-building blocks (wires, rods and tubes) can provide a high accessible specific surface area for dye loading. In addition, the comparable size of the spherical assemblies to the wavelength of visible light can enhance the light utilization efficiency via effective light scattering, which lengthens the light transport path inside the photoanode film and thus increases the probability of photon absorption by dye molecules. Moreover, such spherical structures can densely pack, and have a highly interconnected framework due to crystal intergrowth inside the spheres or the interwoven 1D building blocks to boost the electron transport. Therefore, high J_{sc} and PCEs were obtained when these materials were employed to fabricate photoanodes for DSCs. The dual pore sizes of the resulting bead films with relatively large inter-bead voids and mesopores throughout the whole bead could facilitate pore-filling with a highly viscous gel, polymer electrolytes or solid-state hole-conducting materials, thus making these hierarchical, mesoporous, spherical-particle films highly promising for (quasi-) solid state DSC application [35].

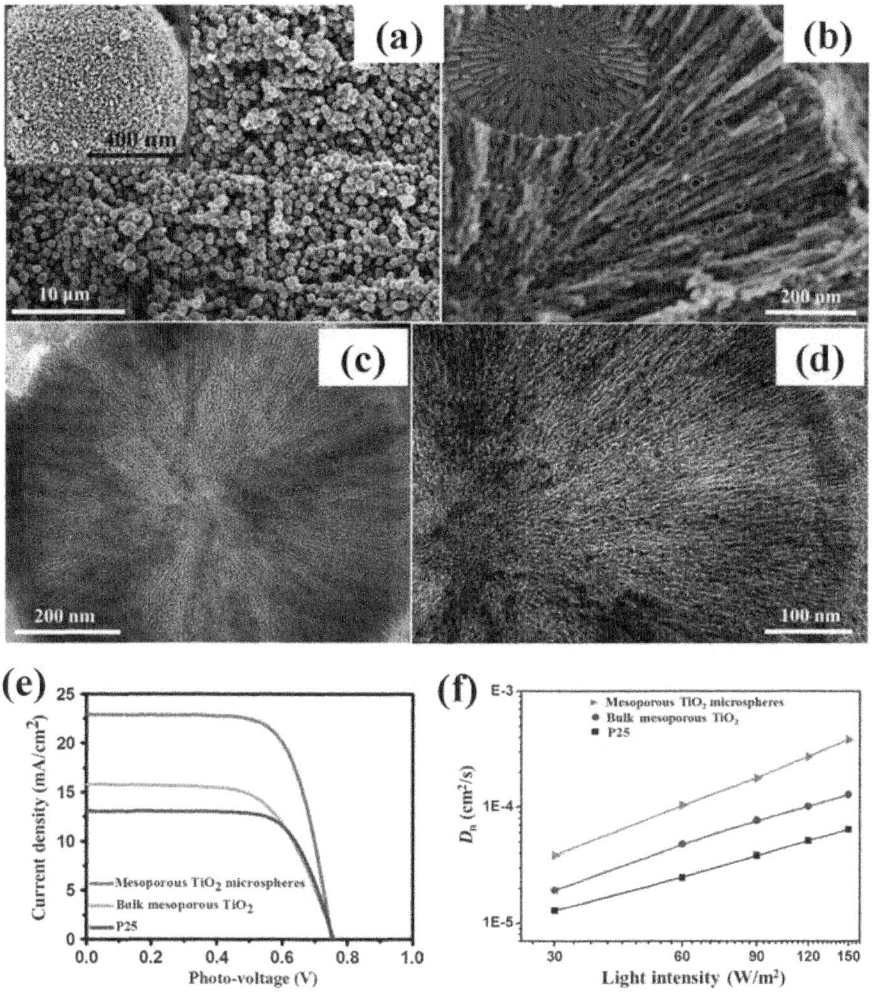

Figure 7.7. SEM images of (a) mesoporous titania microspheres and (b) radially-oriented microchannels within a mesoporous titania microsphere with a large number of interchannel pores. Inset in (b) is a schematic representation of the structure for the radially oriented channels with interchannel pores; (c, d) TEM images of a single ultramicrotomed mesoporous titania microsphere. (e) *J-V* curves and (f) electron diffusion coefficient of the DSCs fabricated with three different titania samples as indicated [34]. Images reproduced with permission [34]. Copyright 2015, Science (AAAS).

3.3 Hierarchically porous 1D structures

To overcome the electron transfer kinetic limitations of nanoparticles, diverse 1D nanostructures including nanorods, nanowires and nanotubes, have been widely investigated due to their advantages of providing direct electron transfer pathways for fast electron transport and efficient charge collection [36, 37]. However, the surface area of conventional 1D nanostructures is low, thus limiting their potential application in DSCs with high efficiency. A common strategy is to decorate 1D structures with nanoparticles to significantly increase the surface area of the photoanode film, thus increasing dye uptake, while at the same time maintaining effective electron transport through the 1D backbone. As shown in Figs. 7.8a and 7.8b, 1D hierarchical titania nanorods and nanotubes were prepared by coating titania nanoparticles onto the surface of pristine nanorods or nanotubes [38, 39]. When applied in DSCs, enhanced performance was observed due to the enlarged specific surface area arising from the hierarchical, porous structure, relatively fast electron transport and long electron lifetimes [38, 39]. However, a large number of grain boundaries still exist within the nanoparticle-coated outer layer, which will adversely affect the charge collection efficiency. Recent work from Kuang and co-workers has demonstrated the fabrication of well-aligned anatase titania core-shell arrays with interconnected 2D nanosheets covering the surface of a 1D nanowire scaffold (Fig. 7.8c). These 2D nanosheet shells on the nanowire backbone could in principle direct and speed up the electron transport, and thus efficiently concentrate the photo-generated electrons from the nanosheet shell to the nanowire core, which would improve the charge collection efficiency for photovoltaic devices [40]. A similar concept was previously reported with 1D hierarchical nanostructures of titania nanosheets on SnO_2 nanotubes for high-performance solid-state DSCs by Kim et al. [41].

Apart from 0D nanoparticles and 2D nanosheets, 1D nanobranches such as nanorods and nanowires were also decorated onto 1D nanostructures to construct "tree-like" or "forest-like" architectures [45]. The trunks offer efficient pathways for fast electron transport, while the branches enhance the specific surface area for dye loading and build a 3D network for efficient light scattering, trapping and utilization. Fonzo and co-workers reported a forest-like titania film consisting of high aspect ratio and tree-like nanostructures via a pulsed laser deposition method. Such films provided a high surface area for dye loading and enhanced light harvesting, favoured the mass transport of tri-iodide in an ionic liquid electrolyte and improved the electron diffusion length, leading to an impressive PCE of 4.9% for a 7 μm thick film [46]. Further hierarchical branched titania nanostructures have since been reported that were typically fabricated via a hydrothermal method, producing PCEs ranging from 4 to 9% when applied as photoanodes in DSCs [46–48]. A family of branched (Fig. 7.8d) or hyperbranched titania nanostructures (Figs. 7.8e and 7.8f) with secondary or even tertiary branches were designed by Kuang and co-workers [42–44]. These tree-like, porous structures offer a number of synergic advantages:

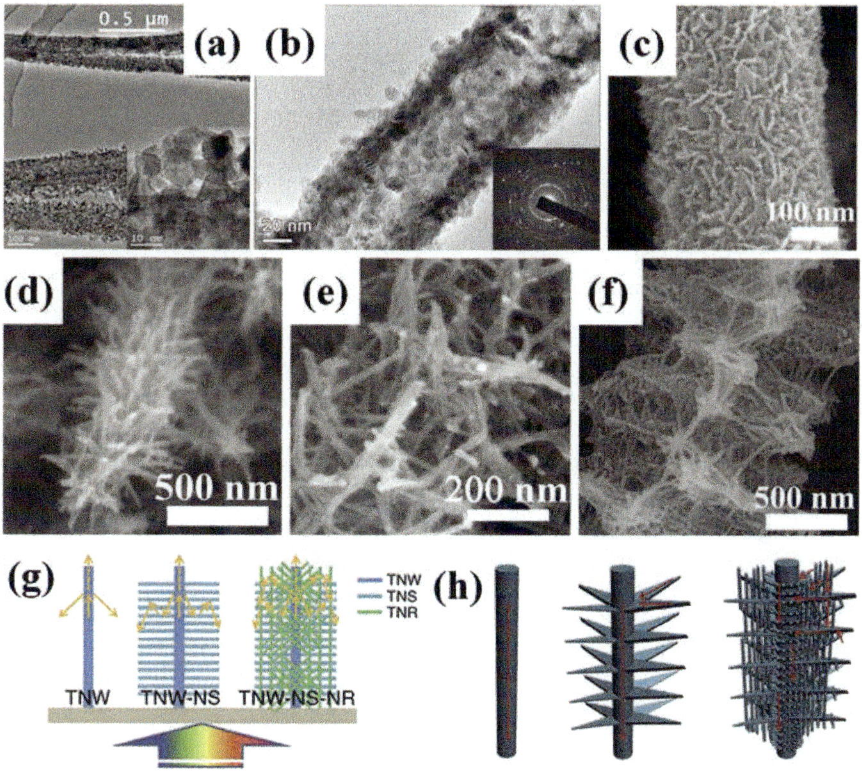

Figure 7.8. TEM images of (a) 1D hierarchical titania nanorods [38] and (b) 1D hierarchical titania nanotubes [39]; SEM images of (c) titania nanowire-nanosheet core-shell structure [40], (d) hierarchical titania nanowires consisting of a long nanowire trunk and numerous short nanorod branches [42]; (e) hyperbranched titania nanoarrays consisting of long nanowire trunks, short nanorod branches as well as tiny nanorod leaves [43]; (f) titania nanowire (TNW)-nanosheet (NS)-nanorod (NR) hyperbranched arrays consisting of nanowire trunks and nanosheet leaves, linked by nanorod branches [44]; and (g) schematic of light reflecting and scattering within the simplified TNW, TNW-NS and TNW-NS-NR architectures and (h) electron transfer pathways for individual TNW, TNW-NS and TNW-NS-NR [44]. Images reproduced with permission [38–40, 42–44]. Copyright 2010, The Royal Society of Chemistry [38], Copyright 2012, Wiley-VCH Verlag GmbH & Co. [39], Copyright 2010, American Chemical Society [40], Copyright 2014, Elsevier [42–43], Copyright 2014, Nature Publishing Group [44].

high surface area for sensitizer anchoring and increased interfacial contact with electrolyte, improved light adsorption (prolonged optical path, increased light trapping and superior multi-scattering capability) for enhancing light harvesting (Fig. 7.8g), excellent electron transfer (direct charge carrier transfer pathway in both trunks and branches (Fig. 7.8h), and long electron lifetime for boosting the charge collection efficiency, as well as the porous network between interlaced and intertwined nanobranches for efficient electrolyte infiltration [44]. As a result, the DSCs based on these oriented nanowire-nanosheet-nanorod hyperbranched porous titania arrays gave an impressively high PCE of 9.09% [44]. Apart from hierarchical 1D titania nanostructures, interesting semiconducting metal oxide stem/branch structures, for instance, ZnO nanoforests [49], caterpillar-like ZnO nanowire arrays [50], titania/ZnO hybrid hyperbranched arrays [51], have been demonstrated for DSCs or quantum dot-sensitized solar cells.

The advantages of tree-like hierarchical porous 1D nanostructured photoanodes are also beneficial when used for solid-state DSCs (ssDSCs). This is because the large accessible mesopores generated by the interconnected nanobranched building blocks would enhance infiltration and pore-filling of the solid electrolyte, while at the same time the films would still retain enhanced light trapping and scattering to ensure higher photon-sensitizer interaction, as well as enhanced electron transport. Kim and co-workers reported long hierarchical pine tree-like anatase titania (PTT) nanotube arrays consisting of a vertically oriented long nanotube stem (diameter from 100 to 300 nm) for facilitating charge transport and a large number of short nanorod branches (lengths from 230 to 430 nm) for high dye loading. The solid-state polymer electrolyte of poly((1-(4-ethenylphenyl) methyl)-3-butyl-imidazoliumiodide) was used for the fabrication of ssDSCs. The ssDSCs assembled with 19 μm long PTT arrays exhibit an outstanding PCE of 8.0%, two-fold higher than that of devices with a photoanode prepared from commercially available Dyesol nanorod paste (4.0%), and is one of the highest PCE values obtained for N719 dye-based ssDSCs [52]. Di Fonzo and co-workers demonstrated improved optical density and PCE in ssDSCs with a photoanode architecture comprising an array of hierarchical tree-like hyperbranched nanostructures ("trees having branches and leaves") self-assembled from pulsed laser deposition methods. The ssDSCs based on arrays of hyperbranched titania nanostructures on FTO glass sensitized with D102 dye and using spiro-OMeTAD as electrolyte showed a maximum PCE of 3.96%, which was 66% higher than the reference mesoporous photoanode [45].

4. Mesoporous Materials as Electron Transporting Layers for PSCs

The functioning of PSCs was initially based on the DSC where the dye was simply replaced by perovskite as the sensitizer. These perovskites have tunable optical properties [53], high absorption coefficients [54], very long charge carrier diffusion lengths [55], and are processable in solution at low-temperature [56]. To date, the record certified cell efficiency has reached 20.1% for a state-of-the-art mesoscopic structured PSC device

[4]. The organometallic perovskite materials possess balanced electron and hole transport properties and thus can be constructed into three main device structures: a mesoscopic structure employing a mesoporous semiconducting metal oxide (i.e., titania, ZnO, WO_3 and $SrTiO_3$) film as the electron transport layer (ETL) [13, 57–59]; meso-superstructured solar cells employing insulating mesoporous Al_2O_3 or ZrO_2 as a scaffold layer [60, 61]; and planar heterojunction structures (*p-i-n* solar cells) without a mesoporous layer [62]. In mesoscopic PSCs, the mesoporous thin films are normally deposited through screen printing or spin-coating of a diluted semiconducting metal oxide paste followed by annealing to remove the polymeric binders. This mesoporous ETL that selectively and efficiently extracts the photogenerated electrons from the perovskite light absorber is of great significance. Obviously, the ETL/perovskite interface affects the optical and photoelectric properties of the devices, hence control over the morphology, thickness and crystallinity of the mesoporous ETL will play a crucial role in determining the cell performance.

4.1 Mesoscopic thin films

In 2009, Miyasaka and co-workers reported on the $CH_3NH_3PbI_3$ and larger band gap analogue, $CH_3NH_3PbBr_3$, perovskite-sensitized mesoporous titania based solar cells with liquid halide electrolytes. A PCE of 3.8% was obtained from $CH_3NH_3PbI_3$ with the photocurrent onset observed from ~800 nm [12]. Park et al. optimized the mesoporous layer thickness, perovskite concentration, electrolyte formulation and the method of depositing the perovskite, resulting in an increased PCE of 6.5% [63]. However, the rapid dissolution of the perovskite in the organic electrolyte solvent is a big challenge to device stability. A key advance in both PCE and stability was subsequently made by replacing the liquid electrolyte with a solid-state hole conductor (spiro-MeOTAD) in 2012. As a result, Park, Grätzel and co-workers reported a PCE of 9.7% (J_{sc} = 17.6 mA cm^{-2}, V_{oc} = 0.888 V and FF = 0.62) for $CH_3NH_3PbI_3$-sensitized titania mesoporous layer (~600 nm in thickness), in conjunction with *ex-situ* long-term stability tests conducted for over 500 hours, where the devices were stored at ambient temperature without encapsulation [13]. The high absorption coefficients and long charge carrier diffusion lengths characteristic of these perovskite sensitizers allows for a significant reduction in film thickness for the mesoporous ETL (normally < 1 μm).

Along with a significant development for solution-based deposition of perovskite materials onto the mesoporous scaffold, Grätzel and co-workers utilized a sequential deposition process for the fabrication of PSCs [64]. This process was carried out by first spin-coating PbI_2 onto the titania layer, which allows sufficient infiltration of precursor within the mesoporous layer (Fig. 7.9a). Subsequently, the yellow coloured films are dipped into a

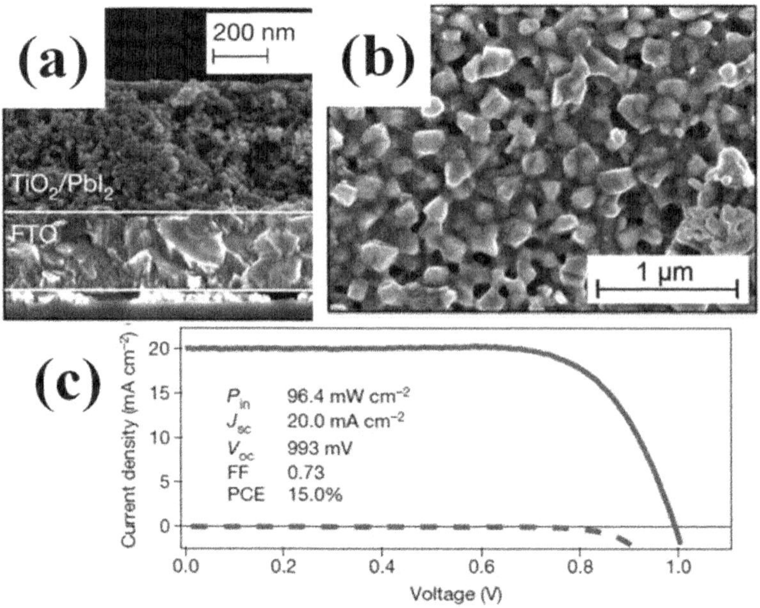

Figure 7.9. (a) Cross-sectional SEM image of a mesoporous titania film infiltrated with PbI_2; (b) Top-view SEM image of the $CH_3NH_3PbI_3$ film obtained using the sequential deposition method; (c) *J-V* curves for the best-performing cell prepared using the perovskite sequential deposition method [64]. Images reproduced with permission [64]. Copyright 2013, Nature Publishing Group.

CH_3NH_3I/2-propanol solution, converting to dark brown $CH_3NH_3PbI_3$ in a few seconds (Fig. 7.9b); thus producing a perovskite sensitizer infiltrated porous titania film. It has been noted that the thickness of the mesoporous layer can significantly affect the PCE. The highest PCE (~15.0%, Fig. 7.9c) for a FTO/Perovskite-infiltrated mesoporous titania/HTM/Au configuration has been obtained with a 350 nm thick mesoporous titania layer [64].

Recent efforts to improve perovskite film surface coverage, increase the crystal size and improve the crystallinity of the grains have resulted in high performance mesoscopic titania based PSCs being achieved. Apart from titania, various n-type semiconducting mesoporous metal oxides, such as ZnO, SnO_2, WO_3 and $SrTiO_3$, have been deposited onto the transparent conducting oxide substrates as the ETL layer for PSCs, providing many options for constructing highly efficient perovskite photovoltaic devices [13, 57–59].

Snaith et al. discoverd a new efficient meso-superstructured solar cell configuration, which employed an insulating mesoporous layer, e.g., Al_2O_3 or ZrO_2, on top of the compact titania layer as a scaffold for perovskite infiltration [60, 61]. For instance, when a mesoporous scaffold made of

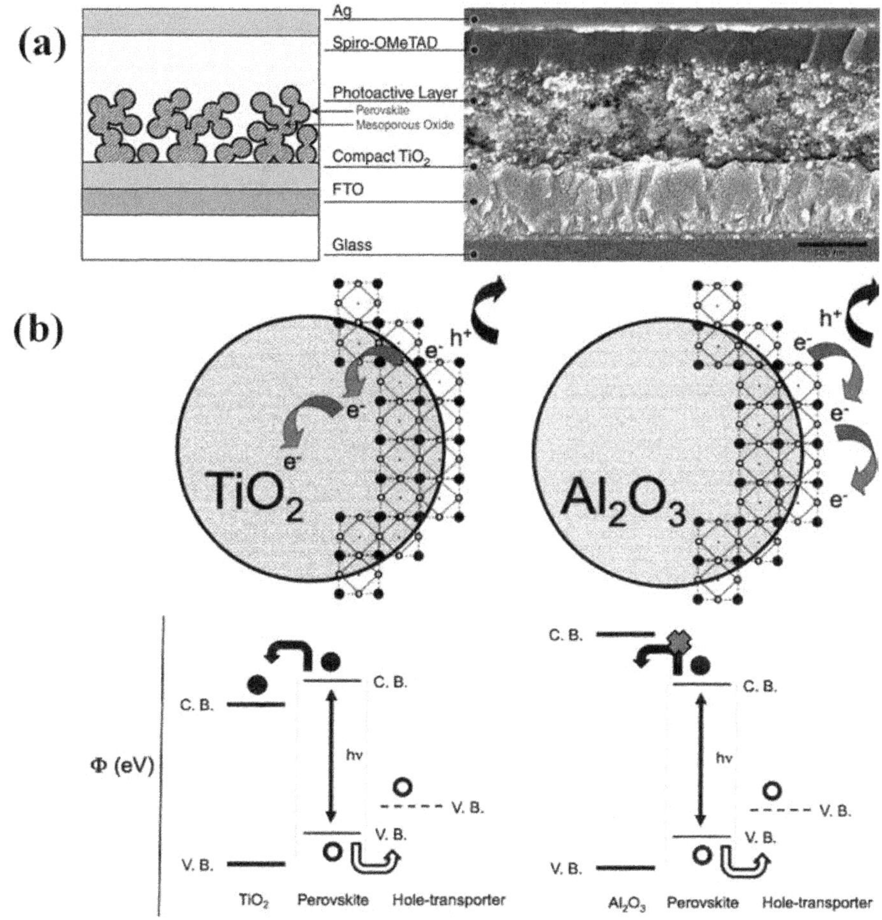

Figure 7.10. (a) Sketch and SEM cross section illustrating the meso-superstructured solar cells where the mesoporous oxide can be an n-type titania, ZnO or an insulation layer such as Al_2O_3 or ZrO_2. (b) Schematic diagram showing the charge transport in a titania based PSC and a non-injecting Al_2O_3 based PSC [60]. Images reproduced with permission [60]. Copyright 2012, Science (AAAS).

Al_2O_3 was used instead of titania (Fig. 7.10a), the Al_2O_3 does not take part in the electron extraction due to its large band gap (Fig. 7.10b), and the PSCs still yielded a similar PCE [60]. This suggested that the perovskite itself had an ambipolar nature and can transport the electrons to the front collector electrode very efficiently and that the perovskite formed within the mesoporous layer is continuous [60].

Han and co-workers reported a mesoscopic HTM-free PSC based on a double layer of titania and ZrO_2 as the scaffold, and carbon as a back contact

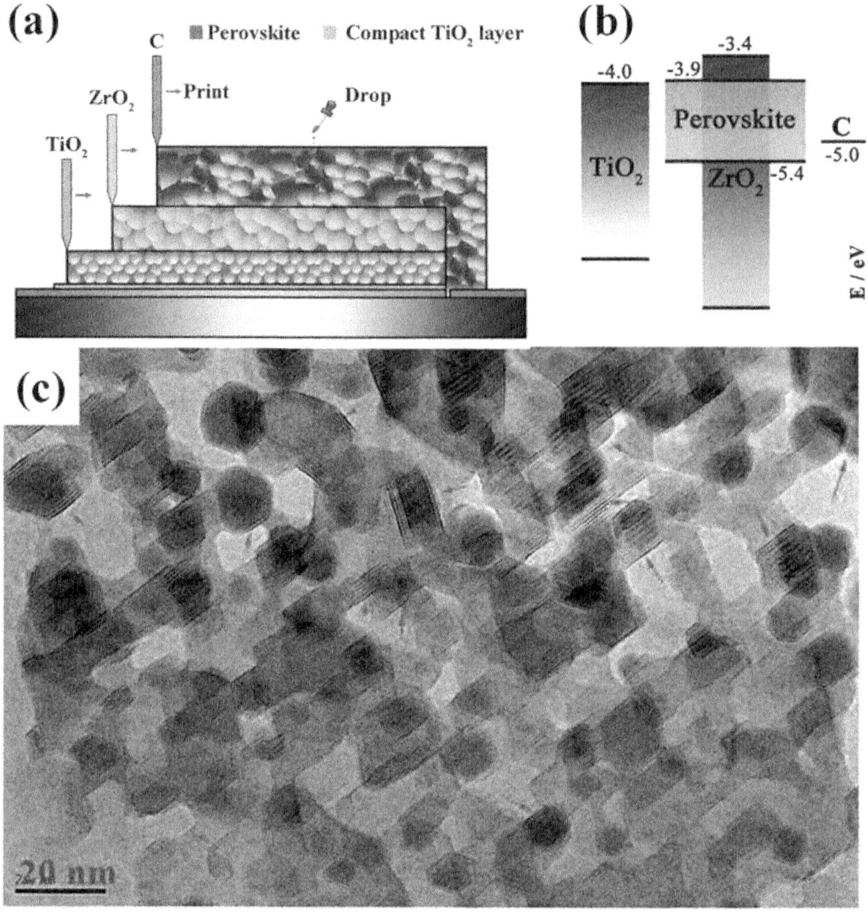

Figure 7.11. (a) Sketch showing the device architecture of perovskite-infiltrated triple titania/ZrO_2/C layer based fully printable mesoscopic cell. (b) Energy band alignment of the triple-layer device. (c) TEM image of titania particles infiltrated with perovskite scraped off from a mesoporous film, arrows indicate dense coverage of the TiO_2 particles with (5-ammoniumvaleric acid)$_x$(methylammonium)$_{1-x}$$PbI_3$ [61]. Images reproduced with permission [61]. Copyright 2014, Science (AAAS).

(Fig. 7.11a). The titania/ZrO_2/C triple mesoscopic layer was fully printable with perovskite infiltrated into the mesoporous layer by drop-casting from solution, densely covering the titania surface (Fig. 7.11c). Due to the presence of the ZrO_2 insulating layer, the electrons injected from the photoexited perovskite into the titania ETL can be prohibited from recombination with the back contact because of the 0.6 eV offset between the titania conduction band and that of ZrO_2 (Fig. 7.11b). PSCs based on this structure reached a certified PCE of 12.8%, exhibiting outstanding stability under long-term light soaking [61]. In their follow-up work, the size effect of the titania ETL on the cell performance was also investigated. The size of the titania particles can control the pore size between the particles and affects the infiltration of the perovskite precursor, and thus significantly influence the contact and charge transfer dynamics at the titania/perovskite interface. PSCs based on titania nanoparticles with a diameter of 25 nm yielded a champion PCE of 13.41% [65]. Fully printable device fabrication provides incentive for future large-scale production and commercialization of the PSCs.

4.2 Ordered mesostructured thin films

The previously discussed mesoscopic thin films constructed of disordered nanoparticle networks slow the electron transport because of the random-walk pathway, which increases the chance of charge recombination and thus inhibits the charge collection [66]. The well-organized and ordered mesoporous materials with interconnected pores, controllable pore sizes and mesochannels, are good candidates for thin film photoelectrodes in PSCs. Mesostructured thin films synthesized using template methods have shown promising features for their application in PSCs.

Seok and co-workers fabricated a well-ordered, crack-free mesoporous titania thin layer (Figs. 7.12a and 7.12b) with a cage-like mesostructure using an inorganic sol-gel process and block copolymers as the sacrificial template; the pore sizes of the film can be adjusted by changing the amount of swelling agent. The ordered mesostructured layers were then employed as photoanodes to construct PSCs. PCEs of 11.7 and 12.8% were obtained for devices containing layers with 10 and 15 nm mesopores, respectively. The better performance observed in the 15 nm mesoporous layers was attributed to the easier pore-filling of perovskite into the bigger mesopores [67].

Snaith et al. demonstrated a scalable technique for fabricating highly ordered metal oxide scaffolds with honeycomb structure (Fig. 7.12c), that can be used to pattern the perovskite thin films on a photonic length scale. The perovskite material fills the large pores of the honeycomb well (Fig. 7.12d) and hence the domain size and film thickness of the perovskite can be controlled and tuned by varying the honeycomb structure as well as perovskite precursor concentration. PSC devices fabricated from the periodic SiO_2 honeycomb and 20 wt% perovskite precursor solution (Figs. 7.12e and 7.12f) were semi-transparent with increased V_{oc} and FF, yielding PCEs up to 9.5% with an average visible transmittance through the active layer of around 37% (Fig. 7.12g) [68].

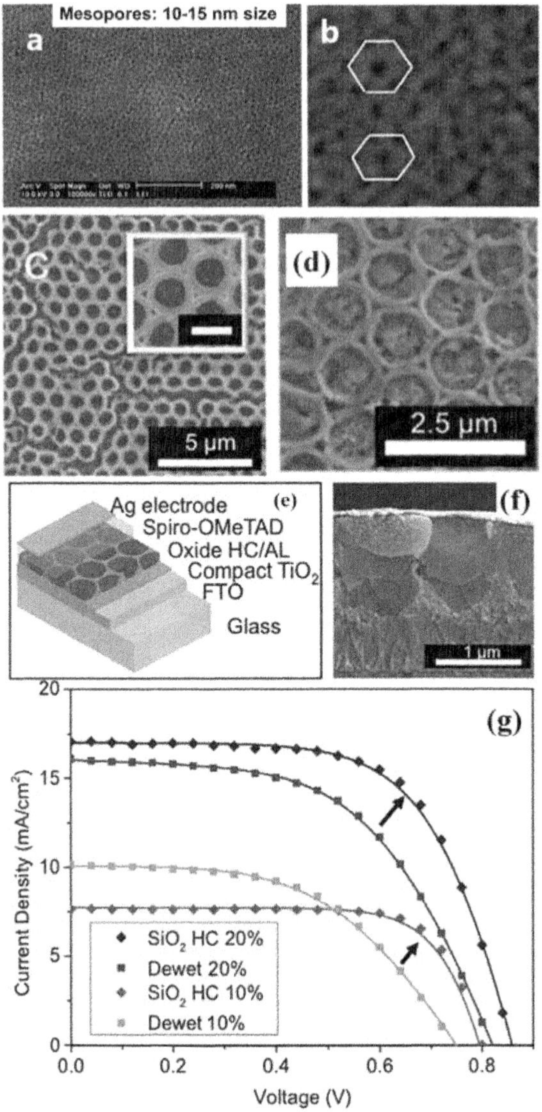

Figure 7.12. (a) Top-view SEM image of the mesoporous titania film prepared with 1, 3, 5-trimethylbenzene. (b) 6-fold arrangement of mesopores [67]; SEM images of (c) porous titania with honeycomb structure and (d) the same titania scaffold after filling with 20 wt% perovskite precursor solution; (e) Sketch of the device architecture with oxide honeycomb scaffold; (f) Colour enhanced cross-sectional SEM image of a device fabricated from the SiO_2 honeycomb with 20 wt% perovskite precursor solution; (g) *J-V* characteristics of devices fabricated with (SiO_2 HC) and without (Dewet) SiO_2 honeycomb scaffolds and covered with perovskite films prepared with 10 and 20 wt% precursor solution [68]. Images reproduced with permission [67, 68]. Copyright 2014, American Chemical Society, and Copyright 2015, The Royal Society of Chemistry.

4.3 Mesoporous oxide single crystals

The key to high efficiency sensitized solar cells is the high surface area electrode that provides highly accessible surfaces for sensitizer loading. However, the conventional mesoporous nanocrystal films require in-film thermal sintering to reinforce electronic contact between particles, otherwise the electron mobility would be significantly decreased because of the high number of grain boundaries and the long electron diffusion paths through the particle network [66]. The Mesoporous Single Crystal (MSC) semiconductor can meet both criteria of fast electron transport and high surface area. Snaith et al. achieved the growth of large anatase titania mesoporous single crystals based on seeded nucleation and growth inside a mesoporous template (close-packed assembly of silica nanospheres) in a dilute reaction solution (Fig. 7.13a). The crystal was grown entirely within the pores of the silica template through hydrothermal treatment in a TiF_4 solution (Fig. 7.13b). The mesoporous crystal possesses a similar surface area for dye adsorption (~70 $m^2\ g^{-1}$ for 20 nm pores in 2 μm^3 crystals) as the conventional nanoparticles (Dyesol 18NR-T, 20 nm average particle size, 75 $m^2\ g^{-1}$). The TEM and electron diffraction characterization indicate the mesoporous particles consist of a single underlying anatase crystal domain (Figs. 7.13c and 7.13d). Such titania MSCs were used as an electron collector in PSCs, and a PCE of 7.3% (Fig. 7.13f) has been obtained due to their high conductivity and electron mobility for fast electron transport (Fig. 7.13e), as well as high surface area [69].

Yang and co-workers have recently reported the synthesis of mesoporous SnO_2 single crystals (SnO_2 MSC, Figs. 7.14a and 7.14b) by a simple silica templated hydrothermal method. The SnO_2 MSCs were coated with a thin layer of titania via $TiCl_4$ treatment and used as photoanodes for PSCs, achieving a PCE of 8.54%, which was higher than that of pure SnO_2 MSC based cells (Fig. 7.14c). An Electrochemical Impedance Spectroscopy (EIS) study revealed that the SnO_2 MSC based PSC had a lower transport resistance than the titania nanocrystal PSCs as the single crystal provides high electron mobility and fewer bulk defects and impurities. Moreover, the thin titania coating layer on the SnO_2 MSC considerably suppressed the charge recombination while maintaining the superior electron transport properties (Fig. 7.14d) [70].

4.4 Thin films of 1D structures

In the mesoscopic PSC, the transfer of photoinduced electrons from perovskite to the ETL, the pore-filling of the perovskite and the HTM infiltration play crucial roles in determining the cell performance. Normally, these three factors are hampered by the highly convoluted porous channels

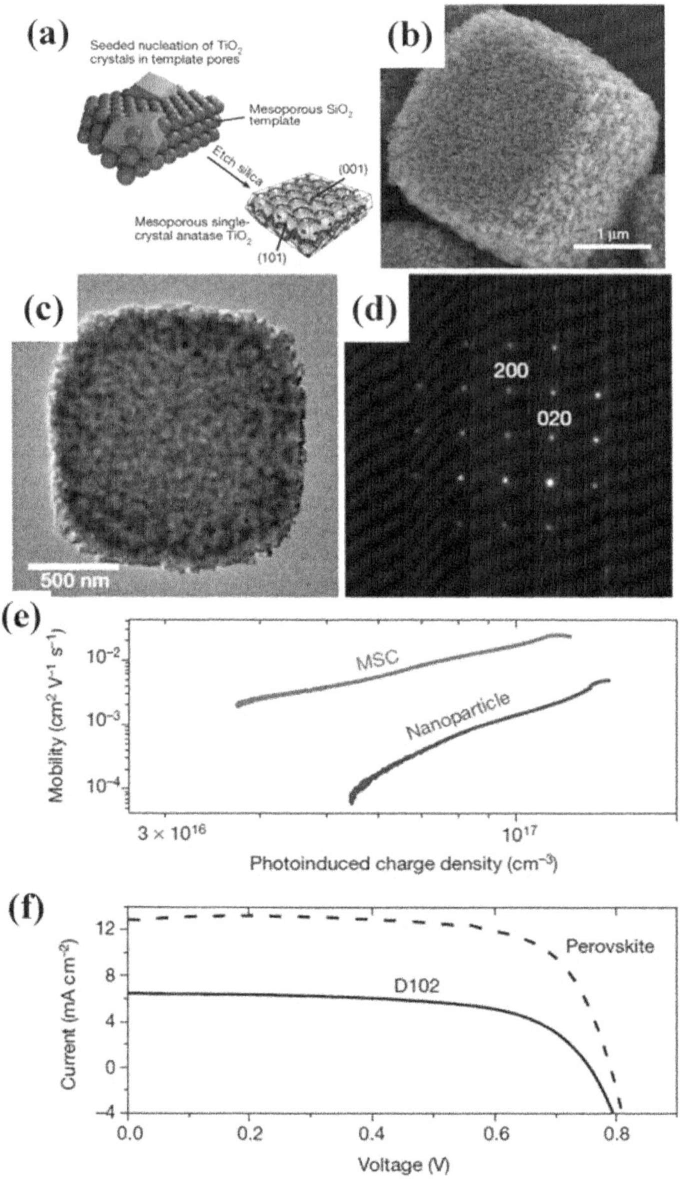

Figure 7.13. (a) Schematic of titania MSC nucleation and growth within a mesoporous silica template; (b) SEM image of fully mesoporous titania single crystal; (c) TEM image and (d) electron diffraction pattern collected from a complete mesoporous crystal; (e) Mobility dependence on photoinduced charge density for titania MSC and nanoparticle films; and (f) J-V curves of solar cells containing titania MSC photoanodes with either the D102 dye or perovskite sensitizer [69]. Images reproduced with permission [69]. Copyright 2013, Nature Publishing Group.

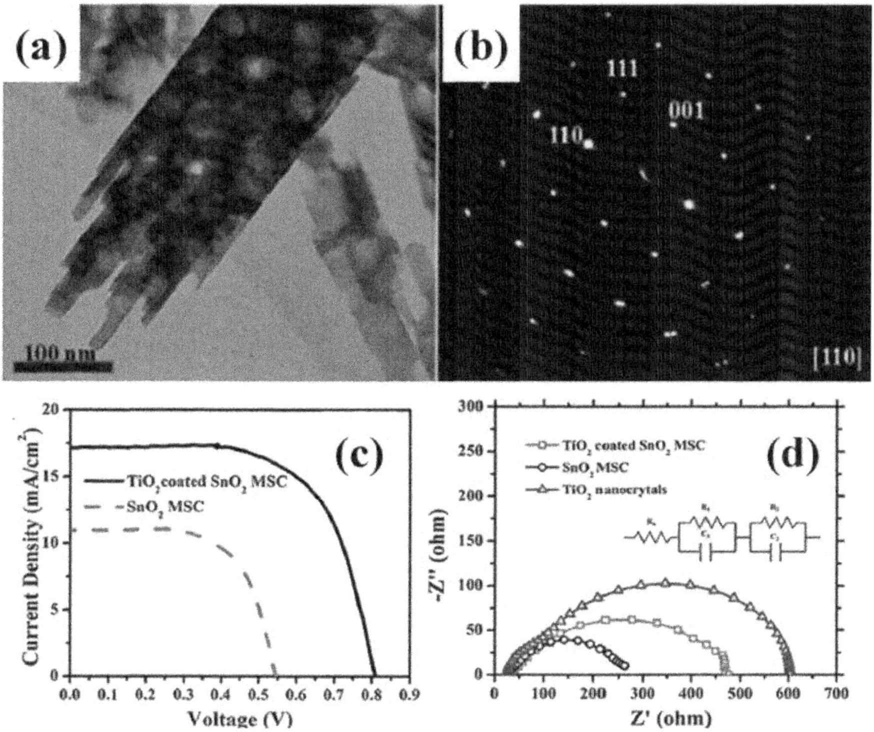

Figure 7.14. (a) TEM image and (b) selected-area electron diffraction of the SnO_2 MSCs; (c) *J-V* curves of the PSCs based on pure SnO_2 MSC and titania-coated SnO_2 MSC; and (d) EIS Nyquist profiles of the PSCs based on pure SnO_2 MSC, titania-coated SnO_2 MSC and titania nanocrystals [70]. Images reproduced with permission [70]. Copyright 2015, The Royal Society of Chemistry.

within the nanoparticulate film [66, 71]. Therefore, thin films composed of 1D nanostructures have been widely employed as efficient ETLs for enhancing the performance of PSCs. For instance, Park et al. reported a perovskite-adsorbed submicrometre-thick rutile titania nanorod film (~500–600 nm) that yielded a PCE of 9.4% [71]. Very recently, Feng and co-workers demonstrated that 900 nm rutile titania nanowires with large voids exhibited PCEs as high as 11.7% in PSCs [66]. Such nanorods or nanowires are excellent for perovskite anchoring, facile infiltration of the HTM, while at the same time being superior in electron transport and suppressing electron/hole recombination compared to nanoparticulate films when applied in PSCs. Similarly, other 1D oxide nanostructures, such as, titania nanofibres, titania nanocones or ZnO nanorods, were also employed as efficient ETLs in PSCs with enhanced photovoltaic performance [57, 72, 73].

Figure 7.15. SEM images of (a) titania hyperbranched nanofibre-nanorod array [74]; (b) 3D titania nanowire architecture (inset is an enlarged SEM image showing the tree-like branched nanowire structure) [75]; (c) dendritic titania nanowire thin film covering the FTO glass; (d) TEM image of these titania nanowires scratched from the FTO glass, clearly indicating their dendritic nanostructure; and (e) schematic of a completed PSC device constructed with anatase titania nanowire (ATNW) charge collection layer on FTO glass [76]. Images reproduced with permission [74–76]. Copyright 2015, Wiley-VCH Verlag GmbH & Co. [74, 76], and Copyright 2015, American Chemical Society [75].

The hierarchical branched titania nanowire architectures coupled with large surface area for high perovskite loading and fast electron transport, as well as suppressed charge recombination are considered as superior electrode candidates for PSCs. Recently, novel hyperbranched titania nanorod-nanofibre arrays (Fig. 7.15a) have been prepared via a multistage electrospinning and hydrothermal route and then applied as ETLs for PSCs. These hyperbranched materials with optimal electron transport and carrier lifetimes lead to highly efficient mesostructured PSC with a maximum PCE of 15.50% [74]. Wang and co-workers also developed branched titania nanowire architectures (Fig. 7.15b) via a ZnO nanowire template-assisted surface-reaction limited pulsed chemical vapour deposition technique. The high PCE of 9.0% was achieved using ~600 nm long branched titania nanowire architectures (additionally the hysteresis effect was largely suppressed) [75]. A mild one-step hydrothermal process to fabricate dendritic anatase titania nanowire thin films (~80–220 nm in thickness) directly on a FTO substrate (Figs. 7.15c and 7.15d) was reported. PSCs constructed using the optimized dendritic titania thin films (Fig. 7.15e) yield PCEs up to 14.2% with a J_{sc} of 20.32 mA cm^{-2}, V_{oc} of 993 mV and FF of 0.70 [76].

Branched titania nanowire architectures with enlarged surface area and unique tree-like morphology were able to achieve larger perovskite loadings, enhanced light harvesting and improved electron transport properties, which enhanced the photovoltaic performance.

5. Conclusion and Outlook

Different mesostructured photoanodes made with nanoparticles, spherical assemblies (beads, hollow spheres, spiky/branched beads, etc.), ordered mesoporous structures, mesoporous single crystal particles, and porous architectures composed of 1D materials, show great capacity for enhancing the performance of solar cells in view of their unique, highly interconnected porous structure as well as excellent optical and electronic properties. In brief, the mesostructured photoanodes enhance the PCE of DSCs and PSCs in four primary ways. Firstly, they provide a relatively high surface area for sensitizer loading (dyes or perovskite pigments). Secondly, the strong light trapping and scattering ability of such materials can enhance the light harvesting efficiency owing to their unique geometric features. Thirdly, relatively fast electron transport and suppressed charge recombination within the mesoporous material-based photoelectrodes with well-connected crystalline building blocks boosts the electron transport and thus high charge collection efficiency can be achieved. Last but not least, the hierarchical multi-scale porous structures are beneficial to the permeation and diffusion of electrolyte in the photoanode.

In the near future, it is expected that further improvement in cell performance will follow materials innovation. Research efforts will focus on the following aspects: (1) developing and enriching synthesis procedures considering simplicity, the environment and low-cost for practical application of such mesoporous materials; (2) synthesizing novel mesoporous materials with more complex nanoscale structures or modifying existing materials (i.e., doping with other functional elements or components) to further improve the optical and electronic properties; and (3) correlating the synthesis parameters to the final material properties to achieve accurate control over the morphology, crystallinity, pore structure and pore size of the resulting materials to optimize for application.

The potential applications of these mesoporous material-based photoelectrodes are not limited to solar cells. These advanced mesoporous materials, or the processes used to form such structures, could be applied in other areas, such as supercapacitors, photocatalysis, water splitting, gas sensors, fuel cells and Li-ion batteries.

Acknowledgement

RAC is the recipient of an Australian Research Council Future Fellowship (FT0990583).

References

[1] Lewis, N.S. 2007. Toward cost-effective solar energy use. *Science*, 315: 798–801.

[2] Grätzel, M. 2009. Recent advances in sensitized mesoscopic solar cells. *Accounts Chem. Res.*, 42: 1788–1798.

[3] Hagfeldt, A., Boschloo, G., Sun, L.C. et al. 2010. Dye-sensitized solar cells. *Chem. Rev.*, 110: 6595–6663.

[4] Yang, W.S., Noh, J.H., Jeon, N.J. et al. 2015. High-performance photovoltaic perovskite layers fabricated through intramolecular exchange. *Science*, 348: 1234–1237.

[5] Zhang, Q.F., Uchaker, E., Candelaria, S.L. et al. 2013. Nanomaterials for energy conversion and storage. *Chem. Soc. Rev.*, 42: 3127–3171.

[6] Guo, M., Xie, K.Y., Lin, J. et al. 2012. Design and coupling of multifunctional TiO_2 nanotube photonic crystal to nanocrystalline titania layer as semi-transparent photoanode for dye-sensitized solar cell. *Energy Environ. Sci.*, 5: 9881–9888.

[7] Chen, D.H., Huang, F.Z., Cheng, Y.B. et al. 2009. Mesoporous anatase TiO_2 beads with high surface areas and controllable pore sizes: a superior candidate for high-performance dye-sensitized solar cells. *Adv. Mater.*, 21: 2206–2210.

[8] Sauvage, F., Chen, D.H., Comte, P. et al. 2010. Dye-sensitized solar cells employing a single film of mesoporous TiO_2 beads achieve power conversion efficiencies over 10%. *ACS Nano.*, 4: 4420–4425.

[9] Bisquert, J., Cahen, D., Hodes, G. et al. 2004. Physical chemical principles of photovoltaic conversion with nanoparticulate, mesoporous dye-sensitized solar cells. *J. Phys. Chem. B*, 108: 8106–8118.

[10] Grätzel, M. 2005. Solar energy conversion by dye-sensitized photovoltaic cells. *Inorg. Chem.*, 44: 6841–6851.

[11] Jung, H.S. and Park, N.-G. 2015. Perovskite solar cells: from materials to devices. *Small*, 11: 10–25.

[12] Kojima, A., Teshima, K., Shirai, Y. et al. 2009. Organometal halide perovskites as visible-light sensitizers for photovoltaic cells. *J. Am. Chem. Soc.*, 131: 6050–6051.

[13] Kim, H.S., Lee, C.R., Im, J.H. et al. 2012. Lead iodide perovskite sensitized all-solid-state submicron thin film mesoscopic solar cell with efficiency exceeding 9%. *Sci. Rep.*, 2: 591.

[14] O'Regan, B. and Grätzel, M. 1991. A low-cost, high-efficiency solar cell based on dye-sensitized colloidal TiO_2 films. *Nature*, 353: 737–740.

[15] Mathew, S., Yella, A., Gao, P. et al. 2014. Dye-sensitized solar cells with 13% efficiency achieved through the molecular engineering of porphyrin sensitizers. *Nat. Chem.*, 6: 242–247.

[16] Bai, Y., Mora-Seró, I., De Angelis, F. et al. 2014. Titanium dioxide nanomaterials for photovoltaic applications. *Chem. Rev.*, 114: 10095–10130.

[17] Chen, H.Y., Kuang, D.B. and Su, C.Y. 2012. Hierarchically micro/nanostructured photoanode materials for dye-sensitized solar cells. *J. Mater. Chem.*, 22: 15475–15489.

[18] Ito, S., Chen, P., Comte, P. et al. 2007. Fabrication of screen-printing pastes from TiO_2 powders for dye-sensitised solar cells. *Prog. Photovoltaics*, 15: 603–612.

[19] Hore, S., Vetter, C., Kern, R. et al. 2006. Influence of scattering layers on efficiency of dye-sensitized solar cells. *Sol. Energ. Mater. Sol. C*, 90: 1176–1188.

[20] Li, Y., Wang, H., Feng, Q.Y. et al. 2013. Gold nanoparticles inlaid TiO_2 photoanodes: a superior candidate for high-efficiency dye-sensitized solar cells. *Energy Environ. Sci.*, 6: 2156–2165.

[21] Fisher, A.C., Peter, L.M., Ponomarev, E.A. et al. 2000. Intensity dependence of the back reaction and transport of electrons in dye-sensitized nanocrystalline TiO_2 solar cells. *J. Phys. Chem. B*, 104: 949–958.

[22] Koo, H.J., Kim, Y.J., Lee, Y.H. et al. 2008. Nano-embossed hollow spherical TiO_2 as bifunctional material for high-efficiency dye-sensitized solar cells. *Adv. Mater.*, 20: 195–199.

[23] Chen, D.H., Cao, L., Huang, F.Z. et al. 2010. Synthesis of monodisperse mesoporous titania beads with controllable diameter, high surface areas, and variable pore diameters (14–23 nm). *J. Am. Chem. Soc.*, 132: 4438–4444.

[24] Kim, Y.J., Lee, M.H., Kim, H.J. et al. 2009. Formation of highly efficient dye-sensitized solar cells by hierarchical pore generation with nanoporous TiO_2 spheres. *Adv. Mater.*, 21: 3668–3673.

[25] Zhang, Q.F., Chou, T.R., Russo, B. et al. 2008. Aggregation of ZnO nanocrystallites for high conversion efficiency in dye-sensitized solar cells. *Angew. Chem. Int. Ed.*, 47: 2402–2406.

[26] Wu, X., Lu, G.Q. and Wang, L. 2011. Shell-in-shell TiO_2 hollow spheres synthesized by one-pot hydrothermal method for dye-sensitized solar cell application. *Energy Environ. Sci.*, 4: 3565–3572.

[27] Dong, Z.H., Lai, X.Y., Halpert, J.E. et al. 2012. Accurate control of multishelled ZnO hollow microspheres for dye-sensitized solar cells with high efficiency. *Adv. Mater.*, 24: 1046–1049.

[28] Qian, J.F., Liu, P., Xiao, Y. et al. 2009. TiO_2-coated multilayered SnO_2 hollow microspheres for dye-sensitized solar cells. *Adv. Mater.*, 21: 3663–3667.

[29] Fang, W.Q., Yang, X.H., Zhu, H.J. et al. 2012. Yolk@shell anatase TiO_2 hierarchical microspheres with exposed {001} facets for high-performance dye sensitized solar cells. *J. Mater. Chem.*, 22: 22082–22089.

[30] Chen, D.H., Huang, F.Z., Cao, L. et al. 2012. Spiky mesoporous anatase titania beads: a metastable ammonium titanate-mediated synthesis. *Chem. Eur. J.*, 18: 13762–13769.

[31] Liao, J.-Y., Lei, B.-X., Kuang, D.-B. et al. 2011. Tri-functional hierarchical TiO_2 spheres consisting of anatase nanorods and nanoparticles for high efficiency dye-sensitized solar cells. *Energy Environ. Sci.*, 4: 4079–4085.

[32] Heiniger, L.-P., Giordano, F., Moehl, T. et al. 2014. Mesoporous TiO_2 beads offer improved mass transport for cobalt-based redox couples leading to high efficiency dye-sensitized solar cells. *Adv. Energy Mater.*, 4: 1400168.

[33] Wu, W.Q., Xu, Y.F., Rao, H.S. et al. 2014. Multistack integration of three-dimensional hyperbranched anatase titania architectures for high-efficiency dye-sensitized solar cells. *J. Am. Chem. Soc.*, 136: 6437–6445.

[34] Liu, Y., Che, R., Chen, G. et al. 2015. Radially oriented mesoporous TiO_2 microspheres with single-crystal–like anatase walls for high-efficiency optoelectronic devices. *Sci. Adv.*, 1: e1500166.

[35] Pazoki, M., Oscarsson, J., Yang, L. et al. 2014. Mesoporous TiO_2 microbead electrodes for solid state dye-sensitized solar cells. *RSC Adv.*, 4: 50295–50300.

[36] Liu, B. and Aydil, E.S. 2009. Growth of oriented single-crystalline rutile TiO_2 nanorods on transparent conducting substrates for dye-sensitized solar cells. *J. Am. Chem. Soc.*, 131: 3985–3990.

[37] Varghese, O.K., Paulose, M. and Grimes, C.A. 2009. Long vertically aligned titania nanotubes on transparent conducting oxide for highly efficient solar cells. *Nat. Nanotechnol.*, 4: 592–597.

[38] Qu, J., Li, G.R. and Gao, X.P. 2010. One-dimensional hierarchical titania for fast reaction kinetics of photoanode materials of dye-sensitized solar cells. *Energy Environ. Sci.*, 3: 2003–2009.

[39] Zhuge, F.W., Qiu, J.J., Li, X.M. et al. 2011. Toward hierarchical TiO_2 nanotube arrays for efficient dye-sensitized solar cells. *Adv. Mater.*, 23: 1330–1334.

[40] Wu, W.Q., Feng, H.L., Rao, H.S. et al. 2014. Rational surface engineering of anatase titania core-shell nanowire arrays: full-solution processed synthesis and remarkable photovoltaic performance. *ACS Appl. Mater. Interfaces*, 6: 19100–19108.

[41] Ahn, S.H., Kim, D.J., Chi, W.S. et al. 2013. One-dimensional hierarchical nanostructures of TiO_2 nanosheets on SnO_2 nanotubes for high efficiency solid-state dye-sensitized solar cells. *Adv. Mater.*, 25: 4893–4897.

[42] Wu, W.Q., Rao, H.S., Feng, H.L. et al. 2014. Morphology-controlled cactus-like branched anatase TiO_2 arrays with high light-harvesting efficiency for dye-sensitized solar cells. *J. Power Sources*, 260: 6–11.

[43] Wu, W.Q., Rao, H.S., Feng, H.L. et al. 2014. A family of vertically aligned nanowires with smooth, hierarchical and hyperbranched architectures for efficient energy conversion. *Nano Energy*, 9: 15–24.

[44] Wu, W.Q., Feng, H.L., Rao, H.S. et al. 2014. Maximizing omnidirectional light harvesting in metal oxide hyperbranched array architectures. *Nat. Commun.*, 5: 3968.

[45] Passoni, L., Ghods, F., Docampo, P. et al. 2013. Hyperbranched quasi-1D nanostructures for solid-state dye-sensitized solar cells. *ACS Nano.*, 7: 10023–10031.

[46] Sauvage, F., Di Fonzo, F., Bassi, A.L. et al. 2010. Hierarchical TiO_2 photoanode for dye-sensitized solar cells. *Nano Lett.*, 10: 2562–2567.

[47] Sheng, X., He, D.Q., Yang, J. et al. 2014. Oriented assembled TiO_2 hierarchical nanowire arrays with fast electron transport properties. *Nano Lett.*, 14: 1848–1852.

[48] Wu, W.Q., Xu, Y.F., Su, C.Y. et al. 2014. Ultra-long anatase TiO_2 nanowire arrays with multi-layered configuration on FTO glass for high-efficiency dye-sensitized solar cells. *Energy Environ. Sci.*, 7: 644–649.

[49] Ko, S.H., Lee, D., Kang, H.W. et al. 2011. Nanoforest of hydrothermally grown hierarchical ZnO nanowires for a high efficiency dye-sensitized solar cell. *Nano Lett.*, 11: 666–671.

[50] McCune, M., Zhang, W. and Deng, Y. 2012. High efficiency dye-sensitized solar cells based on three-dimensional multilayered ZnO nanowire arrays with "caterpillar-like" structure. *Nano Lett.*, 12: 3656–3662.

[51] Feng, H.L., Wu, W.Q., Rao, H.S. et al. 2015. Three-dimensional hyperbranched TiO_2/ZnO heterostructured arrays for efficient quantum dot-sensitized solar cells. *J. Mater. Chem. A*, 3: 14826–14832.

[52] Roh, D.K., Chi, W.S., Jeon, H. et al. 2014. High efficiency solid-state dye-sensitized solar cells assembled with hierarchical anatase pine tree-like TiO_2 nanotubes. *Adv. Funct. Mater.*, 24: 379–386.

[53] Noh, J.H., Im, S.H., Heo, J.H. et al. 2013. Chemical management for colorful, efficient, and stable inorganic-organic hybrid nanostructured solar cells. *Nano Lett.*, 13: 1764–1769.

[54] Xing, G.C., Mathews, N., Sun, S.Y. et al. 2013. Long-range balanced electron- and hole-transport lengths in organic-inorganic $CH_3NH_3PbI_3$. *Science*, 342: 344–347.

[55] Stranks, S.D., Eperon, G.E., Grancini, G. et al. 2013. Electron-hole diffusion lengths exceeding 1 micrometer in an organometal trihalide perovskite absorber. *Science*, 342: 341–344.

[56] Wang, J.T.W., Ball, J.M., Barea, E.M. et al. 2014. Low-temperature processed electron collection layers of graphene/TiO_2 nanocomposites in thin film perovskite solar cells. *Nano Lett.*, 14: 724–730.

[57] Son, D.Y., Im, J.H., Kim, H.S. et al. 2014. 11% efficient perovskite solar cell based on ZnO nanorods: an effective charge collection system. *J. Phys. Chem. C*, 118: 16567–16573.

[58] Mahmood, K., Swain, B.S., Kirmani, A.R. et al. 2015. Highly efficient perovskite solar cells based on a nanostructured WO_3-TiO_2 core-shell electron transporting material. *J. Mater. Chem. A*, 3: 9051–9057.

[59] Bera, A., Wu, K., Sheikh, A. et al. 2014. Perovskite oxide $SrTiO_3$ as an efficient electron transporter for hybrid perovskite solar cells. *J. Phys. Chem. C*, 118: 28494–28501.

[60] Lee, M.M., Teuscher, J., Miyasaka, T. et al. 2012. Efficient hybrid solar cells based on meso-superstructured organometal halide perovskites. *Science*, 338: 643–647.

[61] Mei, A.Y., Li, X., Liu, L.F. et al. 2014. A hole-conductor-free, fully printable mesoscopic perovskite solar cell with high stability. *Science*, 345: 295–298.

[62] Liu, M.Z., Johnston, M.B. and Snaith, H.J. 2013. Efficient planar heterojunction perovskite solar cells by vapour deposition. *Nature*, 501: 395–398.

[63] Im, J.-H., Lee, C.-R., Lee, J.-W. et al. 2011. 6.5% efficient perovskite quantum-dot-sensitized solar cell. *Nanoscale*, 3: 4088–4093.

[64] Burschka, J., Pellet, N., Moon, S.J. et al. 2013. Sequential deposition as a route to high-performance perovskite-sensitized solar cells. *Nature*, 499: 316–319.

[65] Yang, Y., Ri, K., Mei, A. et al. 2015. The size effect of TiO_2 nanoparticles on a printable mesoscopic perovskite solar cell. *J. Mater. Chem. A*, 3: 9103–9107.

[66] Jiang, Q., Sheng, X., Li, Y. et al. 2014. Rutile TiO_2 nanowire-based perovskite solar cells. *Chem. Commun.*, 50: 14720–14723.

[67] Sarkar, A., Jeon, N.J., Noh, J.H. et al. 2014. Well-organized mesoporous TiO_2 photoelectrodes by block copolymer-induced sol-gel assembly for inorganic-organic hybrid perovskite solar cells. *J. Phys. Chem. C*, 118: 16688–16693.

[68] Horantner, M.T., Zhang, W., Saliba, M. et al. 2015. Templated microstructural growth of perovskite thin films via colloidal monolayer lithography. *Energy Environ. Sci.*, 8: 2041–2047.

[69] Crossland, E.J.W., Noel, N., Sivaram, V. et al. 2013. Mesoporous TiO_2 single crystals delivering enhanced mobility and optoelectronic device performance. *Nature*, 495: 215–219.

[70] Zhu, Z.L., Zheng, X.L., Bai, Y. et al. 2015. Mesoporous SnO_2 single crystals as an effective electron collector for perovskite solar cells. *Phys. Chem. Chem. Phys.*, 17: 18265–18268.

[71] Kim, H.S., Lee, J.W., Yantara, N. et al. 2013. High efficiency solid-state sensitized solar cell-based on submicrometer rutile TiO_2 nanorod and $CH_3NH_3PbI_3$ perovskite sensitizer. *Nano Lett.*, 13: 2412–2417.

[72] Dharani, S., Mulmudi, H.K., Yantara, N. et al. 2014. High efficiency electrospun TiO_2 nanofiber based hybrid organic-inorganic perovskite solar cell. *Nanoscale*, 6: 1675–1679.

[73] Zhong, D., Cai, B., Wang, X. et al. 2015. Synthesis of oriented TiO_2 nanocones with fast charge transfer for perovskite solar cells. *Nano Energy*, 11: 409–418.

[74] Mahmood, K., Swain, B.S. and Amassian, A. 2015. Highly efficient hybrid photovoltaics based on hyperbranched three-dimensional TiO_2 electron transporting materials. *Adv. Mater.*, 27: 2859–2865.

[75] Yu, Y., Li, J., Geng, D. et al. 2015. Development of lead iodide perovskite solar cells using three-dimensional titanium dioxide nanowire architectures. *ACS Nano.*, 9: 564–572.

[76] Wu, W.Q., Huang, F.Z., Chen, D.H. et al. 2015. Thin films of dendritic anatase titania nanowires enable effective hole-blocking and efficient light-harvesting for high-performance mesoscopic perovskite solar cells. *Adv. Funct. Mater.*, 25: 3264–3272.

Index

A

Acidity 110, 113, 115, 121, 122, 124, 125
acidity of the material surface 113

B

button cells 132, 133, 139

C

capillary condensation effect 108, 111, 117, 118, 120, 121
Carbon materials 51, 52, 55, 59–61, 65–69, 71, 77, 82–85, 87, 88
Carbon synthesis 66, 82
co-condensation method 114, 119, 120
Collect/collected (relating to charge) 294, 295, 316
composite silicate 115

D

Doped carbon 16–18, 20, 30, 33, 71, 85–89

E

Electrocatalysts 152, 187, 195, 203, 208
Electrochemical capacitors 51, 52, 55
Electrode materials 152
Electrode reaction 152, 208
Energy storage 262, 263, 265
energy storage and conversion 1, 2, 28, 37, 38
Extract/extracted/extraction (relating to charge) 293, 311

F

ferroxane 110
filler content 122, 123
Fuel cells 97, 98, 101, 108, 110, 111, 113, 114, 116, 117, 125, 131–133, 139–141, 152, 153, 166, 185, 189, 200, 202–204, 208
functionalized Hollow Mesoporous Silica 117
functionalized mesoporous silica 106, 113, 114, 117, 121, 122, 124, 126, 129–131, 134, 138, 139

H

Harvest/harvested/harvesting (relating to light) 292, 293, 296–300, 302, 306, 307, 319
Hierarchical/hierarchically 292, 299, 300, 302–308, 318, 319
HPAs-functionalized meso-silica 124
HPW 124–135, 137, 139–141
HPW-meso-silica 127–135, 139

I

Inject/injected/injection (relating to charge) 294, 296, 300, 312
interpenetrating polymer network 117

K

Kelvin equation 121, 122

L

Li-air (O2) battery 281
Li-ion battery 266
Li-S battery 281, 282, 286
lithium-air batteries 240
lithium ion batteries 231, 232

M

Mesoporous alumina 110
Mesoporous materials 221, 222, 234, 238, 240, 243, 251, 252, 261, 263, 270–275, 278, 279, 281, 284, 285
mesoporous Nafion 101–108
mesoporous silica 101, 104–106, 111–114, 116–127, 129–132, 134, 138, 139, 141

Mesoporous structured materials 97, 101, 135, 136, 138, 141
Mesoporous structures 152, 159, 160, 162, 166–168, 170, 172–174, 180, 206, 208
mesoporous TiO_2 108, 109, 140
morphology control 1

N

Nafion 97, 99–108, 110, 114, 116–124, 131, 133, 135, 136, 139
Nanoparticle/nanoparticulate 292–294, 296–300, 302, 304–306, 313, 315–317, 319
Nanoporous electrode 270
Nanorod 302, 305–308, 317, 318
Nanostructures 280
Nanotube 305–308
Nanowire 305–308, 317, 318
nature of the surface groups 111

O

one-pot synthesis 125, 126, 132

P

Periodic Mesoporous Organosilica 120
phosphonic acid functionalized mesoporous-structured silica 114
Polybenzimidazole 97, 100
Pore structure 109, 111, 121
Porous carbon 1–10, 12–26, 28–38, 51, 52, 61, 65, 69, 75, 76, 79, 86, 87
porous silica glass 111–113, 138
post-grafting method 119
post synthesis by impregnation 125
proton conduction 101, 104, 112, 113
proton diffusion mechanism 118
Proton Exchange Membrane 97–99, 124
Proton exchange membrane fuel cells 97, 98
Pseudocapacitance 51, 83

R

Recombination (relating to charge) 295, 316
Reflect 302

S

Scattering 293, 296–302, 304, 306–308, 319
self-assembly 97, 102–104, 111, 120, 123, 126, 127, 129
Shell 300–302, 306, 307
sodium ion batteries 232, 233
sol-gel technique 109
Sphere/spherical 296, 297, 299–304, 319
stack cells 133
sulfonic acid functionalization 113

T

template methods 1, 2, 8, 37, 38

#0002 - 240317 - C36 - 229/152/18 [20] - CB - 9781498747998